Cal Zecca

ABOUT THE AUTHOR

TIMOTHY FERRIS's works include *Seeing in the Dark*, *The Mind's Sky* (both *New York Times* Best Books of the Year), and *The Whole Shebang* (listed by *American Scientist* as one of the hundred most influential books of the twentieth century). A fellow of the American Association for the Advancement of Science, Ferris has taught in five disciplines at four universities. He is an emeritus professor at the University of California, Berkeley and a former editor of *Rolling Stone*. His articles and essays have appeared in *The New Yorker*, *Time*, *Newsweek*, *Vanity Fair*, *National Geographic*, *Scientific American*, *The Nation*, *The New Republic*, *The New York Review of Books*, *The New York Times Book Review*, and many other publications. A contributor to CNN and National Public Radio, Ferris has made three prime-time PBS television specials: *The Creation of the Universe*, *Life Beyond Earth*, and *Seeing in the Dark*. He lives in San Francisco.

ALSO BY TIMOTHY FERRIS

Seeing in the Dark

Life Beyond Earth

The Whole Shebang

The Universe and the Eye

The Mind's Sky

The World Treasury of Physics, Astronomy, and Mathematics (editor)

Coming of Age in the Milky Way

The Practice of Journalism (with Bruce Porter)

SpaceShots

Galaxies

The Red Limit

THE SCIENCE OF LIBERTY

DEMOCRACY, REASON, AND THE LAWS OF NATURE

TIMOTHY FERRIS

HARPER PERENNIAL

NEW YORK • LONDON • TORONTO • SYDNEY • NEW DELHI • AUCKLAND

HARPER PERENNIAL

A hardcover edition of this book was published in 2010 by HarperCollins Publishers.

FIRST HARPER PERENNIAL EDITION PUBLISHED 2011.

Designed by Jaime Putorti

The Library of Congress has catalogued the hardcover edition as follows:

Ferris, Timothy.
The science of liberty / Timothy Ferris. — 1st ed.
p. cm.
Includes bibliographic references and index.
ISBN 978-0-06-078150-7
1. Science—Social aspects. 2. Democracy and science. 3. Science—History. I. Title.
Q175.5.F477 2010
303.48'3—dc22
2009027505

ISBN 978-0-06-078151-4 (pbk.)

11 12 13 14 15 OV/RRD 10 9 8 7 6 5 4 3 2 1

IN MEMORIAM, H.S.T.

This poet had the foretaste of a vision, a great lust for which tormented his soul. From it he derived his great eloquence of desire and craving, lifting his readers above his work and all mere works and lending them wings to soar. . . .

—NIETZSCHE

Great doubt: great awakening.
Little doubt: little awakening.
No doubt: no awakening.

—ZEN MAXIM

CONTENTS

THE SCIENCE OF LIBERTY

CHAPTER ONE

SCIENCE & LIBERTY

Liberty . . . is the great parent of science and of virtue; and . . . a nation will be great in both, always in proportion as it is free.

—THOMAS JEFFERSON TO JOSEPH WILLARD, 1789

Science as subversion has a long history.

—FREEMAN DYSON, 1989

Over the past few centuries, two transformations—one scientific, the other democratic—have altered the thinking and the well-being of the human species. The scientific revolution is still gathering momentum, but has already revealed more about the universe than had been learned in all prior history, while technological applications of scientific knowledge have rescued billions from poverty, ignorance, fear, and an early grave. The democratic revolution has spread freedom and equal rights to nearly half the world's inhabitants, making democracy the preference of informed peoples everywhere.

These two transformations were linked, and remain so today: Every scientific nation in the world at the close of the twentieth century was a liberal, or at least partly liberal, democracy (meaning a state that guarantees human rights to its citizens, who elect their leaders). But how are they linked?

The scenario most of us learned in school presents the transformation in three acts—the Renaissance, the Scientific Revolution, and the Enlightenment. In the Renaissance (meaning "rebirth," from around 1450 to 1600), classical Greek and Roman writings became available to Europeans through trade with the Arab world, producing an outpouring of humanistic art and thought along with a few green shoots of science—as when Copernicus in 1543 demonstrated that the motions of planets in the sky could as readily be explained by the earth orbiting the sun as by the old earth-centered cosmology. The resulting brew of humanistic and scientific thinking eventually produced the Enlightenment, which in turn sparked the democratic revolution: Hence the Enlightenment is often dated as beginning with the English Revolution of 1688 and ending with the French Revolution of 1789. Meanwhile there was for some reason a scientific revolution, and so the modern world emerged.

The traditional scenario works reasonably well as a framework of study, but it portrays the simultaneous rise of science and liberal democracy as little more than a series of coincidences. The situation becomes clearer if we ask what was new: What was the innovative ingredient—the crystal dropped in the supersaturated liquid, suddenly solidifying it—without which the democratic revolution would not have occurred?

This book argues that the new ingredient was science. It maintains that the democratic revolution was sparked—*caused* is perhaps not too strong a word—by the scientific revolution, and that science continues to foster political freedom today. It's not just that scientific creativity has produced technological improvements, which in turn have enhanced the prosperity and security of the scientific nations, although that is part of the story, but that the freedoms protected by liberal democracies are essential to facilitating scientific inquiry, and that democracy itself is an experimental system without which neither science nor liberty can flourish.

To investigate this proposition and its implications, this book attempts to do three things.

First, it explores the historical link between science and liberty from the Renaissance through the end of the eighteenth century, examining science as an ongoing enterprise that requires freedom of speech, travel, and association. It maintains that scientific skepticism is corrosive to authoritarianism, and that scientific experimentation provides a better model for governance than any of the systems that preceded it.

Second, the book traces the development of the democratic and scientific societies from the eighteenth century onward, to see how the ideas and practices of science influenced their social policies. To some extent this amounts to testing the argument by making retroactive predictions about what should have happened if, indeed, science promotes liberty and democracy. Admittedly such a process is fraught with potential for abuse—anyone today, knowing how things turned out, can sift through the evidence for facts which suit his case—but such are the hazards of history.

Finally the world today is examined by the lights of science and liberty—taking into account powerful antiscientific forces that have cast shadows across our times—revealing, amid a welter of problems, surprisingly ample grounds for hope.

The word science comes from the Latin *scientia*, meaning "knowledge." In that broad sense of the word Anaximander of Miletus may be called a biologist for having proposed, in the sixth century BC, that humans were descended from fish, and Aristarchus of Samos an astronomer, since he hypothesized in the third century BC that the earth orbited the sun. The trouble with this approach is that it makes a scientist out of any philosopher who happened to voice a reasonably accurate opinion on a subject that has since *become* a science. To speculate and be proved right is not in itself to do science: As the American philosopher Alfred North Whitehead noted, "Everything of importance has been said before by somebody who did not discover it." The essence of science is experimentation, and scientific experiments were carried out by only a few ancient thinkers, among them Eratosthenes of Cyrene, who made a geometrical measurement of Earth's diameter in the third century BC; Strato of Lampsacus, who experimented with vacuums and compressed air at about the same time; and Galen, who dissected animal and human corpses a bit later. So for the sake of clarity this book uses the term science to mean what is often called modern science—that is, research involving observation and experiment, conducted as an ongoing social enterprise by career scientists working in laboratories and contributing to professional conferences and journals.

Liberty means the observance of human rights and freedoms. In practice the governments that have done so have almost all been liberal democracies, so the rise of liberty is roughly equated with the rise of liberal democracy. This process got off to a slow start, but has accelerated during the last hundred years. In 1900 there was not a single liberal democracy in the world (since none yet had universal suffrage); by 1950 there were twenty-two. As of 2009, despite recent reversals, there were eighty-nine democracies, comprising 46 percent of the world population.

The claim that science flourishes only in liberal-democratic environments rests on five assertions.

First, science is inherently antiauthoritarian. In order to qualify as scientific, a proposition must be vulnerable to experimental testing. If it repeatedly fails such tests it tends to fall by the wayside, regardless of who supported it or how much it may have seemed to make sense. The verdict of experiment has rudely dismissed the pronouncements of great thinkers from Aristotle (who thought that men and women were born with a different number of teeth) to Einstein (who insisted that quantum physics must be deterministic), and has sufficed to unhorse the claims of alchemists who sought to turn lead into gold and the folk wisdom behind a thousand racial, ethnic, and sexual stereotypes. The very process of doing first-rate science—of making important discoveries rather than merely refining old ideas—depends on unfamiliar and sometimes unpopular ideas being freely promulgated, discussed, and in some instances accepted. The fact that millions of people today are open to new ideas and skeptical about political or intellectual authority is largely due to the rise of science.

Second, science is self-correcting. Corrupted data, ill-begotten theories, and instances of outright fraud may not be caught at once, but if significant are unlikely to go undetected for long. When a scientist makes a major discovery, his or her colleagues flock to it, seeking to exploit and expand it as best they can, and such attentions are not entirely benign: Each new generation of scientists seeks to build a reputation by exposing weaknesses in the theories of its elders and replacing them with newer and more commodious theories. In this manner science presents a model for liberal governance, where it is similarly useful—although often frustrating—for plans and proposals to be widely debated and repeatedly altered before being enacted. Indeed a major failing of liberal democracies is that they are not yet sufficiently self-correcting: Programs that fail to accomplish their intended aims frequently survive anyway, by virtue of their popularity among the few who benefit from them or the many who assume that they are working.

Third, science in order to flourish must draw on all available intellectual resources. Nations aspiring to compete in the front ranks of science and technology cannot afford to suppress any element of their society—since none has a monopoly on brainpower—and so are obliged to educate their people and to maximize individuals' opportunities to advance on their merits. Liberal democracy approaches this ideal more closely than any other known system. As Francis Bacon put it, "There is but one state of learning, and that ever was and ever will be the democratic."

Fourth, science is powerful. Knowing things is empowering in itself, and the power of applying science to technology is so much a fact of modern life as to hardly require comment, whether you are using a mobile phone to find work in an African village or the Internet to chart your stock holdings. The power provided by science and technology is obvious in military matters, but it also drives national economies in countless ways. Historically, economic growth has proceeded apace in nations where science has flourished, while the local clocks in the less scientific and technological nations have run more slowly.

Finally, science is a social activity. In the early days when almost nothing was known about how the world worked, an isolated experimenter could make important discoveries, but to make progress today requires the combined talents of many participants. Even a solitary scientist needs to stay in touch with the literature, and in big sciences like experimental high-energy physics, collaborations have become so extensive that the number of contributors listed on the front of a technical paper may exceed that of its readers. The further science progresses down this road, the more it requires freedom of speech, travel, and association. Its resemblance to democratic institutions is clear. As John Dewey maintained, "Freedom of inquiry, toleration of diverse views, freedom of communication, the distribution of what is found out to every individual as the ultimate intellectual consumer, are involved in the democratic as in the scientific method." Authoritarian governments such as Nazi Germany, the Soviet Union, and communist China tried to address this issue by encapsulating scientists in special zones of relative freedom, but such partial measures had only partial success: The local clocks not only slowed down but sometimes stopped or went backward.

These claims may be tested against the verdict of history by applying the journalistic "five *Ws*"—who, when, what, where, and why—to the changes that have created the modern world.

Who implemented the democratic revolution? If its cause was science, disproportionately high numbers of scientifically minded individuals should be found among the instigators of the democratic revolution. And so they are. The Whig insurgents who brought about the English Revolution were largely proponents of science, while their opponents tended to be traditionalists disinclined toward scientific innovation. The American Revolution was incited and carried out in disproportionate measure by amateur scientists like Thomas Paine, Benjamin Franklin, and Thomas Jefferson. And scientists have been found at the forefront of political reforms down to the present day, from dissidents in communist China to planetary scientists struggling to alert governments to the hazards of global warming.

When did they do these things? Science arose to prominence immediately prior to the Enlightenment—as would be expected if, indeed, science was the one indisputably new ingredient in the social and intellectual ferment that produced the Enlightenment and the democratic revolutions that followed. The principal Enlightenment catalyst was the publication, in 1687, of Isaac Newton's *Principia*, which subsumed prior astronomy into a unified theory of gravitation that could be employed to predict natural phenomena ranging from tides to the orbits of comets. This thunderclap caught the startled attention of thinkers throughout the Western world, scientific and otherwise. The *Principia* put teeth in reason, demonstrating with unprecedented power and scope that mathematical analysis combined with careful observations could expose an elegant simplicity underlying the complex motions of the planets across the sky. In exposing laws of nature it promoted the idea that there are natural laws of human affairs as well, and natural human rights. Newton's book also acted as a kind of vaccine against the predations of superstition and faith-based authoritarianism, making them look paltry by comparison to what scientific and political empiricism could accomplish. The empiricist physician John Locke, whose doctrine of natural rights became the polestar of the Declaration of Independence, was Newton's closest friend—aside, perhaps, from the diarist Samuel Pepys, himself enough of a promoter of science to be elected president of the Royal Society—and Locke described his own philosophical accomplishments as subordinate to those of Newton. Other prime movers of the Enlightenment influenced by Newtonian science included Adam Smith, Fontenelle, Bayle, Montesquieu, Condillac, Diderot, Voltaire, La Mettrie, Leibniz, Spinoza, Holbach, and Buffon. The French *Encyclopédie*, a seminal Enlightenment tome, was dedicated to Bacon, Locke, and Newton; Thomas Jefferson in 1789 commissioned a composite portrait of these enthusiasts of science, calling them "the three greatest men that have ever lived, without any exception."

What became of the liberal democracies? If indeed it was to science that they owed their birth they should have become world leaders in science, and so they did. England—precocious in democracy, having established one of the world's first parliaments—wielded a far greater scientific influence than its population would otherwise have indicated. The United States became both the world's oldest constitutional democracy and its scientific and technological leader. Small nations that became democratic early on, such as the Netherlands, made significant contributions to science, while mightier nations that only later became democracies, such as Spain, experienced a retarded progress toward scientific significance.

Where did the first democracies appear? Precisely in those states where science and technology were most advanced—in England, America, and the Netherlands, and (more fitfully) in France, Italy, and Germany.

Why did this happen? Because—or so this book maintains—science demanded liberty and demonstrated its social benefits, creating a symbiotic relationship in which the freer nations were better able to carry on the scientific enterprise, which in return rewarded them with knowledge, wealth, and power. This process continues today. It is difficult to think of any large-scale human activity that has not benefited from science—from feeding the poor to growing a business, from expanding an economy to protecting the environment, from educating the young to improving the welfare of the elderly.

That much constitutes what might be called the positive argument for a link between science and liberty—that on the whole, science has flourished in free societies and fared poorly in nations with illiberal governments. But what about negative examples—illiberal regimes under which science seemed, however briefly, to have risen to imposing heights? This book pays particular attention to the three most powerful totalitarian regimes of the twentieth century—Hitler's Germany, Stalin's Russia, and Mao's China—which, during World War II and the Cold War, were widely viewed by their adversaries as scientific and technological juggernauts. Were this assessment accurate, the case for the science–liberty link would be badly undermined.

All three nations had strong intellectual and creative traditions that included at least some scientific attainments. Twentieth-century Germany, where democratic reforms had been rising and falling like a ship in a choppy sea, had a similarly sporadic career in science, highlighted by the geographical exploits of Alexander von Humboldt, the optical experiments of Joseph von Fraunhofer, and the theoretical physics of Max Planck and Albert Einstein. Russia had an Academy of Sciences where botanical and zoological research was conducted, and could boast as well of Dmitry Mendeleyev's periodic table of chemical elements, Vasily Struve's studies of stars, and Fyodor Bredikhin's work on comets. China could point to a history of technological attainment stretching back thousands of years, including the invention of paper and gunpowder and the preservation of extensive astronomical records. So if these nations failed to make significant scientific progress while under illiberal rule, the fault cannot have been a lack of indigenous talent. Nor could it be blamed solely on a lack of resources, since all three of the repressive regimes that befell these nations funneled vast resources into programs designed to promote scientific and technological progress.

Yet fail they did. The communist ideology espoused by Stalin and Mao talked a great deal about science—indeed it portrayed itself as a scientific form of government, its universal triumph as inevitable as the outcome of a demonstrative experiment in a high-school physics class—but was unable to adapt when social experiments failed. Instead, each was proclaimed a great success by the controlled news media, to be followed by another Five Year Plan or Great Leap Forward based on faith rather than empirical evidence. The Nazis imagined that science could be put to work generating technical advancements while substantiating their weird biological and cosmological notions. All three regimes tried to exploit their most talented scientists, but wound up silencing, imprisoning, or murdering many of them.

The technological achievements of these totalitarian regimes—such as Germany's rocket program and the Soviet space effort—impressed and alarmed many in the liberal nations, but were based on little more than the momentum of earlier science plus the short-term torque of intense government spending: As Nikita Khrushchev wryly described the Soviet space program to his son, in 1961, "We have nothing to hide: We have nothing, and we must hide it." In the end, totalitarian science collapsed in a morass of scandals (the Lysenko affair, sadistic Nazi "medical experiments") and in catastrophes such as the mass famines that resulted from communist agricultural reforms, killing millions.

The spectacular failure of totalitarian science spotlighted the futility of attempting to treat science as a tool that could be divorced from its ethical imperatives. It is often said that science is ethically neutral—that it shows how things are, not how they ought to be—but there is less to this claim than meets the eye. Applied science has placed enormous power in human hands, and power can be used for good or ill, but to exploit that power without accommodating the scientific culture that produced it, as illiberal states have done, is to kill the goose that lays the golden eggs. The more closely science is examined, the more evident it becomes that science has an ethos and perhaps an ethics. The mathematician and polymath Jacob Bronowski was especially perceptive about this. "The society of scientists," he wrote, "is simple because it has a directing purpose: to explore the truth.

> Nevertheless, it has to solve the problem of every society, which is to find a compromise between man and men. It must encourage the single scientist to be independent, and the body of scientists to be tolerant. From these basic conditions, which form the prime values, there follows step by step a range of values: dissent, freedom of thought and speech, justice, honor, human dignity and self-respect.

These are humanistic values, and the social impact of science has been quite the opposite of popular notions about its turning people into unfeeling robots. "Like the other creative activities which grew from the Renaissance, science has humanized our values," Bronowski notes.

> Men have asked for freedom, justice and respect precisely as the scientific spirit has spread among them. The dilemma of today is not that the human values cannot control a mechanical science. It is the other way about: The scientific spirit is more human than the machinery of governments.

On a personal, philosophical, or religious level, ethics may be concerned with what ought to be—with wished-for improvements in human conduct like those inspired by Lao Tzu, Jesus of Nazareth, and Gautama Buddha. But the political art of governing a state means dealing with people as they are, and ethics on this level is primarily an empirical matter of identifying and promoting actions that actually work. Science, too, is about what works, and demonstrates daily the practical value of freedom, dignity, and autonomy. The world is far from fully absorbing these lessons, but it has come a long way in the last few centuries—thanks to science and liberty.

Although scholars have paid little attention to the role of science in the rise of liberal democracy—an omission that this book endeavors to address—they have identified other conditions that affect a nation's chances of establishing and maintaining liberal-democratic governments. One factor is wealth: Since 1950, democracies established in nations with a per capita gross domestic product (or GDP, which may be loosely translated as "income") of under $3,000 have typically failed within a generation, while in those with a per capita GDP exceeding $6,000, democracy has tended to survive indefinitely. In part this is because their growing economies, spurred by free markets and access to scientific and technological progress, may lift them above the $9,000 mark, a threshold beyond which no liberal democracy adopted in the past half century has reverted to illiberal rule. Another factor is distribution of wealth: States that derive most of their income from exploiting a single natural resource, such as oil, tend toward oligarchy. A third consideration is education: Nations with higher levels of education and literacy are stronger candidates for democracy. Interestingly, all these criteria are related to science as well: Creating scientific research institutions requires wealth that is reasonably well distributed plus an educational system capable of training future scientists and engineers. Nor is it sufficient to spend lavishly on laboratories and schools in boom times, then turn off the spigot when times are hard: The dues required to remain a player in the science-and-liberty club is an ongoing investment of at least 2 percent of GDP in scientific research and development.

Much of the world today is concerned about the future of the Middle East, which at first glance would seem to provide poor soil for the cultivation of science and democracy. Its twenty-two nations constitute the most illiberal block on earth, characterized by official state religions, sanctions against religious and political dissent, and extreme concentrations of wealth in the hands of the few. Human rights abuses are rampant, gender inequality high, and investment in scientific research one-seventh the world average: Israel, the region's only liberal democracy, publishes more scientific papers annually than the rest of the Middle East combined. The region's educational systems are so paltry that many children can learn to read only by enrolling in politically reactionary religious academies—where they are taught, as a Syrian scholar put it, that "the role of thought is to explain and transmit . . . and not to search and question." Add to this mix the vexations felt by once-great peoples at finding themselves left behind in an increasingly scientific world, and little mystery remains as to the sustenance of Islamist radicalism.

Many thinkers, both inside and outside the region, recognize that the way out is through liberal democracy. But the choice is too often couched in terms of culture: Muslims are asked to decide whether to cling to Islamic culture or instead adopt "Western" democracy and science. Science and democracy did happen to originate in the West, but that does not mean that adopting them requires embracing Western culture. Science is practiced by persons of all races and religious beliefs, speaks a universal language, and evaluates results on their merits rather than on their place of origin. Democracy is no more inherently Western than science is. It is the property of no particular culture, but belongs to everyone willing to plant and cultivate it: The world's largest democracy is India. The decision whether to adopt democracy and open the door to indigenous science, whatever else it may involve, is not about abandoning one's culture.

Nor is the potential for a democratized Middle East as unlikely as it seems at first glance. Half the region's nations already stand at or above the $6,000 GDP tipping point, and all are wealthier, per capita, than India—which sports many science and technology endeavors, among them the world's second-largest software company. Herculean efforts will be required to reform and improve public education in some Middle Eastern states, but they can draw on a deep-rooted Islamic intellectual tradition that includes a love of books and learning and an ancient tradition of tolerance for unorthodox philosophical and religious opinions. What's more, most Muslims express enthusiasm for democracy. In a recent poll, more Muslims than Western Christians agreed with the statement "I approve of democratic ideals."

Let's say you are a citizen of Iran. Your government is a strange mix of theocracy and democracy: a democratically elected legislature constrained by a band of religious "guardians" who can bar political candidates as they please. Like many Iranians—70 percent of whom say the nation they most admire, after their own, is the United States—you aspire to full democracy. What must your nation do?

First, the key is not just democracy but *liberal* democracy, one that safeguards fundamental human rights. Since the observance of these rights permits lowly as well as lofty practices, you will have to accept that many Iranians may participate in unseemly activities such as dancing to lewd music or watching pornographic videos. But your ancient culture can survive such affronts, and anyway these practices are already popular in Iran, where their current suppression only adds to their allure.

Second, it must be a *secular* democracy: no official state religion. This may be the bitterest pill of all, especially at the outset, since it almost certainly means that some will stray from the path of Islam. But other faiths have thrived under liberal governments. Christianity, the world's largest religion, has lost some of its faithful in the secular democracies—indeed, has lost some to Islam—yet remains very much a going concern, with more than a third of American Christians saying they attend church services at least once a week. State-sponsored religion is bad not only for science but for religion as well, depriving it of the free, open discussion without which all systems of thought desiccate into deadwood. Secular democracy is also the only way past the sticking point, endlessly cited by Arab princes to justify their rule, that free elections would sweep religious fundamentalists into power. Perhaps they would, for a time, but voters in a functioning secular democracy are free to sweep them out again—as almost certainly would happen in Iran, where theocrats show up as unpopular in the polls.

Third, you must spend what it takes to promote universal public *education*. Ignorance is poverty, not bliss, and the way out is through investment. The United States has above-average public schools and the world's best universities—not because Americans are especially clever, but because they spend nearly as much on their public schools as on national defense, and more on their universities than any other nation in the world.

Finally, you will need to allot at least 2 percent of your GDP to scientific *research and development*. That's a lot of money, but given a growing domestic science and technology sector, free markets, and free government, you'll be making a lot more money, too.

It's your choice. You can join the club—not as second-class Westerners, but as proud Shia and Sunni Muslims, Zoroastrians, Baha'is, Christians, and Jews—or you can continue to fall behind. As the physician Lewis Thomas remarked, the greatest discovery of modern science was of the dimensions, not of cosmic space and time, but of human ignorance.

Widening the circle of firelight in this deep darkness is a noble task, and it could use your help.

Arrayed against this promise are the imposing forces of antiscience. Reactionary politicians who assume that the only way we can have more is for others to have less, and fundamentalist clerics who insist that everything worth knowing is contained in this or that religious text, are to be found in every society. Beneath them stand intellectual mandarins eager to deny all hope of progress. On the popular level lurks the Faust myth—the notion that science must be reined in lest it go too far. Eternal vigilance in the face of such misapprehensions is the price of science and liberty alike.

Modern science and liberal democracy are novelties of recent vintage—considering that, as Whitehead used to say, it takes a thousand years for a genuinely new concept to engrain itself in a culture—so it is not entirely surprising that many today dismiss science as a new priesthood or a scythe wielded by the privileged, and deride democracy for its inefficiency, its ceaseless demands for compromise, and its many glaring faults, blunders, and absurdities. These are not necessarily majority opinions—about half of all Americans support science and vote in at least the occasional election—but their persistence indicates that even many citizens of the democratic, scientific nations do not yet comprehend the twin revolutions that made their world.

It helps to consider that whereas prior systems dealt in claims of certitude, such as philosophers' allegedly airtight reasoning and monarchs' god-given right to rule, science and democracy are steeped in doubt. Both start with tentative ideas, go through agonies of experimentation, and arrive at merely probabilistic conclusions that remain vulnerable to disproof. Both are bottom-up systems, constructed more from individual actions in laboratories and legislatures than from a few allegedly impervious precepts. A liberal democracy in action is an endlessly changing mosaic of experiments, most of which partially or entirely fail. This makes the process frustratingly inefficient, but generates its strength. Democracy is like the "well-tempered" keyboard tuning system championed by Johann Sebastian Bach: By spreading the disharmony, it makes everything imperfect yet produces robust results. Fisher Ames, who helped frame the American Bill of Rights, remarked that "a monarchy is a merchantman which sails well, but will sometimes strike on a rock, and go to the bottom; a republic is a raft which will never sink, but then your feet are always in the water."

This book does *not* make several claims that might be conflated with its argument. It does not claim that governance is a science, or that governments ought to be run by scientists. Scientific information can be valuable to those who govern—economics, for instance, is a science, and nations ignore economics at their peril—but the pipe dream of scientific experts handing down rational dicta to eager multitudes has rightly been rejected by every sensible citizenry to have run afoul of it, and nothing of the sort is being advocated here. Rather, this book favors the messy, selfish, and often foolish and greedy push-and-pull of democracies as they are—neither rational nor expert but experimental—as better attuned to the spirit of science than are enchantments with authoritarian expertise and top-down planning. Scientists play a role in politics, as do all citizens, but this book is not out to pretend that scientists, if empowered, would make any less of a muddle of things.

Nor does it claim that science or democracy is perfect. On the contrary, the very notion of perfection is inimical to both institutions. As the political scientist Stein Ringen writes, "Perfect democracy is an illusion, as is the idea of perfection generally." The scientist who claims that his theory is perfect is a crank; the politician who claims that his administration is perfect is a tyrant. The operations of free societies are inescapably fraught with flaws and mistakes, as are scientific experiments. To expand upon Winston Churchill's famous remark, democracy and science are the worst systems of governance and inquiry, except for all the others. There is a pernicious mode of argument that consists of calling attention to the shortcomings of scientific and political endeavors and claiming that they are therefore bogus. If a liberal democracy conducts itself in ways inconsistent with its professed values—as they all do from time to time—it is indicted as a sham. If a few scientists commit fraud, act cruelly or stupidly in making experiments, or announce a discovery that turns out to be false, then science is said to have feet of clay. All such arguments consist of comparing a real system with an imagined ideal that does not exist and (the deadly dreams of idealists notwithstanding) almost certainly never will.

Finally, this book disavows the position that science is to be valued solely insofar as it produces new technologies that add to the power, specifically the military power, of scientific nations. Science certainly contributes to technological progress—just as technology contributes to scientific progress; you cannot have one without the other—and there are grounds for optimism that insofar as science thrives in free societies, which it empowers with military and other technological might, then the future may belong to the free. But science must comprehend before it can control—as Francis Bacon wrote, "Nature is only subdued by submission"—and power divorced from knowledge is a Moloch. Fortunately the two cannot long be separated before the befuddled monster lumbers to a halt.

CHAPTER TWO

SCIENCE & LIBERALISM

The spirit of liberty is the spirit which is not too sure that it is right.

—JUDGE LEARNED HAND, 1944

The method of freedom is the method of science.

—SOCIOLOGIST LYMAN BRYSON, 1947

Liberalism shares many of the qualities of science, but since both are widely misunderstood it may be useful to examine their relationship. Liberalism nourishes science by fostering a free and flexible milieu in which scientific creativity can flourish, which in turn increases the knowledge, power, and wealth of liberal societies. In doing so, science helps demonstrate that liberal governance works; and so the cycle continues.

Liberalism (from the Latin for "freedom") is based on what John Stuart Mill in 1859 called

> one very simple principle. . . . that the sole end for which mankind are warranted, individually or collectively, in interfering with the liberty of action of any of their number, is self-protection. That the only purpose for which power can be rightfully exercised over any member of a civilized community, against his will, is to prevent harm to others.

This thesis is admirably easy to test: Just increase personal freedoms and see what happens. Originally such experiments took the form of putting limitations on the powers of monarchs, as with the English Bill of Rights of 1689. Today, every day, liberalism is tested through the functioning of the world's liberal democracies. Its accomplishments to date include the abolition of slavery (seven million slaves freed in a single century, an unprecedented accomplishment), the extension of legal rights to women and minorities, the maintenance of free speech and a free press, and an unprecedented global growth in knowledge, prosperity, and power.

Liberalism arose when, with the advent of science, a relatively static Western civilization began to become more creative, dynamic, and free. The word *liberal* entered the English language through education, where it signaled a shift from the narrow preparation of students for the vocations to which their social status had predestined them (usually law or the priesthood) to a "liberal sciences" (later called liberal arts) approach aimed at arming students with the tools needed to thrive in an unpredictable world. Andrew Abbott of the University of Chicago used to tell incoming freshmen, "There are no aims of education. The aim *is* education." Liberally educated men and women will have learned how to keep learning throughout their lives, so that they can participate in a dynamic, changing society. Liberalism's acceptance of unpredictable change distinguishes it from conservatism, which concentrates on the lessons of the past, and from progressivism, which plans for a future about which it knows rather less than it sometimes claims.

Liberalism is inherently nonpartisan: It means freedom for all, or it means nothing at all. It maintains that everyone benefits from everyone's freedom, and that all are diminished whenever one individual or group is not free. This precept can contort liberals into the uncomfortable posture known as tolerance. Some think that tolerance means treating all opinions as equally deserving of respect, but the point of liberalism is not that all views are equally valid. It is that society has no reliable way to evaluate opinions other than to let everybody freely express and criticize them—and, if they can garner sufficient support, to try them out.

It was difficult even for the founders of liberalism to fully embrace tolerance. John Locke would have denied equal rights to atheists: "Those are not at all to be tolerated who deny the being of a God," he declared, in his *Letter Concerning Toleration,* since the "promises, covenants, and oaths, which are the bonds of human society, can have no hold upon the atheist." Many otherwise liberal thinkers today recoil from the prospect of granting homosexual couples the same legal benefits that heterosexual couples enjoy, or affording legal rights of due process to those accused of terrorism. Such concerns—essentially the nagging worry that something terrible will happen if too much freedom is extended to people who do not closely resemble ourselves—have so far prevented societies from becoming entirely liberal. But each step taken to extend equal rights to those previously denied them has in retrospect been seen to benefit not only the group in question but the society as a whole.

As an empirical, experimental philosophy that accommodates error and uncertainty, liberalism rejects all absolutist political claims, including absolute faith in religion at one extreme and in rationalism at the other. Liberalism does not oppose religion—it is a staunch defender of religious freedom—but it demands that the state grant special status to no religion; as Machiavelli observed, religion plus politics equals extremism. Rationalists are apt to imagine that they can reason their way to a political scheme so self-evidently superior that its implementation justifies at least a temporary suppression of opposing views; liberals will make no such concessions, because they appreciate that nobody is prescient enough to justifiably sacrifice present liberties for imagined future gains. That is the sense of Judge Learned Hand's suggestion that Oliver Cromwell's injunction, "I beseech ye, in the bowels of Christ, think that ye may be mistaken," be inscribed over the door of every church, school, and courthouse in the nation.

The ideal of liberalism is universal peace and mutual aid. "The starting point of liberal thought is the recognition of the value and importance of human cooperation," wrote the economist Ludwig von Mises, "and the whole policy and program of liberalism is designed to serve the purpose of maintaining the existing state of mutual cooperation among the members of the human race and of extending it still further. . . . Liberal thinking always has the whole of humanity in view and not just parts." Liberals are opposed to war not only for the usual reasons but also because wars tend to aggrandize governments, ballooning their budgets and emboldening them to draft conscripts. Similarly, liberalism opposes imperialism, colonialism, racism, and every other form of oppression.

There is, however, one inherent problem with liberalism. Since absolute liberty would be anarchy, liberalism must sanction some form of coercion to prevent the strong from abridging the freedoms of the weak. The troubled history of American race relations is replete with examples of such illiberalism on the march: The white-robed Klansmen who bombed churches and burned crosses might be said to have been exercising their freedom of speech and association, as were the white American shipyard workers who harassed their black coworkers during World War II. To prevent such injustices, liberals concede to government a monopoly on coercive force. The police, the national guard, and the military are entitled to employ force, while corporations, vigilante groups, and self-appointed militias are not.

This governmental monopoly on coercive force calls forth two further liberal mandates. The first is equal protection under the law. (Locke: "Where there is no law there is no freedom.") Liberty is abridged whenever the government places itself above the law. The other mandate is that government be kept small, lest the citizenry be snared in growing tendrils of laws and regulations backed by ever-increasing powers of enforcement and intimidation. Each such measure may be well intended—as when voters seek to save the jobs of steelworkers or stockbrokers, curb hate speech, or comb millions of e-mail messages in search of threats to national security—but their proliferation increases the power of government to confiscate your wages and property, put you in jail, and send you off to war. This battle liberalism has been losing. In every liberal democracy today the government's share of the national wealth is increasing with each passing decade. In the United States, a more laissez-faire nation than most, the government currently claims a third of the wealth, and is growing rapidly; some of the socialistic democracies of Scandinavia collect more than half. Elected officials find the trend difficult to reverse. President Ronald Reagan and Prime Minister Margaret Thatcher attempted to shrink the size of government, yet Thatcher never managed to do more than to slow the rate at which the British government grew, and Reagan presided over major increases in federal spending that swelled the national debt from 23 percent to 69 percent of GDP. The only recent presidents to have significantly shrunken federal spending as a percentage of GDP were Harry Truman (–8.6 percent, in the aftermath of World War II) and Bill Clinton (–1.8 percent). This is not a new story; liberalism seems always to be creating wealth and fighting against its confiscation. "Opinion tends to encroach more on liberty," complained J. S. Mill in 1855, "and almost all the projects of social reformers in these days are really liberticide."

To the extent that liberalism is scientific, it is obliged to judge its success in terms of measurable quantities. Conservatives may argue that people in the old days were healthier in spiritual terms—that for instance they were happier when everyone knew his place—but liberals prefer to measure what *can* be measured, in a quantitative way. (As Lord Kelvin said, in words now chiseled on the University of Chicago's Social Science Research Building, "When you cannot measure, your knowledge is meager and unsatisfactory.") One quantitative measure of national status is productivity—the creation of wealth—and there the verdict is favorable: People have most prospered where they have been most free. Nor has this just been the prosperity of a few. Even though the United States grew more economically vertical from 1950 to the early twenty-first century, all five quintiles of the population—from the bottom fifth to the top—increased substantially in wealth during that same period.

The emphasis placed by liberals on economic productivity has prompted critics to dismiss them as materialistic. Liberals plead guilty to this charge. The quantitative measurement of material results may ignore spiritual considerations, but even when people are asked ethereal questions about how they *feel*, those living in liberal states say they prefer it that way, while those living elsewhere—if they have any reasonable access to uncensored information—say that they, too, would prefer liberal governance. It is on such bases that von Mises could call liberalism "the application of the teachings of science to the social life of man."

What this book calls liberalism is the philosophy more often known in the United States as "classical" liberalism—the original liberalism, dating from Thomas Hobbes and John Locke and the American founders. That is not, however, what liberalism has come to mean for most Americans. In the United States today the word *liberal* is more apt to be applied to leftists or progressives—those who value equality over liberty, and are willing to put the force of government behind efforts to create greater political and economic equality even if personal freedoms are abridged in the process.

Consider the term *liberal democracy.* If you're a classical liberal, what matters most to you is the liberal side of the concept. You may be enthusiastically democratic, doubting that any form of government other than a democracy can *be* liberal, but you know that democracies can behave in illiberal ways—as when millions of Germans voted for the National Socialist Party, one of whose leaders declared, "We are socialists, we are enemies of today's capitalistic economic system for the exploitation of the economically weak, with its unfair salaries [and] evaluation of a human being according to wealth and property instead of responsibility and performance." Alert to such dangers, you feel that liberalism is more likely to rescue a wayward democracy than the other way around. In this your position resembles that of Locke and other Enlightenment philosophers, and of the many Whigs and other liberals who, well into the nineteenth century, regarded liberalism as a given but democracy as an ongoing experiment.

If, on the other hand, you are an American "liberal" of leftist inclinations, what matters most to you about liberal democracy is that it's democratic. Democracy is mainly about equality—one man, one vote—and when you look around and see people living in steeply unequal conditions, you're apt to feel that the system isn't working. If, say, African-Americans are substantially poorer and less well educated than are white Americans, you are not likely to be satisfied with (classically) liberal talk about universal freedom and equal protection under the law. You may be willing to vote for higher taxes and a bigger, more intrusive government if the result will be greater equality of outcome, even at the expense of reducing equalities of opportunity. Such views are best described not as liberal but as, say, progressive.

The tension between liberals and progressives—between those who stress liberty and those who stress equality—has long persisted; as Reinhold Niebuhr observed in 1943, "Whether democracy should be defined primarily in terms of liberty or of equality is a source of unending debate." The use of one word to describe them both needlessly confounds American political discourse. Only 20 percent of Americans describe themselves as liberals, even though a majority hold both liberal views (supporting equal rights for all and specifically for homosexuals) and progressive views (60 percent favoring universal health care, and 70 percent agreeing that the government has an obligation "to take care of people who can't take care of themselves"). As a result many classical liberals are reluctant to be called liberals, since the term is so often conflated with big-government progressivism.

Mathematicians are familiar with the distortions that result when a complex structure is collapsed into too few dimensions. A snowflake, smashed into two dimensions, becomes a botched asterisk; squeezed into one dimension it is a line. That is precisely the problem with employing a one-dimensional political spectrum that puts conservatives at the right and so-called liberals at the left, and pretends that all other political philosophies reside between. This crude paradigm, which originated in the French

National Assembly circa 1789, when the Jacobin faction sat to the left of the president's chair and the Girondins to his right, has long caused confusion. Better to add another dimension, forming a triangle:

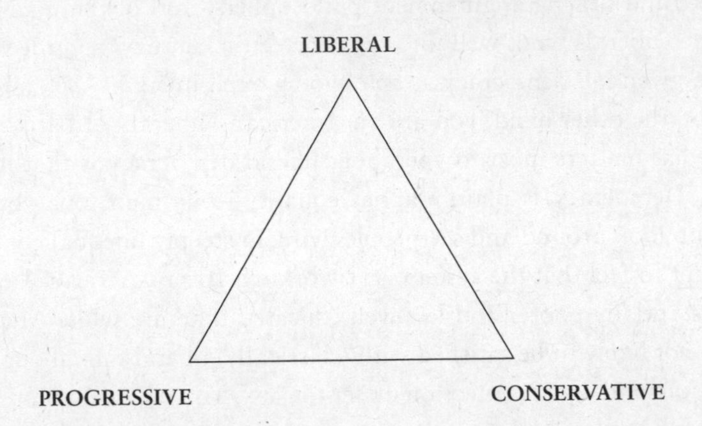

In one corner of the triangle stand the conservatives—upholders of tradition, captains of a sea anchor that steadies the ship of state, comfortingly reliable if sometimes awkward in the face of change. There will always be conservatives, because there will always be traditions worth conserving. If you're a young person who detests conservatism and manages to enact radical reforms which in later years you are obliged to defend, in that respect you shall have *become* a conservative.

Occupying the second corner are the liberals—defenders of liberty, embracers of change, and friends of science.

In the third corner are those who emphasize equality of outcome over freedom of choice—the progressives, socialists, or social democrats. Their standing is made clearer in nations that have three major parties rather than America's two—progressives in Finland and Denmark tend to be Social Democrats, while in Canada they are known as New Democrats—but even there the labels can be confusing. By any name, progressives can boast of many honorable accomplishments, some of their own instigation (universal health care and state pension plans such as Social Security) and others instigated by liberals (women's suffrage and the thwarting of racial and sexual discrimination).

As the triangle illustrates, the distance from conservatism to liberalism is no greater than from conservatism to progressivism. This helps explain why a conservative can be liberal in certain respects, such as by upholding free markets or opposing the jailing of drug abusers, and why a classical liberal may be otherwise conservative ("neoconservative") or progressive ("neoliberal"). It also clarifies the previously puzzling fact that many hard left progressives have borrowed doctrines and terminology from the hard right. As many demagogs have demonstrated, it is possible to oscillate between the left and the right without ever approaching liberalism.

The political triangle is useful as well for visualizing the process of political consensus building. Consider the question of how to find sufficient votes in Congress to pass a bill aimed at reducing carbon emissions to combat global warming. One option is to take a command-and-control approach, empowering the Environmental Protection Agency to regulate limits on carbon emissions, and fining or shutting down power plants that exceed the mandated limits. That appeals to progressives, since it uses the government's power to get rapid results, but is unpalatable for conservatives, since it would be bad for business, and for liberals, since it reduces personal freedom and enhances the power of government. More likely to gain support is a cap-and-trade approach that sets limits while establishing a carbon market in which the greener factories earn credits which they can sell to firms whose dirtier operations put them over the cap. Such a scheme has a reduced impact on business (which appeals to conservatives), and a smaller government role (appealing to liberals). Hence the triangular model predicts that the cap-and-trade bill is more likely to be enacted.

Liberals are deaf to the popular opinion that everything would be fine if everybody would just do what one individual thinks they should do. They oppose that stance on principle and also on the basis of historical fact. The modern world has extensive experience with all-powerful leaders—among them Hitler, Stalin, Mao, and the Kim dynasty in North Korea—who got their way, promising wonderful results in return for the sacrifice of their subjects' liberties. These promises went unfulfilled. The divisions imposed on Korea and China more than a half century ago created the nearest thing to scientific experiments to be found in human affairs on such a large scale. In North Korea, decades of strongman control—*he* gets his way, *we* do not—produced a per capita GDP circa 2008 of $1,700, while the figure for South Korea was $26,000. Communist China's per capita GDP that year was a sixth that of neighboring Taiwan. The communist regimes were unable even to consistently feed their own citizens, with millions dying of famine while North Korean boys at age seven stood, on average, five inches shorter than their South Korean counterparts.

It can be objected that the results of social experiments are contaminated, inasmuch as no nation has ever had a purely liberal or a purely progressive government. That certainly is the case: Even the most stringently socialist states have found *some* room for free enterprise, if only in the form of flourishing black markets operating under the radar of the controlled economy, while the United States—a wild-and-wooly bastion of free enterprise, where the word *Liberty* appears on every coin—has Social Security and a vast federal bureaucracy. But it is not necessary to have unsullied examples of conservative, socialist, and liberal societies in order to make rough yet reasonable determinations of what does and doesn't work. The modest proposal of those who favor liberalism and science is not to subject people to involuntary experimentation of any sort, but simply to assess the empirical successes and failures of the innovations wrought by free peoples.

In practice, democratic nations constantly slosh around within the triangle, tending toward conservatism, liberalism, or progressivism at various times and for various reasons. Their short-term motivation may amount to little more than disenchantment with the status quo and a desire to "throw the bums out," but cultural differences play a major long-term role. High taxes and socialistic programs are more popular in nations that have ethnically homogeneous populations, such as Sweden, than in ethnically diverse nations such as the United States. (It may be easier to think of your compatriots as one big happy family if they all *look* like members of the same family.) Within a given nation, sociologists find that city dwellers tend to be more socialistic than those in the countryside—and that it's not a matter of voting one's pocketbook. In the United States, the rural red counties that vote Republican are net consumers of federal dollars, while the blue cities that vote Democratic contribute more federal taxes than they consume; if the so-called urban elites were to change course, voting to cut taxes and slash federal government spending, the people worst impacted would be the rugged individualists in the red counties.

Liberal democracy normalizes such differences by holding elections, but when political parties are swept from office the programs they created frequently survive. The resulting proliferation of costly programs that have outlived their usefulness (or never had any to begin with) can be reduced by building evaluative feedback loops into the legislation that creates them—as the U.S. Congress and other legislatures have been trying to do, with the Department of Education, for example, spending $100 million a year to see which learning programs work best. It remains to be seen whether improved social-scientific evaluations can prevail against political pork, but governments do at least now have better tools for determining whether programs are doing what they were chartered to do. Statecraft used to be like driving at night with the headlights off; today legislators have headlights that, though flickering and dim, are better than nothing.

When liberalism is evaluated quantitatively, via the social science of economics, it comes off rather well. By fostering individual creativity and free markets, liberalism has created enormous prosperity—but it seems that a dash of government involvement helps, too. Party politics may be a crude metric, but the United States in the past half century has experienced faster GDP growth, lower unemployment, and higher corporate profits during Democratic than during Republican administrations. The stock markets performed better, too, with annualized returns on investment averaging almost 9 percent when Democrats were in office against less than 1 percent for the Republicans. It is not yet possible to identify exactly which conditions best bolster a free-market economy's production of wealth—given the internal differences in the geography, history, ethnicity, and other conditions of various nations, there may *be* no one best way to set the dials—but it is clear that almost everyone is economically better off in the liberal-democratic, free-market states. Eighteen of the world's twenty largest economies circa 2007 were liberal democracies—the exceptions being China, a communist state which boosted its economy into double-digit growth rates by admitting free-market practices, and Russia, which was foundering between democracy and oligarchy on an ocean of petroleum profits. That same year, all forty of the world's cities said to offer the "highest quality of life" (measured in terms of thirty-nine factors ranging from recreational opportunities to political stability) were in liberal-democratic nations. It also appears that liberalism really does move humanity toward universal peace: Seldom do liberal democracies make war on one another, and as the world has become more liberal it has also become more peaceful.

But how, exactly, does liberalism facilitate scientific inquiry, and science benefit liberalism? Primarily through *change* and *creativity*.

Liberalism stands out among political philosophies in its readiness to embrace change. Liberals do not pretend to know what the future will bring, and so are skeptical about planning. They stress the importance of individual creativity, noting that humans are profoundly ignorant and so must be free to keep learning. As the liberal economist Friedrich Hayek declared, "All institutions of freedom are adaptations to this fundamental fact of ignorance, adapted to deal with chances and probabilities, not certainty. Certainty we cannot achieve in human affairs."

Prior to the rise of science and liberal democracy, people had few choices other than to put their faith in a regnant belief system (which was girded against dissent by the power of church and state) or reason their way to an idiosyncratic system of their own (which unless done discretely could land them in jail). Science opened up a third option. Scientific research, whether conducted by an eccentric loner or a club of propertied gentlemen, could adduce facts of universal validity—facts objectively verifiable from every point of view. Such phenomena are called *invariant*, meaning that they are the same regardless of the perspective from which they are examined. All scientific laws are statements of invariance, and as such transcend both official dogmas and individual subjectivities. Liberalism's claim that all humans have equal rights is therefore mirrored in the universal validity of scientific facts. "Far from being relative truths, scientific results tend to make everyone's truth property the same across cultures," noted the American philosopher Robert Nozick. "In this sense, science unifies humanity."

Some thinkers accept that science finds facts but imagine that science can say nothing about values. Nozick summarizes their argument in terms of a syllogism: (1) science is objective, (2) values are not objective, so (3) values have no place in science. But is it true that values are nonobjective? The fact that the majority of cultures, while differing in many particulars, take quite similar positions on fundamental moral and ethical values like prohibiting murder and incest, suggests that there may be a universal, evolutionary basis for human ethics and values. As Nozick notes, "Science could contain values, and even make value assertions, and still be objective *if* values themselves are objective."

The rapid changes in knowledge, power, and preference produced by the rise of science and technology have spotlighted the value of creativity. Back when the pace of change was glacially slow, the average person had scant opportunity to imagine, much less live, a life greatly different from that of his parents and grandparents: If you were the child of a peasant or a serf you almost certainly were going to remain in a similar station. Science and liberalism changed this dynamic. They opened up economic opportunities—if you could invent an improved clock or steam engine, or discover the universal laws of gravitation and inertia, you could become rich and famous regardless of whether your father was a carpenter, as was James Watt's, or a farmer, as was Isaac Newton's—and created a dynamic of progress that benefited not just the inventors but the general public. "In an advancing society," noted the mathematician H. B. Phillips, "any restriction on liberty reduces the number of things tried and so reduces the rate of progress. In such a society freedom of action is granted to the individual, not because it gives him greater satisfaction but because if allowed to go his own way he will on the average serve the rest of us better than under any orders we know how to give." This wasn't clear in the old days, when few individuals enjoyed many benefits from creativity beyond hearing a stirring sermon from the pulpit or a new song produced by a bard from a faraway court. Today, the value of individual creativity is much more evident: Starting with the rise of literacy and libraries, and now expanding in a world of mobile phones and the Internet, people can appreciate that their well-being is enhanced by the creativity of others, and that the world's total expertise far exceeds the personal understanding of any one individual. Hence the benefits of everyone's being free to come up with new ideas and inventions become increasingly clear, even if most of their work is often too specialized for the rest of us to comprehend. Liberalism fosters science, which expands the intellectual and material universe, and liberalism can best cope with the changes that it and science have engendered.

Although thinkers long assumed otherwise, freedom is efficient. Perhaps it was the machine age, with its picture of society as resembling a factory run by a boss from a windowed aerie high above it all, that misled so many on this point, but by now it has become clear that the world is far too complex to be run by individuals or by committees of experts. No leader can assimilate enough information to accurately price tomatoes, much less chart the course of scientific or social advance. So the liberal ideal of peoples being free to decide matters for themselves turns out to have practical value—and the more complicated society becomes, the more apparent become the benefits of liberty.

Democracy has always been less popular than liberalism—if liberalism is a gift, democracy is the rattletrap truck that delivers it—and the two have distinctly different origins. Liberalism arose as a matter of principle, while democracy arose more for sociological than for intellectual reasons, and was always more about power than ideals. As Mill observed in 1859, the movements of societies toward democracy "are not the work of philosophers, but of the interests and instincts of large portions of society recently grown into strength." Prominent among them were what the English called the gentry, a class whose wealth and influence came largely from scientific and technological innovations and free-market economies. They were trades and crafts people who gained power by gaining wealth, clawing their way to political prominence on the votes of previously disenfranchised multitudes who saw them as opening up fresh opportunities for all—which may explain why the poor, though impatient with snobs, demonstrate little animus toward the self-made rich. The process was not pretty, but it worked. "It could be said of democracy," writes the historian Roland Stromberg, "that all theory was against it and all experience for it."

Experience with democracy having been in short supply for most of human history, the great mass of humankind was long thought too ignorant, stupid, or distracted to make sound judgments. Intellectuals holding antidemocratic views took their cues from Plato, who regarded the people as a "beast," and were inclined to agree with Sir Thomas Browne that "the multitude" is the "great enemy of reason." Such opinions were thick on the ground for many centuries. "The multitude is always in the wrong," sniffed Wentworth Dillon, the Earl of Roscommon, in 1684. "The public has very crude ideas," opined Stendhal. Locke feared that the people were mired in "passion and superstition," and apt, as Voltaire wrote, to "speak without thinking." Jean-Jacques Rousseau asked rhetorically how "a blind multitude" could "carry out for itself so great and difficult an enterprise as a system of legislation?" The susceptibility of mobs to passionate irrationality, detailed in Charles Mackay's 1841 bestseller, *Extraordinary Popular Delusions and the Madness of Crowds*, was thought to demand the steely leadership celebrated by the poet Alfred Tennyson as "civic manhood firm against the crowd." Education wouldn't be much help if, as Henry David Thoreau assured his readers, "The mass never comes up to the standard of its best member, but on the contrary degrades itself to a level with the lowest." The historian Thomas Carlyle spoke for the majority of thinkers in declaring, "I do not believe in the collective wisdom of individual ignorance."

They had a point. The public *can* be crude, and sullen in the face of change. (Creative artists frequently encounter this problem; as Bob Dylan remarked, "You're nobody if you don't get booed sometimes.") But there were at least three serious flaws in the traditional derogation of the multitude. First, the intellectuals were educated and the masses were not. It is misleading to compare the ignorant to the learned unless the ignorant have a choice in the matter—which is why liberalism stresses the importance of universal education. Second, to object to the liberal precept that all persons are created equal, on the grounds that everybody is not equally intelligent, strong, or creative, is to miss the point. Liberalism does not maintain that everyone is equal *in any particular regard*. What is meant is that everyone has—or ought to have—equal standing as citizens, because the strength of the society resides in the very diversity of their abilities. If a nation required only a few talents, then it might perhaps be justified in conferring a superior standing to those who were the best accountants, carpenters, brewers, or whatever else was needed. But no such nation exists, and even if it did it could not be sure that the talents it valued today would be the most useful tomorrow; in diversity there is strength, both at present and in the unknowable future. Third, the empirical evidence acquired with the spread of democracy has failed to support the dire views of democracy's critics. Democracies have, on the whole, compiled better performance records, by just about every agreed-upon quantitative measure of such matters, than any other system of governance.

But why is this so? Why does democracy work so well?

Social scientists are beginning to glimpse part of an answer, in what is known as the wisdom of crowds.

That story begins with an event in 1906 that may assume a place in the annals of scientific folklore, although only one among the many persons present saw anything remarkable in it. The scene was a weight-guessing competition held at the annual West of England Fat Stock and Poultry Exhibition, in Plymouth. Wandering through the crowd was a snowy-haired polymath named Francis Galton. This not-quite-eminent Victorian had done a number of things, several of which showed sparks of originality although none was likely to earn him a place in scientific history. A prodigious child—he was reading and doing arithmetic by age two and memorizing Shakespeare by age six—Galton never quite settled down. He studied a little medicine and took up mathematics at Trinity College, Cambridge, but abandoned these studies in 1844 when his father died and left him financially independent. A decade of gentlemanly exploration followed, the young Galton traveling in Egypt and filling his notebooks with advice on how to pitch a tent, steer by the stars, fire a rifle from horseback, and sleep safely when being pursued (tie your horse's reins to your wrist, so that if the horse stirs you will be awakened). Thereafter he settled in South Kensington to study heredity, meteorology, psychology, and criminology. (His was the first proper study of fingerprints.) A repressed, eccentric, and rather solitary man, Galton belonged to various scientific organizations but never held any professional post. He admired beautiful women from afar—literally so, using a sextant to calculate their measurements. ("I surveyed them in every way . . . and tabulated the results at my leisure.") His marriage—to Louisa Butler, the daughter of Erasmus Darwin, who was Charles Darwin's grandfather—produced no children.

Galton was fascinated (if also from afar) by breeding. Poring over his in-law Charles Darwin's *Origin of Species,* he concentrated on its many accounts of animal husbandry. Most readers were unexcited by these parts, which contain discursions on "peculiarities in the silkworm," "the many admirable varieties of the strawberry," and "the former and present state of carrier and tumbler pigeons in Britain, India, and Persia," but Galton saw in them a blueprint for social engineering. He feared that democracy might not work, given that "the stupidity and wrong-headedness of many men and women [is] so great as to be scarcely credible." His solution was to isolate what he called "degenerates"—of whom, in his view, there were a great many—to prevent their having children. "I think that stern compulsion ought to be exerted to prevent the free propagation of the stock of those who are seriously afflicted by lunacy, feeble-mindedness, habitual criminality, and pauperism," he wrote, proposing that this be done in "ways yet to be devised that are consistent with a humane and well-informed public opinion." It was his conviction that "a democracy cannot endure unless it be composed of able citizens; therefore it must in self-defense withstand the free introduction of degenerate stock." Galton coined a word for this proposed practice; he called it *eugenics*, from the Greek for "well born." Stillborn as a science and self-evidently illiberal, eugenics rose to notorious practice in the twentieth century, when it was taken up by British socialists, German Nazis, and a strange combination of American progressives and racists. (The U.S. Supreme Court in 1927 upheld a Virginia law *requiring* the sterilization of those declared mentally defective, while southern laws prohibiting interracial marriage were not struck down until 1967, Chief Justice Earl Warren writing for the Supreme Court that "the freedom to marry is one of the 'basic civil rights of man.' ") But if eugenics today is little more than a disreputable memory, for Galton it was a bright prospect.

Such was the background of the eighty-five-year-old amateur scientist who found himself at the West of England Fat Stock fair that fall day in 1906. There he watched as nearly eight hundred people paid sixpence each to guess the weight, slaughtered and dressed, of an ox on display. The winner came pretty close. "The sixpenny fee deterred practical joking," Galton noted, "and the hope of a prize and the joy of competition prompted each competitor to do his best." Intrigued, Galton obtained the entry tickets and subjected them to statistical analysis. Adept at quantitative thinking, he was open-minded enough—though his scientific career seemed to be over and his memoirs had already been published—to make a discovery that ran counter to his beliefs.

Galton found that while no one competitor had come up with the exact weight of the slaughtered and dressed ox, and while many were wildly off, the mean value of all the entries was astonishingly accurate. The ox's weight was 1,198 pounds; the crowd as a whole had estimated it to be 1,197 pounds. "The vox populi"—the voice of the people—was "correct to within one percent," Galton reported. "This result is, I think, more creditable to the trustworthiness of a democratic judgment than might have been expected."

Many other examples of the wisdom of crowds have since been adduced. Ask fifty students to guess the number of jellybeans in a jar, and you'll usually find that the class as a whole not only comes quite close to the right answer but is closer than any one of the students. On the television quiz show *Who Wants to Be a Millionaire?*, the purportedly expert consultants telephoned by contestants got the right answer 65 percent of the time, while the studio audience—"random crowds of people with nothing better to do on a weekday afternoon than sit in a TV studio," as the journalist James Surowiecki writes in his book *The Wisdom of Crowds*—got the right answer 91 percent of the time.

Predictions markets—Web sites where thousands of people invest in predicting election results and other unambiguous future events—routinely outperform the experts. The Iowa Electronic Markets bested 596 opinion polls in predicting presidential elections from 1988 to 2004, 74 percent of the time. The Dublin-based Intrade.com accurately forecast not only the 2004 and 2008 American presidential elections but congressional results in all fifty states. Futures markets for orange juice concentrate predict Florida weather better than the National Weather Service does. Such markets have proved to be so useful that corporations literally capitalize on them. Arcelor Mittal, the world's largest steel maker, uses internal markets to predict the price of steel. Best Buy invites its employees to bid on which video game consoles will sell best. Invested crowds perform as well or better than experts in predicting the gross ticket sales of feature films and in categorizing craters on Mars. Encouraged by such results, corporations put puzzling problems on the Web and offer prizes to anyone who can solve them. When Eli Lilly created a Web site called InnoCentive to "crowd-source" problems that had stumped its vast R&D teams, more than a third of the problems were solved; since spun off, the site has overseen the invention of a method for dealing with oil spills, identification of an obesity biotarget, and a low-cost way of processing spices in developing countries. Everybody knows how hard it is for bettors to beat the odds on professional football games, but that is because the odds are established by the bettors themselves: The Las Vegas bookies establish only the initial line, a task they generally outsource, and thereafter adjust it to keep equal numbers of bettors on either side of the predicted outcome. So the individual bettor is up against not the experts, but the combined predictive power of all the other bettors. As Galton's discovery suggested, the public not only *has* the right to make decisions but very often *is* right.

A great deal remains to be learned about how crowds can demonstrate superior predictive power, but a few precepts are already emerging. First, the outcome must be unambiguous, as when predicting election results or guessing the number of jellybeans in a jar. Second, the group members should function independently of one another, making their own determinations rather than being influenced by others; mob rule doesn't work. Third, it helps if the group members are invested in the process, as happens when people purchase calls on Internet prediction markets; this was the point Galton grasped, noting that the price of tickets at the ox-weighing competition "deterred practical joking." Finally, and quite interestingly, the accuracy of group predictions improves with its diversity: The more socially, ethnically, sexually, and intellectually diverse the composition of the group, the better it performs.

The discovery of the wisdom of crowds tends to validate liberal faith in free democratic processes, providing clues as to why liberal-democratic systems of governance outperform the competition. Liberalism may have inspired stirring rhetoric ("Give me liberty or give me death!"), but in essence it was an experiment—one that many skeptics predicted would fail. Today, with a growing body of evidence suggesting that liberal democracy makes people content, and their nations prosperous and peaceful, it is no longer necessary to rely upon moral and ethical grounds to argue, say, that the United States Senate would do better if it contained more women and minorities. The scientific data suggest that a more diverse Senate would be a more intelligent and aware Senate.

To appreciate how differently things look from this new perspective, consider Francis Galton's eugenics. While Galton was wandering around livestock festivals, a number of younger scientists and social reformers were pushing his eugenics. Their motive was understandable: Darwin had shown that heredity plays a significant role in human behavior, yet reformers out to ameliorate the plight of the poor tended to focus entirely on environmental concerns such as poor housing and inadequate medical care. Eugenicists like Karl Pearson, who took over Galton's Eugenics Records Office and renamed it the Francis Galton Laboratory for National Eugenics, argued that environmental improvements were a waste of money if the problem was due to genetics. Why build public housing and fund public hospitals if the suffering of the poor, indeed the very existence of poverty, could be remedied by isolating, sterilizing, or euthanizing the "defectives" who were, through no fault of their own, genetically responsible for their own suffering? This campaign eventually foundered in the face of liberal distaste at the prospect of the poor and helpless being treated like cattle, but questions have persisted as to whether there might have been something to it.

It is now possible to understand why the answer is no. At the time, one clear objection to eugenics was that nobody could know which genetic characteristics ought to be suppressed. The example often offered was alcoholism: Given a prevalence of alcoholic actors, poets, and painters, who could be sure that breeding out alcoholism might not also reduce a society's artistic creativity? To this hesitation may now be added a stronger argument, drawn from the importance of diversity in the wisdom of crowds. Even if scientists and politicians could agree on which putatively undesirable characteristics ought to be eradicated, the effect would be to reduce the diversity of the whole—and that would reduce the overall intelligence of a necessarily diverse multitude. Selective breeding is not appropriate to human beings because the strength of humanity resides in its diversity rather than in any specific characteristics that a breeder might select for. Liberalism was right about that after all.

The legacies of science and liberalism include our sense of what it is to be modern. The historian of science Gerald Holton defined a modern individual in terms of four criteria: "Being an informed participant citizen," "having a marked sense of personal efficacy (being able to control one's own destiny and events in the world)," "being highly independent and autonomous," and "being open to new ideas and experiences." Despite opposition from many philosophies and almost all ideologies, science and liberalism have made real progress in all these regards, as measured against prior experience rather than an imagined ideal: The Earth is a planet, not a paradise. Humankind is much closer to the beginning than to the end of this endeavor; in many quarters, ignorance and superstitions continue to hold sway over knowledge and freedom. A flame has, however, been struck, the fire is being fed, and the cold and the dark are receding.

CHAPTER THREE

THE RISE OF SCIENCE

Only the educated are free.

—EPICTETUS, FIRST CENTURY AD

Works, not Words;
Things, not Thinking . . .
Operation, not merely Speculation.

—CHEMIST GEORGE THOMPSON, 1660s

Prior to the rise of science there were, as Aristotle noted, only two valid ways to evaluate the merit of an idea—by critiquing its internal logic, or by comparing it to other ideas. Science offered a third option, that of testing ideas through controlled experimentation. A telling experiment might put an end to disputation, by obtaining answers directly from nature.

The conducting of scientific experiments takes time, money, and (since most experiments fail) perseverance. To support it requires of society a degree of affluence plus an appetite for innovation and change. These conditions most often arose where states vigorously competed against one another through trade, which created wealth and promoted technological and financial innovation. Scientific experimentation should therefore be expected to have first appeared in a region that combined diversity and competition, even to the point of strife, with a geography favorable to trade—a region of long coastlines, natural seaports, and proximity to foreign lands from which its people acquired a taste for the exotic and the unfamiliar. The European region that best fit those criteria was Italy.

The Italian Renaissance, generally dated from the 1400s, was kick-started by the fall of Greece and the rest of Byzantium to the Turks in 1453, an upheaval which sent thousands of Byzantine art treasures and books—notably ancient Greek and Latin manuscripts—westward by ship to the markets of the Italian peninsula. Italy was not yet a nation but a set of city-states, renowned for their vitality and their enthusiasm for liberty and independence. In Ferrara it was being said as early as 1177 that "our liberty, which we have inherited from our forefathers, we can under no circumstances relinquish, except with life itself." The watchwords of Florence—whose celebrated poet Dante Alighieri called freedom "God's most precious gift to human nature"—were *libertas* and *libertà*. These stirring slogans meant that citizens (a small minority of the population) had certain legal rights, and that each city-state asserted a right to govern itself rather than to be dominated by the Spanish, the French, or any other distant power. Although no Italian city-state would be called a democracy by modern standards, the circumstances within and among them produced a profusion of social experimentation, with one city-state after another oscillating between republicanism, limited democracy, and reversions to various authoritarian forms of government. Pisa, Milan, Arezzo, Lucca, Bologna, and Siena were all at some point governed by elected officials, while Florence flirted with republicanism and Venice tried various parliaments and councils to sufficient effect that William Wordsworth, ruefully surveying the fall of the Venetian Republic to Napoleon in 1797, called her "the eldest child of liberty." Innovation and trade created new wealth and propelled merchants and artisans into the ranks of the aristocracy. This economic mobility lent a meritocratic cast to even the Kingdom of Naples—of which the prominent seventeenth-century attorney and amateur scientist Francesco D'Andrea declared, "There is no city in the world where merit is more recognized and where a man who has no other asset than his own worth can rise to high office and great wealth . . . without having to depend on either birth or money to get there." Amid such social tumult, the scions of old wealth took care to at least outwardly observe an "equality of aristocrats" that embraced the newly wealthy on equal terms; even the Doge of Venice found it advisable to dress like a merchant and be bowed to only in private. The author and libertine Giacomo Casanova, whose personal preferences ran to luxury, was amused by the egalitarian appearance affected by the Venetian nobles but understood its utility:

> It is impossible to judge of equality, whether physical or moral, except by appearances; from which it follows that the citizen who

> wants to avoid persecution must, if he is not like everyone else or worse, bend his every effort to appearing to be so. If he has much talent, he must hide it; if he is ambitious, he must pretend to scorn honors; if he wants to obtain anything, he must ask for nothing; if his person is handsome, he must neglect it; he must look slovenly and dress badly, his accessories must be of the plainest, he must ridicule everything foreign; he must bow awkwardly, not pride himself on being well-mannered, care little for the fine arts, conceal his good taste if he has it; not have a foreign cook; he must wear an ill-combed wig and be a little dirty.

The city-states' vigorous competition with one another often escalated into war and rebellion. Their hardball politics was exemplified in the works of Niccolò Machiavelli, a career diplomat who negotiated the surrender of Pisa to Florence but wound up being tortured on the rack by the Medici. In his book *The Prince*, which he wrote upon retiring to the country after having lost one political battle too many, Machiavelli advises that "it is far safer to be feared than loved," adding that "being unarmed . . . causes you to be despised" and that "a prince ought to have no other aim or thought . . . than war." Small wonder that Baldassarc Castiglione, in his internationally popular *A Manual for Gentlemen* of 1528—a book praised for its "civilizing" influence—observed that "the principal and true profession of a courtier ought to be in feats of arms."

But Castiglione added that "the principal matter . . . is for a courtier to speak and write well." To minimize crippling losses of blood and treasure, the Italian city-states did two things. First, they trained some of the world's first professional ambassadors, nurturing staffs of political advisors to guide their leaders and advertise their civic virtues—especially "civic humanism," the idea, prominent in liberal democracies today, that citizens should devote time to government service. Second, they promoted their ongoing competition in technology and the arts as a peaceful and profitable alternative to war.

The artists of the Italian Renaissance were often involved in such political competitions. Michelangelo's larger-than-life sculpture *David* was commissioned, in 1501, to celebrate the establishment of a republican government in Florence, which had just freed itself from rule by the anti-Renaissance book burner Fra Girolamo Savonarola. Leonardo da Vinci, the original Renaissance man, secured a position in Milan by writing a long letter to the mechanically inclined Duke Lodovico Sforza that stressed his abilities as an inventor and military engineer, mentioning only in passing that he was also an artist. Leonardo was being candid: The painter of the *Mona Lisa* (which he never finished) did not in fact much like to paint, and was impatient with the fine arts generally. He preferred to work on plans for his many mechanical inventions—a pile driver, an automobile, a helicopter, a parachute, a diving bell, a robot, and various military defense systems. Leonardo's penchant for experimentation extended to his artwork, sometimes with unfortunate results. His fresco *Battle of Anghiari*, done with a high-gloss oil technique of his own invention, reportedly failed to dry and ultimately slithered off the wall, vanishing entirely within fourteen years. His inventions ran so far ahead of existing technology that few could be built, much less tested, while he was alive. As the historian John Herman Randall Jr. remarked, Leonardo's "thought seems always to be moving from the particularity of the painter's experience to the universality of intellect and science, without ever quite getting there."

In short, the Italy that created the Renaissance was a fragmented and disputatious gaggle of city-states that contended using every weapon they could lay their hands on, from daggers and cannons to frescoes and carved marble, their experimentation producing a torrent of political, artistic, and ultimately scientific creations. The Republic of Venice in the fifteenth century built the world's best ships and was known for its fabrics, leather goods, and glass—its glassmakers tellingly if falsely advertising that their elegant goblets would shatter on contact with poison. Florence, hampered by a landlocked geography except when it controlled Pisa, created limited-liability partnerships for investment in the manufacture and trade of leather goods and wine and became an international center of banking and finance. Bologna invented hydraulically powered silk mills and operated more than a hundred of them, producing a million pounds of raw silk annually before the trade secret got out and was adopted by competitors. Milan, a major exporter of silk, velvet, wool, brocade, and military armor, impressed visiting Londoners who noted with astonishment that the Milanese sported over a thousand horse-drawn coaches elegant enough to turn heads back home. The papermakers of Genoa, expanding production to meet the growing appetites of an increasingly literate public, were operating a hundred paper mills by the end of the sixteenth century. By 1680, when Antonio

Stradivari set up his violin-making shop in Cremona, the trade boom was slackening. Portuguese and Spanish navigators had opened up water routes to Asia around Africa, blunting the Italian advantage and putting Venice into decline, while in the Mediterranean the trading companies of Holland and England were offering fast maritime transport at competitive prices. Nevertheless, the skills acquired by Italian craftsmen kept them in the high-end luxury market, where their descendants continue to thrive today.

Many of the great Renaissance artists were men of common origins who resented being condescended to by aristocrats and by the scholars who aped aristocratic manners. Complained Leonardo, the illegitimate son of a liaison between a lawyer and a young lady said to have been a household servant:

> I am fully aware that the fact of my not being a man of letters may cause certain presumptuous persons to think that they may with reason censure me, alleging that I am a man without learning. Foolish folk! Do they not know . . . that my subjects require for their exposition experience rather than the words of others? . . . Though I have no power to quote from authors as they have, I shall rely on a far bigger and more worthy thing—on experience, the instructress of their masters. . . . And if they despise me who am an inventor, how much more should they be blamed who are not inventors but trumpeters and reciters of the works of others.

Michelangelo—a brusque, muscular workman who went for months without bathing or taking off his dogskin boots, and whose conspicuously broken nose resulted from a fistfight—was similarly quick to take offense. When Leonardo made fun of him, joking with friends who were disputing some lines of Dante in front of the Palazzo di Gavina that "Michelangelo will explain it to you," Michelangelo's response was to question not Leonardo's art but his technological ability: "You designed a horse to be cast in bronze and, as you could not cast it, you abandoned it from shame—and those stupid Milanese believed in you." The peerless goldsmith Benvenuto Cellini was a hooligan who boasted in his autobiography of killing four men and of reducing wealthy noblemen to near-groveling as they vied to get him to accept their commissions. Although Cellini's tales smack of exaggeration, they capture the frustration felt by many artists and artisans whose dexterity brought them money and recognition but inadequate social status.

The making and accounting of money helped advance quantitative modes of analysis without which science might otherwise have stalled at a merely descriptive level. Florentine bankers, Milanese merchants, and Venetian traders created important innovations such as double-entry bookkeeping—which their successors learned from Luca Pacioli's authoritative *Summa de Arithmetica,* first published in Venice in 1494—while technological improvements pioneered by the Genoese, Florentine, and Venetian mints gained their coins an international reputation for reliable weight and purity. The minting of money became a subject of scientific interest, with Nicolaus Copernicus publishing a treatise on coinage in Poland in 1526 and Isaac Newton taking control of the London mint from 1699.

Once the printing press had radically reduced the cost of books, scientific and technological innovators found that there was money to be made by writing for the public, in the vernacular rather than Greek or Latin. Their popular books made an end run around the authorities, setting the stage for a dramatic confrontation between tradition and the emerging forces of innovation and creativity. It came in Italy. Its martyr was Galileo.

A born scientist, lifelong experimenter, and compelling writer, Galileo Galilei presented a triple threat to the professors and priests whose careers were invested in the proposition that everything worth knowing could be found in ancient books. (Galileo said that they studied "a world on paper," whereas his experiments revealed "the real world.") His iconoclasm appears to have been inherited from his father, the merchant, musician, and mathematician Vincenzio Galilei, who wrote papers disputing scholarly opinions about music that conflicted with experience. When his former teacher Gioseffo Zarlino asserted that the semitone cannot be divided into two equal parts—because to do so meant invoking an irrational number, which is to say a number that cannot be expressed as the ratio of two integers, a Christian taboo—Vincenzio countered that since any capable musician could hear such tones, a mathematical theory that denied their existence must be flawed. To test the prevailing claim that two strings of equal lengths which sound an octave apart must differ in tension by the ratio 2:1, Vincenzio suspended various weights from lyre strings and found that the ratio actually is 4:1. So the young Galileo had plenty of opportunity to see how experimentation could overturn the opinions of scholars.

Sent to study medicine at the University of Pisa, Galileo soon lost interest and resorted to tinkering on his own. When a concerned Vincenzio came down from Florence and found his son immersed in experiments that had little to do with his schoolwork, he bowed to the inevitable and let the boy come home. A period of free investigation followed, during which Galileo paid particular attention to pendulums—a natural step, given his father's penchant for investigating music by hanging weights from strings. These experimentations eventually led to several important discoveries. One was "isochronism"—that a pendulum takes the same time to complete each swing, whether moving rapidly in a long arc or slowly in shorter arcs. (Isochronism suggested to Galileo that pendulums could be employed to make more accurate clocks, as they were from 1656, when Christiaan Huygens patented one.) Another finding opened the door onto gravitational physics. Galileo suspended pendulum bobs of differing weights from equal lengths of wire, and discovered that the light ones (made of cork) swung at almost the same rate as heavy ones (made of lead). Even when one ball was "a hundred times heavier" than the other, he reported, they swung in step. To test the matter further, he rolled balls down parallel inclined planes and got the same result: Regardless of their weight, they all rolled to the bottom at the same rate.

This seemingly minor observation had epochal implications. The Aristotelian professors taught that heavy objects fall faster than light objects, as everyday experience indicates. Owing to air resistance, coins dropped here on earth fall faster than feathers do. Galileo's experiments minimized the effect of air resistance, allowing him to glimpse the truth—that gravity accelerates all objects at the same rate, regardless of their mass. With that, Galileo the college dropout opened a door for future scientists from Newton to Einstein and beyond.

Galileo's lack of a degree frustrated his initial efforts to secure a teaching position—he was turned down by the universities of Siena, Bologna, and Florence—but he eventually landed a mathematics lectureship at Pisa, and then a higher-paying post at Padua. Padua was part of the free republic of Venice, and Galileo's indifference to authority and vivid rhetorical style made him a favorite among the students there. The advancement of military technology was a priority in the Venetian republic, and Galileo soon supplemented his income by making inventions. He frequented the Venice arsenal, one of the world's most advanced centers for the construction and outfitting of ships, and garnered a commission to study the physics of oars. He invented a compass for aiming cannons, a horse-driven water pump (patented in 1594), and a thermometer, eventually setting up a small business for the manufacture of scientific and military instruments.

In May 1609, having learned that telescopes were being fashioned in the Netherlands, Galileo began making telescopes of his own, turned them on the sky, and embarked on observations that demolished the Aristotelian universe. That model, fashioned in the fourth century BC by Aristotle and the Greek astronomer Eudoxus and refined by Claudius Ptolemy of Alexandria in the second century AD, placed the earth at the center of the universe and depicted the sun, moon, planets, and stars as waferlike discs attached to revolving crystalline spheres. The terrestrial realm had four elements—earth and water, whose gravity made them go downward, and air and fire, whose levity inclined them to rise. Everything above was made of a fifth element, ether, which obeyed a physics all its own. The first four elements could change, but ether could not: Hence the earth was dynamic and the heavens immutable. Flaws in this official cosmology had been observed in the behavior of comets, whose elliptical orbits required that they fly right through the crystalline spheres, and novae—"new" stars, which are actually old stars that explode and become bright enough to be seen for the first time with the unaided eye. When a nova appeared in 1604, Galileo gave three public lectures about it, pointing out that any new apparition in the heavens violated the Aristotelian proscription against change among the heavenly spheres. Such occasional anomalies could be explained away by learned Aristotelians in the Vatican and at the colleges, but Galileo's telescopic observations ended the argument—or should have. The moon displayed rugged mountains, like the earth's, only craggier, and in no way resembled an ethereal wafer. Venus displayed phases like the moon's, indicating that it pursues an orbit around the sun, inside Earth's orbit, just as depicted in the Copernican cosmology.

Galileo's book presenting these observations, *Sidereus Nuncius* ("Messages from the Stars"), created a sensation. With the Italian city-states competing for geniuses in something like the way that its cities today vie for football stars, a newly famous Galileo could take his pick of academic appointments. But he assumed, as scientists sometimes do, that politics was simpler than science, and so made a series of political miscalculations.

His first mistake was to underestimate the value that liberalism afforded him as a citizen of the Venetian Republic. The University of Padua offered him tenure at double his previous salary, but he instead negotiated an appointment with Cosimo II, the Grand Duke of Tuscany, whom he had once tutored in mathematics—in effect bartering away his liberty for the glamour of a court position and the luxury of no longer having to teach undergraduates. "It is impossible to obtain wages from a republic, however splendid and generous it may be, without having duties attached," he mused. "I can hope to enjoy these benefits only from an absolute ruler." His friend Giovanni Sagredo, a career diplomat, warned Galileo about leaving a free republic for "a place where the authority of the friends of the Jesuits counts heavily"—the Jesuits having been banished from the Venetian Republic. "I have seen many cities," Sagredo added, "and truly it seems to me that God has much favored me by letting me be born in [Venice]. . . .

> Here the freedom and the way of life of every class of persons seem to me an admirable thing, perhaps unique in the world. . . . Where will you find the freedom and sovereignty you enjoy in Venice? In the tempestuous sea of a court, who can avoid being . . . upset by the furious winds of envy?

But Galileo was already sailing into political seas about which he remained blithely ignorant. "I deem it my greatest glory to be able to reach princes," he wrote fawningly to Cosimo's secretary of state, Belisario Vinta. "I prefer not to teach others."

In the end it was the Dominicans rather than the Jesuits who persecuted Galileo, but by that time he had managed to provoke them both, and many other clerics besides. Having railed against the professorial "fools" and "blockheads" who refused to look through his telescopes, Galileo now trained his withering scorn on Vatican officials who clung to the old cosmology—an amateur mistake that left them no honorable path of retreat. He put too much faith in personal contacts: Cosimo was his former pupil and the Pope a personal friend, but rulers, like the states they govern, act on their interests rather than their friendships. Part of it was pride. Unlike Einstein—who could joke that "to punish me for my contempt for authority, Fate made me an authority myself"—Galileo hardened into the position that all his opinions, even flawed ones like his mistaken theories of tides, ought to be accepted on the strength of his personal authority alone. Believing in the ultimate rationality and goodwill of the Church, he thought that earnest churchmen, once acquainted with the facts, would cease to mislead their parishioners. When the Vatican banned his books and summoned him to the Inquisition, Galileo wrote to the Pope's nephew Cardinal Barberini that

> the prohibition of the printing and sale of my *Dialogues* has been a cruel blow to me [and] depresses me to such an extent as to make me curse the time I have devoted to these labors—yes, I regret having given to the world so much of my results. I feel even the desire to suppress, to destroy forever, to commit to the flames, what remains in my hands. Thus I should satisfy the burning hate of my enemies.

This bluff the hardened veterans of Vatican politics were prepared to call.

All the world knows the rest of the tale: Threatened with torture, a seventy-year-old Galileo knelt before the Inquisition, recanted such "errors and heresies" as having maintained the "false opinion that the sun is the center of the world and does not move and the earth is not the center of the world and moves," and spent the rest of his life under house arrest—where, nevertheless, he continued to experiment with pendulums. But the reason the world knows the story is, of course, that Galileo's side—the side of free speech and unencumbered investigation—won. His research demonstrated that nature is more intriguing, and its study more rewarding, than had been thought—a lesson that would not be lost on Isaac Newton, born the year Galileo died. By publishing in the vernacular, so that ordinary people could "see that just as nature has given to them . . . eyes with which to see her works, so she has also given them brains capable of penetrating and understanding them," Galileo helped establish the liberal, antiauthoritarian ethos of modern science. "The modern observations deprive all former writers of any authority," he declared, "since if they had seen what we see, they would have judged as we judge." In this he resembled the English physicist and physician William Gilbert, whose work he admired. Gilbert's treatise on magnetism, published in 1600, stressed the efficacy of experimentation:

> In the discovery of secret things and in the investigation of hidden causes, stronger reasons are obtained from sure experiments and demonstrated arguments than from probable conjectures and the opinions of philosophical speculators. . . . Men are deplorably ignorant with respect to natural things, and modern philosophers, as though dreaming in the darkness, must be aroused and taught the

uses of things, the dealing with things; they must be made to quit the sort of learning that comes only from books, and that rests only on vain arguments from probability and upon conjectures.

Following Galileo's persecution by the Church, the scientific epicenter shifted from the Catholic south to the Protestant north. The poet John Milton, who visited Galileo at the villa outside Florence where he was being confined, wrote that the "tyranny" of the Inquisition had "dampened the glory of Italian wits." By the close of the seventeenth century, when important scientific works were being published in the north and Anglican ministers were preaching that science could be reconciled with religion, scientists in Naples were being arrested and tried for maintaining "that there had been men before Adam composed of atoms equal to those of other animals."

Every age since has produced its own martyrs to science. The chemist Joseph Priestley, skeptical about certain religious dogmas, was driven out of Birmingham in 1791 by rioters who sacked his house and destroyed his laboratory. Charles Darwin long suppressed his theory of evolution rather than face the religious indignation that indeed greeted its eventual publication. More than a hundred members of the Soviet Academy of Sciences were imprisoned, and some of them executed, for failing to toe the party line. The American mathematician Chandler Davis, a member of the Institute for Advanced Study at Princeton, spent six months in jail for refusing to divulge names of so-called subversives to the House Committee on Un-American Activities. Scientists are being persecuted today, in China and Iran and elsewhere, but science lives on wherever people are willing to tolerate uncertainty, to admit how little they know, and to learn.

Science has evolved along two consilient paths, the Baconian and the Cartesian, named after Francis Bacon and René Descartes. Bacon stressed induction, an approach that starts with observation and adduces hypotheses from them: "All depends on keeping the eye steadily fixed on the facts of nature, and so receiving their images as they are," he wrote, "for God forbid that we should give out a dream of our own imagination for a pattern of the world." Descartes stressed deduction, reasoning from first principles and rejecting any precept about which he could conjure any doubt. "I showed what the laws of nature were," he claimed, "and without basing my arguments on any principle other than the infinite perfections of God, I tried to demonstrate all those laws about which we could have any doubt, and to show that they are such that, even if God created many worlds, there could not be any in which they failed to be observed."

It is tempting to associate induction and deduction with experimental and theoretical science respectively, although either without the other would be the sound of one hand clapping. Politically, it is said that the legacy of (mostly British) induction fostered liberalism, while (mostly French) deduction promoted socialism. Be that as it may, Bacon is a good place to start.

Bacon is often called the prophet of the scientific revolution, an apt title in that prophets make predictions by methods unknown even to themselves. His biography is a shattered mirror in which can be read everything from the criminality of a wrecked lawyer to the integrity of a poet sufficiently talented to be credited with writing Shakespeare's plays. Bacon moved in aristocratic circles—he was made lord chancellor in 1618, and was titled Baron Verulam—yet saw the future as meritocratic. He suffered from ill health but accomplished staggering amounts of work in law, government, literature, and philosophy. He was repelled by the Aristotelian doctrines taught at university (comparing his Cambridge professors to "becalmed ships; they never move but by the wind of other men's breath") but disliked disputation, saying that he came as a guest and not as an enemy. Prescient about the future rise of science, he cared little about the actual scientific research going on around him: He dismissed Copernicus' cosmology, discounted Galileo's telescopic discoveries, remained indifferent to William Harvey's discovery of how blood circulates in the human body, and ranked Gilbert's experiments with magnets little higher than the alchemists' vain attempts to turn base metals into gold. The Marquis de Condorcet observed:

> Bacon, though he possessed in a most eminent degree the genius of philosophy, did not unite with it the genius of the sciences. The methods proposed by him for the investigation of truth, consisting entirely of precepts he was unable to verify, had little or no effect in accelerating the rate of discovery. . . . Yet the very temerity of his errors was instrumental to the progress of human thought. He gave activity to minds which the circumspection of his rivals could not awaken from their lethargy. He called upon men to throw off the yoke of authority, and to acknowledge no dogma but what reason sanctioned; and his call was obeyed by a multitude of followers, encouraged by the boldness and fascinated by the enthusiasm of their leader.

For all his limitations, Bacon foresaw that science would transform not only knowledge but politics, economics, and society. "I have held up a light in the obscurity of philosophy which will be seen centuries after I am dead," he wrote.

> It will be seen amidst the erection of temples, tombs, palaces, theaters, bridges, making noble roads, cutting canals, granting multitudes of charters and liberties . . . in the foundation of colleges and lectures for learning and the education of youth; foundations and institutions of orders and fraternities for nobility, enterprise, and obedience; but above all, the establishing [of] good laws . . . and as an example to the world.

And so it did.

When Bacon studied philosophy at Trinity College, Cambridge, he found the teachers underpaid, excessively devoted to sterile exercises in logic and rhetoric, ill equipped to conduct scientific experiments and demonstrations, "shut up in the cells of a few authors, chiefly Aristotle their dictator," insufficiently rewarded "for inquiries in new and unlabored parts of learning," and therefore apt to "spin cobwebs of learning, admirable for the fineness of thread and work, but of no substance or profit." "In the universities, all things are found opposite to the advancement of the sciences," he wrote. He came away impressed by the sheer enormity of human ignorance. To assuage it, a process that he compared to diverting fresh spring water into a stagnant swamp, Bacon argued for a new method of learning based on observation, experiment, and inductive reasoning.

Formal scholarship in Bacon's day consisted mainly of theology and philosophy. Both looked backward—theology to the sacred texts of old, philosophy to the Greeks and Romans—and saw the present as a pale shadow of past glories. It was evident that Europe had no philosophers comparable to Plato and Aristotle, while for theologians humankind had abided in sin since Adam and Eve were expelled from the Garden of Eden. Neither camp saw much reason to study nature. Bacon accused his philosophy professors of teaching that the ancients knew things they never actually knew, thus producing students who in turn thought that *they* knew things they did not know. The theologians, especially those rooted in Pauline and Augustinian thought, regarded nature as depraved. In their view the purpose of life, as a modern scholar puts it, was "to escape upwards from this Satan-ridden earth . . . beyond the planetary spheres with their disastrous influences, into the divine empyrean."

Bacon stood this hierarchy on its head. He judged the ancient Greeks to be worthy within their limits, but far too preoccupied with logical formalities—what he called "the ostentation of dispute"—to be of much use in comprehending how nature works. "The sciences we possess have been principally derived from the Greeks," he writes, "but the wisdom of the Greeks was professional and disputatious, and thus most adverse to the investigation of truth." A legendary Egyptian priest had described the Greeks as childlike. Bacon: "They certainly have this in common with children, that they are prone to talking, and incapable of generation, their wisdom being loquacious and unproductive of effects. Hence the external signs derived from the origin and birthplace of our present philosophy are not favorable." Far from being eternally indebted to the Greeks, we are hobbled by our devotion to them: "The reverence for antiquity, and the authority of men who have been esteemed great in philosophy, and general unanimity, have retarded men from advancing in science, and almost enchanted them." Books should follow science, not science books, if "the discovery of things is to be taken from the light of nature, not recovered from the shadows of antiquity."

"Our only hope," Bacon asserted, "is in genuine induction"—the building up, from observations and experiments, through "progressive stages of certainty" to larger conclusions. Induction is quite distinct from the deductive logic of syllogisms:

> The difference between it and the ordinary logic is great, indeed immense. . . . Hitherto the proceeding has been to fly at once from the sense and particulars [that is, sense perceptions] up to the most general propositions, as certain fixed poles for the argument to turn upon, and from these to derive the rest by middle terms: a short way, no doubt, but precipitate, and one which will never lead to nature, though it offers an easy and ready way to disputation.

If instead you work patiently upward from observations, the most general propositions "are not reached till the last, but then when you do come to them, you find them to be not empty notions but well defined, and such as nature would really recognize as her first principles, and such as lie at the heart and marrow of things." Bacon's induction might look grubbily workmanlike when compared to the "pure reasoning" of the scholastics, but it promised to bring human ideas about the material world into closer concordance with that world, to aid in "cultivating a just and legitimate familiarity between the mind and things."

Those of us who live in today's liberal-scientific societies may find it difficult to appreciate just how odd Bacon's arguments must have seemed to his contemporaries. The material world was with them all right—they heard it and smelled it and scraped it off their boots at the door upon arriving home—but was hardly a fit subject for study. The material world was where your workday lasted twelve hours in winter and fourteen hours in the summer, where half of your children died before reaching age thirty, where a grinning highwayman would murder you for your clothes on a road adorned with the rotting corpses of his hanged colleagues, and where the plague could kill a fifth of all Londoners in a single year. A Venetian visitor reported that England's greatest city was covered by "a sort of soft and stinking mud which abounds here at all seasons, so that the place more deserves to be called *Lorda* [filthy] than *Londra* [London]." Drunken lorry drivers fought openly in streets with such frankly descriptive names as Stinking Lane, one Thomas Dekker observing in 1606 that

> in every street, carts and coaches make such a thundering as if the world ran upon wheels: at every corner, men, women, and children meet in such shoals, that posts are set up of purpose to strengthen the houses, lest with jostling one another they should shoulder them down.

Criminals were so abundant that new words were coined to describe their enterprises: Priggers stole horses, nips sliced open purses on the streets, curbers hooked clothes they found hanging out to dry. Public executions, conducted by axe-wielding killers wearing white butcher's aprons to show off the blood, were reliably popular amusements. The common people were so xenophobic that foreigners risked being spit on or beaten if they dared venture forth wearing their native attire; Jews were banished from England altogether for 365 years before being admitted by Cromwell in 1655. The majority of English subjects were illiterate and poor and so had no voice in history, but when the journalist Henry Mayhew began interviewing them in the 1800s he recorded testimonies like this one, from a middle-aged woman:

> I don't think much of my way of life. You folks as has honor, and character, and feelings, and such, can't understand how all that's been beaten out of people like me. I don't feel. I'm used to it. . . . I don't want to live, and yet I don't care enough about dying to make away with myself. I aren't got that amount of feeling that some has, and that's where it is.

One did not willingly embrace this world. If possible one escaped from it, into the university and the Church. (The two were closely linked, since, as Joseph Priestley noted, "None but the clergy were thought to have any occasion for learning.") It was the genius of Francis Bacon to foresee that this would not always be the case.

Bacon's vision of the future, although necessarily imprecise, arose from a realization that scientific knowledge—combined with better tools and techniques that he called "instruments and helps"—could elevate the human condition to unimaginable heights. "Neither the naked hand nor the understanding left to itself can effect much," he wrote. "It is by instruments and helps that the work is done, which are as much wanted for the understanding as for the hand. And as the instruments of the hand either give motion or guide it, so the instruments of the mind supply either suggestions for the understanding or cautions." In Bacon's day it was assumed that anyone out to control nature must subdue it, much as a militia might subdue a riot, by pitting power against power. Bacon saw that the key was not force but comprehension ("Nature to be commanded must be obeyed"), and that knowledge would bring dominion over nature: "Knowledge itself is power."

Bacon's unfinished novel *New Atlantis* hinted at a forthcoming age of ingenuity. The Atlantans boast of having refrigerators, "new artificial metals," desalinization plants, air purifiers, agriculturally engineered fruit "greater and sweeter, and of differing taste" than those found in nature, experimental zoological parks where they "make a number of kinds of serpents, worms, flies, [and] fishes," beverages that last unspoiled for decades, sophisticated pharmaceuticals, "perspective houses, where we make demonstrations of all lights and radiations and of all colors," advanced telescopes and microscopes, synthetic magnets and crystals, "sound houses" where "the voices and notes of beasts and birds" are mechanically produced, "trunks and pipes" that act as telephones, "engine houses, where . . . we imitate and practice to make swifter motions than any of you have," flying machines and submarines, and a kind of cinema "where we represent all manner of feats of juggling, false apparitions, impostures and illusions."

Bacon was fond of material things ("He wrote like a philosopher and lived like a prince"), a predilection that may have contributed to his dismissal as lord chancellor on charges of accepting bribes. Such reversals did little to blunt his influence in the centuries that followed. Black magic, based on the erroneous assumption that nature could be commanded by casting spells without understanding how or why they worked, devolved from a fearsome instrument of satanic power to a set of magician's props. The Royal Society was founded "to provide instruction to sailors and merchants in useful arts, especially practical mathematical techniques." Machinists, metallurgists, navigators, and chemists inched upward in social status, and the old world of lifelong class distinctions began to slip its moorings.

Bacon was the prophet of the liberal-scientific world, not its practitioner. He condemned superstition but credited reports of miracles, thinking it self-evident that "all knowledge is to be limited by religion." He envisioned future societies dedicated to peace and progress, yet displayed little enthusiasm for political reform. Voltaire said that Bacon "did not yet understand nature, but he knew and pointed out the roads leading to it," building "the scaffold with which the new philosophy was raised; and when the edifice was built, part of it at least, the scaffold was no longer of service." The poet Abraham Cowley wrote that

> Bacon, like Moses, led us forth at last
> The barren Wilderness he past,
> Did on the very Border stand
> Of the blest promis'd Land,
> And from the Mountains Top of his Exalted Wit,
> Saw it himself, and shewed us it.

Bacon said as much himself: "I have only taken upon me to ring a bell to call other wits together."

Meanwhile René Descartes, a generation younger than Bacon, was working the other side of the street, attempting to rebuild philosophy through pure deduction. Born near Tours into a reputable family—his father was a member of Parliament—Descartes was a hothouse flower who suffered from debilitating respiratory ailments. As a boy he was permitted to linger in bed for as long as he liked in the mornings, a habit that his fellow mathematician Blaise Pascal thought ideal for mathematical reasoning. He had an aversion to cold, overheating his living quarters so much that he called them "stoves." (Many modern readers are confused by this reference, but the residents of dank seventeenth-century homes often slept close to their stoves or even on top of them.) Educated in the Jesuit college of La Fleche, an experience that he always recalled with respect and delight, Descartes enlisted in the Dutch army of Maurice of Nassau—an odd choice for a sickly young man, but one that befit his family's tradition of military valor and afforded him ample time to think. While dreaming one night in his "stove," Descartes had a vision of a unified system of knowledge based solely on reason. Recognizing that such ideas would be viewed with hostility by the clergy, he moved in 1628 to the Dutch Republic, Europe's freest and most tolerant state, where he remained for twenty years. Even there he found it imprudent to publish his first book, *Le Monde*, which advocated the Copernican cosmology. (It was completed in 1633, just as Galileo was being punished by the Inquisition for taking the same position.) In September 1649, under attack for his alleged blasphemies, Descartes boarded a Swedish warship and departed for Stockholm.

He was responding to an invitation to tutor Queen Christina of Sweden, a young scholar of insatiable intellectual appetites. (Still in her twenties, she had already written two treatises on ethics and established Sweden's first newspaper. In later years she would abdicate the throne and move to Rome, where she built an astronomical observatory and underwrote the construction of Rome's first opera house.) Kristina took her learning seriously, requiring that the two meet at 5 a.m. for classes lasting four to five hours. The deepening Swedish winter afflicted Descartes, who by January was complaining of being "out of his element" in cold so severe that "men's thoughts are frozen here, like the water." He contracted pneumonia, and died on February 11, 1650, aged fifty-four.

As John Stuart Mill remarked, Descartes had a "purely mathematical type of mind." This was the strength of his many contributions—such as the Cartesian coordinate system, which scientists have used ever since—but also their liability, in that Descartes thought of science as completely deducible *from* mathematics. "The only principles which I accept, or require, in physics are those of geometry and pure mathematics," he wrote. "Those principles explain all natural phenomena."

Well, yes and no. Science has since demonstrated that all phenomena can be described mathematically, but while some mathematics is scientifically applicable to the observed world, much of it is not. We do not, for instance, live in a world of ten, rather than three, spatial dimensions—unless, as string theory posits, the higher dimensions exist but are for some reason imperceptible. The trick is to find which mathematical tools apply to which aspects of the observable universe, a matter more often determined by observation and experiment than by deductive reasoning. Some theories survive experimental testing, but all are colored by a tincture of doubt. Even when a theory makes exquisitely accurate predictions (as does, say, quantum electrodynamics), a better one may come along that takes in a wider range of phenomena—which is why scientists today are trying to find a unified account of electrodynamics and gravitation. Because Descartes did not see this, his conception of science was in some ways sterile; it could explain things but predict next to nothing.

A man of considerable charm, Descartes in his *Discourse on Method* makes the egalitarian claim that "good sense is of all things in the world the most equally distributed," adding, "I have never ventured to presume that my mind was in any way more perfect than that of the ordinary man." He modestly recounts "the increasing discovery of my own ignorance," going on to note, with robust skepticism, that "there is nothing imaginable so strange or so little credible that it has not been maintained by one philosopher or other." All quite modern, that, but the book remains almost entirely confined within Descartes' ample but solitary skull. Exhibiting little of the collegiality that would come to characterize the scientific community, it resembles the lone lamp of a solitary ship, searching dark oceans for an unassailable truth. In this sense Descartes belongs among the ancients, confined by what Bernard Williams of Oxford calls his "preoccupation with the indubitable."

Beneath the beacons of brilliant philosophers like Bacon and Descartes toiled budding scientists like the decidedly unphilosophical Irish chemist Robert Boyle. Born into a titled family (his tombstone bears the epitaph, "Father of Chemistry and Uncle of the Earl of Cork"), Boyle put Bacon's dicta into action although he acknowledged neither Bacon nor anyone else as a teacher, preferring to stress that his learning was adduced directly from nature. Boyle was well rounded (he fenced and played tennis at Eton), and well traveled (he visited Switzerland, Holland, and France, and was in Florence when Galileo died there), but was happiest in his laboratory. Working at Oxford with Robert Hooke (who did pioneering work in microscopy and coined the biological term *cell*), Boyle invented an air pump and demonstrated that charcoal and sulfur would burn only in the presence of air. (Nor, he learned, can mice or kittens survive in an airless environment, although he was too tenderhearted to persist with such experiments.) This was an important step toward the discovery by Joseph Priestley in 1774 of the component of air that supports combustion, which Antoine Lavoisier named oxygen. Boyle also established what has been known ever since as Boyle's law—that at a constant temperature, the volume of a given mass of air or any other gas varies inversely with its pressure—and he measured the density of air, using an instrument he named the barometer. Ceaselessly inventive, Boyle built a portable camera obscura that focused images on a piece of translucent paper, developed a hydrometer to measure the densities of liquids, and fashioned the first match, from heavy paper and phosphorus.

Boyle gave short shrift to any notion that did not emerge from experimental evidence or at least answer to it. Descartes had denied the possibility of a vacuum, preferring to view space as pervaded by an ether. Boyle found that the sound of a rung bell would not propagate through an evacuated chamber, indicating that there was nothing left in the chamber to carry the sound, so he dismissed ether as superfluous. Like Descartes, he saw the world as composed of matter and motion; unlike Descartes, he never let such conceptions take him very far from the laboratory. His contemporaries regarded him as perhaps the first true scientist among them, so much so that the empirical, mechanistic, and decidedly nonmystical world view that ascended with the Enlightenment was often referred to, in England at least, as the work "of Newton and Boyle."

Newton's *Principia* drew the disparate threads of Renaissance science together into a quantitative theory of gravitation that made accurate predictions of the motions of planets and comets in the sky. Copernicus had suggested that the earth orbits the sun; Newton calculated that the sun's gravitational force holds the earth in its orbit by counteracting the inertia that would otherwise send us careening off into space. Kepler's study of

Tycho Brahe's astronomical observations had revealed that the orbits of the planets are ellipses rather than circles; Newton demonstrated mathematically why this must be so. The planets have gravity, too; hence their orbits cannot be perfect circles, which have but a single "focus" at their centers, but must be ellipses, the two foci representing the fact that two gravitating objects are involved. Galileo had inferred from pendulums and rolling-ball experiments that were it not for air resistance, all bodies would fall at the same rate; Newton proved that the planets' orbital velocities are determined by their distance from the sun, not by their mass: A pebble in the orbit of Saturn moves at the same velocity as does mighty Saturn itself. Bacon and Boyle had championed experimentation; Newton experimented constantly with chemistry, metallurgy, and optics, inventing and building a new sort of telescope that remains the world's most popular today. Descartes (whom the young Newton studied closely) had stressed mathematical analysis; Newton invented the calculus in order to do the analysis required to complete his *Principia*, the full title of which was *Mathematical Principles of Natural Philosophy*. That one book buried the Ptolemaic cosmology, demonstrated that the solar system works according to mathematically precise rules, showed that these same rules apply on earth as in the heavens, and provided the world with a handbook useful in calculating a million practical matters, from the strength of bridge girders to the amount of rocket fuel required to dispatch astronauts to the moon.

Newton also set an example by acknowledging what he did *not* know. He volunteered that he did not know, for instance, what gravity was, or how it could propagate through empty space. "I have explained the phenomena of the heavens and of [the tides] by the force of gravity," he wrote, "but I have not yet assigned a cause to gravity . . . [as] I have not yet been able to deduce from phenomena the reason for [its] properties." (That would remain for Einstein to solve.) Echoing Boyle's sentiment that it is better to establish part of the truth than to pretend to know it all, Newton wrote that "although the whole of philosophy is not immediately evident, still it is better to add something to our knowledge day by day than to fill up men's minds in advance with the preconceptions of hypotheses." By "hypotheses" Newton meant different things at different times (one scholar has counted nine Newtonian definitions of the word), but his point here was to distinguish the scientific method from that of philosophers like Descartes, who thought he could build an entire universe from a single, allegedly indubitable, hypothesis. In that sense, Newton's famous *hypotheses non fingo*—"I do not fabricate hypotheses"—is very nearly the opposite of Descartes' *cogito ergo sum.*

Stylistically severe, crammed with mathematics, and written in Latin rather than the vernacular, the *Principia* is a forbidding work. How is it, asks the historian of science I. Bernard Cohen, "that a largely unintelligible book—its pages closed to all but the most skilled and dedicated mathematicians—would dominate the intellectual character of the Enlightenment and become the most generally influential work of its age"? One answer is that it gave traction to the scientific enterprise, by combining many experiments and observations into a unified mathematical account of gravitation, which along with light constituted the only two forces known at that time. (Discovery of the strong and weak nuclear forces was the work of twentieth-century physicists.) Newton dropped the other shoe with his *Optics*, published in 1704, a work brimming with experiments so edifying that students still perform them today. These two books reinforced both avenues of scientific research—one primarily mathematical, like the *Principia*, and the other, arising from the *Optics*, stressing experiment. The historian Hunter Crowther-Heyck describes Newton "as a conqueror rather than a rebel; if Galileo is the Scientific Revolution's Tom Paine, Newton is Jefferson and Washington rolled into one." The result was the Enlightenment, and the democratic revolution.

CHAPTER FOUR

THE SCIENCE OF ENLIGHTENMENT

The season of fiction is now over.

—JEREMY BENTHAM, 1776

Fact is superior to reasoning.

—TOM PAINE, 1791

Historians agree that the democratic revolution grew out of the Enlightenment, but if you ask what caused the Enlightenment, you are likely to be presented with a list of philosophers—Bacon, Locke, Hume, Berkeley, Descartes, Voltaire, and Diderot among them—who for some reason suddenly began singing in unison on behalf of reason and rights. Unless this was a coincidence, something must have happened—something of lasting consequence, capable of shifting the balance away from traditional beliefs and toward the enlightenment values of inquiry, invention, and improvement.

Science fits the bill. Science discovered new facts that were interesting in themselves but also rewarding—literally so, since they proved effective in curing disease, reducing toil, and making money. Science worked in new ways, rewarding merit rather than social station, challenging aristocratic traditions with a force mustered by no prior system of thought, and underscoring the liberal claim that humans have natural rights. Other influences were at work as well, but the Enlightenment without science would have been a steamship without steam.

The Enlightenment is customarily dated from England's Glorious Revolution of 1688, but a more appropriate date would be one year earlier, with the publication in 1687 of Isaac Newton's *Principia*, which was taken up by thinkers across Europe almost overnight. "Newton rose at once to the highest pinnacle of glory," wrote Thomas Thomson in his history of the Royal Society, and "has stood ever since in the front of the philosophic world." David Hume declared Newton to be "the greatest and rarest genius that ever rose for the ornament and instruction of the species." Alexander Pope celebrated him in a poem: "God said 'Let Newton be,' and all was light." The French philosopher Jean d'Alembert observed in the mid–1700s that Newton's account of gravitation was "so generally accepted, that people were beginning to dispute their author the honor of having discovered it." Newtonian precepts were soon being tried out, with mixed results, on everything from medicine and ethics to government. "We are all his disciples now," said Voltaire in 1776.

Newton was obviously a great scientist, but he was not only a scientist. He was also a dedicated radical Whig—what today would be called a classical liberal—who stood up for liberty in the face of direct threats from the highest authorities. Reclusive, paranoid, and peculiar even by the standards of his fellow dons, Newton was nobody's idea of a professional politician, yet he enjoyed enough political success to be described today as "one of the architects of our civil liberties." His were politically tumultuous times; to appreciate just how tumultuous calls for a bit of background.

When the *Principia* was published, England had been suffering through a bloody half century that saw seven years of civil war; the beheading of a reigning monarch, Charles I, in 1649; a turbulent interregnum during which the monarchy was restored; and the Glorious Revolution itself, in which King James II was deposed. Pressing for further change were the gentry and other self-made traders, merchants, and artisans, who by 1628 comprised the majority of the House of Commons and controlled triple the wealth of their social superiors in the House of Lords. The ship of state seemed near to capsizing.

An adept monarch might have been able to keep it aright for a while, and the Tudor queen Elizabeth I was fondly remembered for having managed to do just that, from her coronation in 1558 until her peaceful death on March 24, 1603, shortly before her seventieth birthday. But the crown thereafter devolved to the Stuart regents James I (who ruled from 1603 until

1625) and his son Charles I (crowned in 1625 and beheaded in 1649). Both were the sort of leaders who, unable to master the intricacies of the job, take refuge in an unblinking adherence to unwavering precepts of religious and secular faith. (The historian G. M. Trevelyan says of James that "he was perpetually unbuttoning the stores of his royal wisdom for the benefit of his subjects, and as there was none who could venture to answer him to his face, he supposed them all out-argued.") When such a ruler gets into trouble he is less apt to question his beliefs than to repeat them with heightened vehemence, as if bluster were a magical spell capable of revitalizing his ebbing power. The worse things get, the less he listens and the more he talks.

James and Charles shared an arrogant disdain for Parliament as well as a belief in the divine right of kings, James informing Parliament during a two-hour harangue in March 1610 that "kings exercise a manner of resemblance of divine power on earth." These traits angered many British subjects, but when Charles finally heard the waterfall's roar of imminent disaster he steered straight for it, assembling an army and declaring war on Parliament. Four years of conflict ended in his defeat, trial, and condemnation, but he remained resolute. Standing on the scaffold on January 30, 1649, Charles said of his subjects, "I must tell you that their liberty and freedom consist . . . not for having share in government, sirs; that is nothing pertaining to them; a subject and a sovereign are clear different things."

Unpopular though Charles may have been, his execution aroused revulsion among those who associated monarchy with the rule of law (an Aristotelian precept still taught at all the universities), and a decade later his eldest son was invited back from exile in France to be crowned King Charles II, in 1661. But hopes for the Restoration were disappointed, Charles II signing a secret treaty with the king of France promising to publicly declare his Catholic faith in return for a bribe and such French troops as might be required to put down any resultant rebellion, then denying to Parliament that any such treaty existed. His brother James II, crowned in 1685, pursued a similar course, further alarming his Protestant subjects by ordering Catholics appointed to high posts at the universities of Oxford and Cambridge. He seems to have underestimated the universities' prestige; as Thomas Babington Macaulay writes in his *History of England*:

> Literature and science were, in the academical system of England, surrounded with pomp, armed with magistracy, and closely allied

> with all the most august institutions of the state. To be the chancellor of an university was a distinction eagerly sought by the magnates of the realm. To represent an university in Parliament was a favorite object of the ambition of statesmen. Nobles and even princes were proud to receive from an university the privilege of wearing the doctoral scarlet. . . . Any attack on the honor or interests of either Cambridge or Oxford was certain to excite the resentment of a powerful, active, and intelligent class scattered over every county from Northumberland to Cornwall.

James II ordered Cambridge to name as Master of Arts a Benedictine monk who refused to take the oath of obedience required by parliamentary law. University officials, objecting that they could not legally obey the edict, dispatched eight distinguished professors to plead their case before an ecclesiastical commission at Westminster in April 1687. Newton was among them. "He was the steady friend of civil liberty and of the Protestant religion," Macaulay writes of Newton, "although his habits by no means fitted him for the conflicts of active life. He therefore stood modestly silent among the delegates, and left to men more versed in practical business the task of pleading the cause of his beloved University." It hardly mattered. The chief inquisitor was Sir George Jeffreys—a manically abusive judge notorious for unleashing torrents of invective on the trembling miscreants brought before his bench—and he dismissed the Cambridge delegation without a hearing. "Go your way and sin no more," he thundered, "lest a worse thing happen to you." Oxford fared little better. When Jeffreys was unable to expel the elected president of Magdalen College and replace him with a bishop of the king's choosing, James himself turned up with a full glittering retinue (having paused along the way at Portsmouth to soothe scrofula victims with "the king's touch") to personally intimidate the dons. "Go home," he roared, tossing aside a faculty petition without reading a word of it. "Get you gone. I am King. I will be obeyed. . . . Let those who refuse look to it. They shall feel the whole weight of my hand. They shall know what it is to incur the displeasure of their Sovereign."

The king prevailed, dispatching troops with loaded carbines to install the bishop and begin the process of transforming Magdalen into a Catholic seminary, but his affronts became the fodder of revolutionary chatter in coffeehouses and newspapers across the land. The birth of a male heir to the previously childless monarch in 1688 aroused fears that yet another generation of Stuart rulers would continue turning back the clock. Within a year James had been deposed, driven into exile by an army that sailed over from Holland under the command of William of Orange at the invitation of seven English peers. (Related by marriage to the English royal family, William had made himself popular with the British by his skillful conduct of Dutch defenses against the French.) No sooner were William III and Queen Mary II enthroned than Parliament, with Newton its representative from Cambridge, enacted the Bill of Rights of 1689. It forbade British monarchs from independently raising money or creating armies in times of peace, infringing on freedom of speech, interfering with due process of law, or suspending acts of Parliament. England had become a constitutional republic, albeit one without a written constitution. Viewed in terms of a dialectic between Whigs (believers in science and material progress, apt to have supported Parliament in the civil war) and Tories (mostly monarchists and royalists, more concerned with stability than with social or scientific innovations), the Whigs had prevailed.

It was in these circumstances that Newton met John Locke, probably in the spring of 1689, the two forming a friendship that thrived until Locke's death fifteen years later. Both were radical Whigs and devotees of science, but they were personally quite distinct. Locke had many friends, the prickly Newton few. Locke was voluble, Newton more apt to reserve his thoughts for the printed page. Newton was the most famous Englishman alive; Locke's celebrity was limited to intellectual circles until publication of three long-suppressed works—his *Essay Concerning Human Understanding*, *Treatises of Government*, and *Letters on Toleration*, arguing that governments are obliged by a "social contract" to protect "natural rights," including right to "life, liberty, and the ownership of property"—made him a luminary. Locke was eager to know not only Newton the preeminent scientist, whose *Principia* had recently dawned like a second sun, but Newton the political activist, who in addition to joining the Cambridge delegation mustered against the intrusions of James II had recently denounced the king's mandates as violations of common law. In that document Newton urged his academic colleagues to "be courageous therefore and steady to the laws," adding that "an honest courage in these matters will secure all, having law on our sides" and reporting to his Cambridge colleagues that "fidelity & allegiance sworn to the King, is only such a fidelity & obedience as is due to him by the law of the land. For were that faith and allegiance more than what the law requires, we should swear ourselves slaves & the King absolute; whereas by the law we are free men."

Locke, like Newton, was a man of modest origins whose early career had been deceptively quiet. Born in a Somerset village on August 29, 1632, to middle-class parents of evident moral standing (his father was sensitive enough to apologize to the adult Locke for once having struck him when he was a child, and Locke described his mother, reputedly a great beauty, as unfailingly loving), Locke was admitted to Christ Church, Oxford, as a scholarship student in 1652. He got off to an unpromising start, complaining that he was not cut out to be a scholar—due most likely to the fact that, as the encyclopedist Jean LeClerc noted, "the only philosophy then known at Oxford was the peripatetic [that is, Aristotelian], perplexed with obscure terms and useless questions." Locke was rescued by his discovery of science, which arrived at Oxford in the person of John Wilkins, a burly, unbookish man who formed a club for the purpose of conducting experiments. Under Wilkins' tutelage, Locke finally found his footing. He learned astronomy from a royalist, Seth Ward, and geometry from a Whig, John Wallis, who, moonlighting as a cryptographer, decoded intercepted Royalist messages for the Parliamentarians in the English Civil War. Eventually settling on medicine, Locke worked with Dr. Thomas Sydenham, called "the English Hippocrates," a founder of epidemiology and enemy of dogma who taught that physicians should trust only "the faithful product of observation," supported by "the only true teacher, experience." Locke pursued science all the rest of his days. He kept meteorological records, studied the atmospheric effects of the great London fire of 1666 (popularly regarded as the satanic apocalypse predicted in the Book of Revelations), observed Jupiter and Saturn from the Paris Observatory, and counted among his friends the Danish astronomer Olaus Rømer, the Irish astronomer William Molyneux (whose pamphlet demanding equal rights for Irish citizens under English rule was burned on order of Parliament), and Robert Boyle, Britain's greatest chemist.

Locke's scientific and medical studies pleased him sufficiently that he remained at Oxford for seventeen years. He never obtained a medical degree, but at one point supervised an operation to remove a suppurating liver cyst that was troubling Lord Ashley, later the first Earl of Shaftesbury, a radical Whig depicted by the royalist poet John Dryden as a Satan whose goal it was

> That Kingly power, thus ebbing out, might be
> Drawn to the dregs of a Democracy.

Against all odds the procedure actually cured Shaftesbury, who became Locke's patron. Thus supported, Locke wrote a number of liberal political essays that he prudently kept to himself. One of them, the *Essay Concerning Toleration*, is a classic statement of the case for religious freedom, and a masterpiece of clearheaded reasoning and soaring rhetoric: "Now, I appeal to the consciences of those that persecute, torment, destroy, and kill other men upon pretence of religion, whether they do it out of friendship and kindness towards them or no?" When Shaftesbury was named lord protector of Carolina, Locke wrote and published the "Fundamental Constitution of Carolina" of 1670, providing that "no person whatsoever shall disturb, molest, or persecute another for his speculative opinions in religion or his way of worship." Its liberalism made Carolina a magnet for refugees fleeing religious persecution.

As one of Shaftesbury's pilot fish, Locke rose high when the great man became the most powerful figure in the court of Charles II, and descended to the depths when, by 1682, Shaftesbury's growing opposition to the king's high-handed fiats put his life in danger. Accused of plotting against the crown along with Lord William Russell, Algernon Sidney, and the Lords Essex and Salisbury—all of whom were soon imprisoned in the Tower of London, and soon thereafter dead—Shaftesbury fled to the Netherlands, where he too promptly died, just two years later. The relatively inconspicuous Locke might have survived in England, but he had by this time drafted his *Two Treatises of Government*, a seditious manuscript that endorsed human rights as natural. Fearing that he was being spied upon (rightly so, the Crown having dispatched lip-readers to transcribe his private conversations in the university dining hall), Locke in 1683 followed his patron's example and decamped for Holland.

The Dutch were the architects and beneficiaries of what would be known as their Golden Age (1609–1713), an era during much of which Holland was pretty nearly the freest—and per capita the wealthiest and most creative—nation in the world. The Dutch dominated world maritime trade, owing in large measure to the fact that their ships, built to suit the practical needs of merchants rather than the vanity of regents, were broad-beamed, commodious, and easily maintained. (The other side of the coin was that Dutch warships tended to be converted merchant vessels, a liability that contributed to their defeat by British ships of the line in the three Anglo-Dutch wars of 1652–1674.) Their maritime empire stretched from Japan, where they enjoyed a trade monopoly that would last until 1853, to New Amsterdam, founded in America in 1625 as a private corporation and granted self-government in 1652, thereafter to become the West's most commercial city. Their robust, republican, and relatively egalitarian society stood up to the mighty military forces of two retrograde regents, Philip II of Spain and Louis XIV of France. They counted among their philosophers immigrant freedom-seekers like Descartes; the liberal skeptic Pierre Bayle, whose *Historical and Critical Dictionary* of 1697 would find its way onto Thomas Jefferson's list of the one hundred books vital to establishing the American Library of Congress; and now John Locke.

The greatest of Holland's indigenous thinkers was Benedict de Spinoza, who like many philosophers professed the virtues of self-sufficiency and tranquil contemplation, but unlike most of them actually lived that way. A lens-grinder by trade, Spinoza contributed to the development of the microscope (and died of glass dust in his lungs), lived modestly (pipe-smoking was his only vice), worked assiduously (a friend said he once went three months without leaving the house), and shunned all honors, declining a professorship at the University of Heidelberg in 1673 "not in the hope of some better fortune, but from love of tranquility." Recognizing science as "the only certain and reliable criterion of truth we possess," Spinoza equated God with the universe and thus regarded scientific investigation of nature as doing God's work—an outlook which at the time seemed so radical that Spinoza was excommunicated by his fellow Jews and narrowly survived an assassination attempt. (Seeing the gleam of the knife thrust at him by a stranger on a dark street, he turned aside and evaded it, keeping his slashed overcoat for the rest of his life as a kind of memento mori.) He was a liberal ("The purpose of the state is freedom") and a democrat. "Democracy," he wrote, "is of all forms of government the most natural and most consonant with individual liberty." He regarded miracle-mongering theocrats as akin to illiberal politicians, since both enhanced their power by encouraging the irrational fears of the multitude:

> The supreme mystery of despotism, its prop and stay, is to keep men in a state of deception, and with the specious title of religion to

cloak the fear by which they must be held in check, so that they will fight for their servitude as if for salvation.

For such efforts—and for denying human immortality and the existence of a personal god who wields supernatural powers and takes an interest in human affairs—Spinoza was called an atheist, notwithstanding his having composed a fourteen-part proof of the existence of God.

The Dutch innovated oil painting, mixing earthen pigments with cold-pressed linseed oil in technologically advanced shops like the Old Holland Oil Color Factory, established in 1644 and still in business today. They created new styles of painting as well, influenced in part by the use of optical devices such as the pinhole camera, the refracting or "Galilean" telescope, and lenses which could project an inverted image onto a blank canvas. One such projector is thought to have been developed by the naturalist and microscopist Antoni van Leeuwenhoek and employed by Johannes Vermeer, both of whom worked in Delft. Art historians today debate whether Vermeer and other painters were in some sense cheating if indeed they used optical projection to sketch the outlines of their paintings, but it is perhaps more to the point to observe that the Dutch were quick to apply the results of scientific experimentation to the fine arts. Dutch paintings celebrated freedom, industriousness, and prosperity—as in Vermeer's *The Lacemaker*, Jacob van Ruisdael's *View of Haarlem with Bleaching Grounds*, and Rembrandt's *The Sampling Officials of the Clothmakers' Guild, Amsterdam*. In a break from the spiritual themes of the past, painters memorialized everyday scenes of housework, shopping, farming (the coming agricultural revolution in England would spring largely from experimental innovations developed by the Dutch), and technology (especially maritime technology; Dutch painters depicted sailing ships so reliably that maritime archeologists today use their paintings to re-create the lost rigging of recovered vessels). Science was celebrated in canvases such as Rembrandt's *The Anatomy Lecture of Dr. Nicolaes Tulp* and Vermeer's *Astronomer*, who studies a celestial globe while bathed in light streaming through a window so radiantly as to suggest an intellectual awakening. The beauty of these canvases indicated that scientific inquiry, technological advancement, and free-market commerce could be carried out in harmony with nature, an outlook later jeered at by romantics who accused science of draining the mystery from existence (Wordsworth's "We murder to dissect") and technology of dehumanizing humankind (William Blake's "dark satanic mills"). Such paintings became so popular that Dutch painters developed "wet-on-wet" overlaying techniques to complete their canvases faster and meet the rising demand, while their agents invented the practice of selling art at auction to maximize profits. As one modern historian puts it:

> Here were products of a major European civilization—the very best products of that civilization—created by burghers for burghers. The existence of these works and their quality proved that no social elite was required for the creation of great art. The aristocracy was superfluous, the church irrelevant. Bourgeois artists attuned to the basic conditions of life around them, trusting their sensibilities and powers of observation alone, were capable of capturing the ultimate meaning of their culture.

Following the death of Rembrandt in 1669, the artistic inclinations of the Netherlands increasingly were expressed through natural history illustrations. Some of these works took the form of tropical landscape paintings, a genre invented by the Dutch that combined geographical and botanical depictions of Brazil, North Africa, and the East Indies. Many appeared in sumptuous books like Maria Sibylla Merian's *Metamorphosis Insectorum Surinamensium* of 1705, a study of the insects of the Dutch colony Surinam; Hendrik Adriaan van Reede tot Drakenstein's *Hortus Malabaricus*, a twelve-volume botanical work containing over six hundred plates; George Rumpf's monumental book on Moluccan shells, *The Ambionese Curiosity Cabinet*; Simon de Vries's two-thousand-page *Great Cabinet of Curiosities*, which ranged from the wampum of American Indians to the marital customs of the Laplanders; and the *Hortus Cliffortianus* of Carl Linnaeus, the "father of taxonomy," who was born in Sweden but developed the basis of his system for classifying plants while studying at the universities of Harderwijk and Leiden. Dutch collectors accumulated such vast inventories of the world's biological and anthropological wonders that many of today's methods of scientific classification had to be developed just to keep track of them.

Dutch society, like Dutch art, evoked a harmonious interplay between liberalism and science. As the historian Johathan I. Israel notes:

> Visitors continually marveled at the prodigious extent of Dutch shipping and commerce, the technical sophistication of industry and finance, the beauty and orderliness, as well as the cleanliness, of the cities, the excellence of the orphanages and hospitals, the limited character of ecclesiastical power, the subordination of military to civilian authority, and the remarkable achievements of Dutch art, philosophy, and science. . . . Until the late seventeenth century many were appalled by the diversity of churches which the authorities permitted and the relative freedom with which religious and intellectual issues were discussed. Others disapproved of the excessive liberty, as it seemed to them, accorded to specific groups, especially women, servants, and Jews, who were invariably confined, in other European countries, to a lowlier, more restricted existence.

The Dutch were tolerant of such criticism; they were tolerant of just about everything.* They could afford to be. Having hit upon the magic triangle of science, technology, and liberal trade policies, they inhabited a seventeenth-century version of Tomorrowland, a state that demonstrated daily how a free, scientific world could work.

Locke thrived during his five years in the Netherlands, maturing into the accomplished, if still almost entirely unpublished, author who would become an architect of the Enlightenment and a preeminent exponent of science and liberalism. He quietly toured Dutch universities and studied local medical and technological improvements while conspiring with fellow Whig exiles to keep rebellion boiling back home in Britain. As a fugitive from justice—a warrant for his arrest and extradition having been issued in England—Locke was obliged to live in hiding, receiving his mail under assumed names and laboriously conducting his research from books hidden in secreted caches. Such a life is easy for no scholar, but Locke was accustomed to conflict—having grown up during the English Civil War, he

---

*There were of course exceptions to the general liberalism, in law if not always in practice. Dutch universities were banned from teaching the philosophy of Descartes but most evaded the ban, and it was said that virtually all the philosophy students at Leiden by the mid-seventeenth century were Cartesians. Rembrandt's housekeeper was called into court in 1654 for behaving like a "whore" in that she shared the painter's bed, yet Amsterdam's brothels ranked with those of Paris as demarking the gold standard of commercial libidinousness.

recalled, "I no sooner perceived myself in the world but I found myself in a storm"—and the Dutch authorities were disinclined to betray him. William of Orange heeded the letter of the law in ordering enforcement of King James' arrest orders—James was, after all, his father-in-law—but he did so with a wink and a nod, issuing fair warning before dispatching the police to search any of the private homes known to shelter Locke. In the preface to his *Essay Concerning Human Understanding*, Locke coolly refers to its having been completed while he was "in a retirement where an attendance on my health gave me leisure," without mentioning that the principal threat to his health was that of his being dragooned back to the Tower of London and the execution block.

Locke read Newton's *Principia* soon after its publication. Its mathematics were beyond him so he asked the great Dutch mathematician Christiaan Huygens for help. He had picked the right man. Huygens was a diplomat, lute player, creator of a thirty-one-tone music temperament, inventor of a cycloidal pendulum employed in wristwatches and marine chronometers, a telescope builder who discovered a satellite of Saturn, and a talented physicist with ample experience in applying mathematics to the study of nature. It was Huygens who had first derived the law of centrifugal force for uniform circular motion, a step that incited Halley to hypothesize that gravitation obeys an inverse-square law of attraction and to ask Newton about it—which in turn led to Newton's composing the *Principia*. Huygens took exception to some aspects of Newton's theory, regarding it as "absurd" to imagine gravitational force somehow propagating itself across a vacuum of space, but he assured Locke that the work was mathematically trustworthy. Locke then devoted himself to absorbing Newton's work, and emerged understanding better than most how and why it changed the world.

It was one thing to speculate, as many scientists had, that nature was built on a logical, mathematically intelligible basis, but Newton had shown nothing less than that the behavior of all objects influenced by gravity—from thrown stones on earth to the orbits of the moon and planets—could be predicted by means of mathematical formulas. Locke was so impressed that in the "Epistle to the Reader" prefacing his *Essay Concerning Human Understanding*, he referred to Newton's "never enough to be admired book" as constituting "the greatest exercise and improvement of human understanding" yet accomplished. Locke added modestly of his own work:

> In an age that produces such masters as the great Huygenius [i.e., Huygens] and the incomparable Mr. Newton, with some others of that strain, it is ambition enough to be employed as an under-laborer in clearing the ground a little, and removing some of the rubbish that lies in the way to knowledge.

Voltaire said of Locke, "He everywhere takes the light of physics for his guide."

With the advent of the Glorious Revolution, Locke returned to England as Princess Mary's personal escort. They made landfall on February 13, 1689, and she was crowned queen the following day. Shortly thereafter her husband offered to name Locke the British ambassador to Frederick III, Elector of Brandenburg, an honor Locke refused on grounds of health—what he called "weak lungs," probably asthma—and sobriety; Frederick was a drinker for whom Locke, "the soberest man" in England, might be poor company. Then Locke went off to find Newton.

Locke described Newton as a "nice" friend in the old sense of the word: "not able to endure much; tender, delicate." When Newton in 1693 had a nervous breakdown—owing perhaps to overwork, lack of sleep, psychotropic poisoning by the toxic chemicals he handled in his alchemical experiments, or to the sexual repression of being a Puritan's Puritan who evidently never had sex with anybody—and apologized to Locke for "being of opinion that you endeavored to embroil me with women," Locke stuck by him, replying, with his customary steadfastness:

> I have been, ever since I first knew you, so entirely and sincerely your friend, and thought you so much mine, that I could not have believed what you tell me of yourself, had I had it from anybody else. And though I cannot but be mightily troubled that you should have had so many wrong and unjust thoughts of me, yet . . . I receive your acknowledgement of the contrary as the kindest thing you could have done me, since it gives me hopes that I have not lost a friend I so much valued. . . . I shall always be ready to serve you to my utmost, in any way you shall like, and shall only need your commands or permission to do it.

By this time Locke had published, almost simultaneously, his *Letter Concerning Toleration*, *Essay Concerning Human Understanding*, and

Two Treatises of Civil Government. They were met with considerable enthusiasm, many agreeing with Lady Mary Calverley that Locke was now to be regarded as "the greatest man in the world."

He had come to philosophy through science rather than the other way round, learning medicine, chemistry, and astronomy before encountering the likes of Descartes and Pierre Gassendi. (Of the two philosophers, Locke felt closer to Gassendi, who as an astronomer, chemist, and physiologist viewed the world empirically and was critical of Descartes.) While working on *Human Understanding*, probably in the winter of 1670–71, Locke attended meetings of the Royal Society and initiated small, informal gatherings to discuss science, philosophy, and religion. At one meeting, with about a half-dozen friends, something unusual happened. In the course of their discussion, as Locke recalled, they became "puzzled" and "perplexed" until "it came into my thoughts that we took a wrong course; and that before we set ourselves upon inquiries of that nature, it was necessary to examine our own abilities, and see what objects our understandings were, or were not, fitted to deal with." This was not a traditional step—philosophers were more apt to use reason to explain everything under the sun than to consider its limits—but it answered to the spirit of science.

Scientists circa 1670 were mostly amateurs whose findings, although often of considerable interest, did not yet add up to anything like the imposing system that would come with Newton's *Principia*. The odd nature of their inquiries—measuring quantities and counting things, mixing smelly chemicals in basement laboratories—was unprecedented and easily ridiculed. Samuel Butler's "Satire upon the Royal Society" lampooned scientists' desire

> To explicate by subtle hints
> The grain of diamonds and flints;
> And in the braying of an ass
> Find out the treble and the bass

Yet these early scientists had hit on something extremely important. The night sky viewed through a telescope revealed countless stars too dim to be seen with the naked eye, scattered endlessly into the depths of space. The microscope showed incredibly rich complexities in the details of every insect, leaf, and drop of pond water. (Even gold, Locke accurately speculated, might cease to look golden if inspected through a strong enough microscope.) Miners digging shafts and naturalists gathering seashells on mountaintops were finding clues that the earth had a longer and more varied history than anyone had imagined. The exotic botanical specimens being brought back from Africa and South America hinted at unanticipated biological extravagances. The effect of all this learning, preliminary and taxonomic as it may have been, was to make the vaunted presumptions of old look puny by comparison. As Richard Feynman would put it in the twentieth century, speaking of biblical accounts of the origin of the universe and the intervention of God in human affairs, "The stage is too big for the drama."

The fact that logical inquiries by generations of capable thinkers had failed to anticipate the richness of the world now being unveiled by science raised the question, for Locke, of how those thinkers had gone wrong. Part of the problem clearly was that reasoning—thinking deductively to arrive at supposedly universal truths, as philosophers had done in the past—was of less use than had been assumed. The most sweeping logical claims, the ones that seemed the loftiest precisely because they lay the farthest from ordinary human experience, amounted to little more than shuffling empty abstractions, inviting the modern jape that "philosophy is the misuse of a terminology which was invented for just this purpose." So Locke attempted a fresh start:

> We should not then perhaps be so forward, out of an affectation of an universal knowledge, to raise questions, and perplex ourselves and others with disputes about things to which our understandings are not suited. . . . Our knowledge being so narrow . . . it will, perhaps, give us some light into the present state of our minds, if we look a little into the dark side, and take a view of our ignorance: which being infinitely larger than our knowledge, may serve much to the quieting of disputes, and improvement of useful knowledge; if discovering how far we have clear and distinct ideas, we confine our thoughts within the contemplation of those things, that are within the reach of our understandings, and launch not out into that abyss of darkness (where we have not eyes to see, nor faculties to perceive any thing,) out of a presumption, that nothing is beyond our comprehension.

His long inquiry into the nature of knowledge led Locke to formulate the philosophy known today as empiricism—the assertion that knowledge is based on experience alone, and that reason is but a secondary quality that arises, for instance, when we compare ideas that have arisen out of our experience.* (Thomas Henry Huxley: "All true science begins with empiricism.") Locke went so far as to argue that the human mind at birth is a blank slate—a "white paper void of all characters, without any ideas"—on which sensations combine to form ideas in something like the way words combine to form sentences. Since human knowledge is limited to sensations and to the ideas formed by comparing and combining sensations, it is specious to claim that knowledge can be attained through reasoning alone, as Descartes did, or to reason from essences rather than appearances, as did Aristotle.

Intellectually, Locke's *Essay* is remarkable for its modesty and its tolerance for ambiguity, two qualities that arose directly from science. Whereas most prescientific philosophers (and many modern and postmodern philosophers as well) sought to construct vast systems that pretended to explain just about everything, Locke was content, as scientists are, to establish facts about specific things. Ultimately the facts may be connected to form large patterns, but the big picture comes later. Thinkers impatient with the empirical approach, Locke noted, usually have failed to take the enormity of human ignorance into account. (The words "ignorant" and "ignorance" appear more than 140 times in Locke's *Essay.*) If the range of empirical knowledge seems narrow by comparison with the grand claims of old, within that range much more may be learned: Locke writes that it would be childish to "undervalue the advantages of our knowledge, and neglect to improve it to the ends for which it was given us, because there are some things that are set out of the reach of it." Aware that, as Feynman would

*The term "empiricism" was not employed by Locke but came into use later on. As a physician he probably wouldn't have liked the word anyway, since in his day the word "empiric" referred to the worthless patent medicines prescribed by quacks. The standard shorthand history is that Locke was one of a trio of empiricists, the others being Berkeley and Hume, who were opposed by a trio of rationalists—Descartes, Spinoza, and Leibniz—with the two strains uniting in Immanuel Kant's view of knowledge as comprised of both innate (i.e., rational, a priori) and acquired (empirical, a posteriori) elements. Spinoza, however, kept at least one foot on the empiricists' side.

put it, "What is not surrounded by uncertainty cannot be the truth," Locke was good-humored about the limitations of his own philosophy:

> The extent of knowledge, or thing knowable, is so vast, our duration here so short, and the entrance by which the knowledge of things gets into our understanding so narrow, that the whole time of our life is not enough to acquaint us with all those things, I will not say which we are capable of knowing, but which it would be not only convenient but very advantageous to know. It therefore behooves us to improve the best we can our time and talent in this respect, and, since we have a long journey to go, and the days are but short, to take the straightest and most direct road we can.

If human knowledge is based on our interactions with a material world that obeys natural laws, it follows that natural law is the proper basis for human governance. It was through this line of reasoning that Locke identified natural rights such as liberty, which he defined as "the power a man has to do or forbear doing any particular action according as . . . he himself wills it." Liberty is rooted in human learning, since people can educate themselves, and act on what they have learned, only insofar as they are free.

> Our understanding and reason were given us, if we will make a right use of them, to search and see, and then judge thereupon. Without liberty, the understanding would be to no purpose: and without understanding, liberty (if it could be) would signify nothing.

Impatient with philosophical arguments against free will, Locke argued that governance should be based on how people act and appear to be, rather than on theories about the inner workings of their minds: "As far as this power reaches, of acting or not acting, by the determination of his own thought preferring either, so far is a man free."

Humans form social contracts with their governments to protect their natural rights. When a government fails to keep its part of the bargain, its citizens are justified in redressing their grievances, either through their elected officials or, if all else fails, through civil disobedience: "It is lawful for the people . . . to *resist* their king." But if the people are to function capably as the ultimate source of political power they must be educated; hence Locke wrote at length about child rearing and public schooling. Although he never married, he evinced great sympathy and warmth toward children, advising parents to let their children get plenty of sleep, ample time to play in the open air, "plain and simple" diet and clothing (no straitlacing or otherwise overdressing of girls), and freedom from excessive discipline: "I have seen parents so heap rules on their children, that it was impossible for the poor little ones to remember a tenth part of them, much less to observe them. . . . Let therefore your rules to your son be as few as possible."

Locke had his limitations, of course. He championed religious tolerance, but not for atheists and Catholics. He celebrated empiricism, but insisted that the existence of God could be established through reason alone. His "blank slate" theory of the human mind finds few adherents today. But he more than any other philosopher fashioned the kit of what was to become the democratic revolution—liberal governance based on natural rights and the consent of the governed; free intellectual, religious, and scientific inquiry; equal justice under law; and universal education designed to nourish individual capacity rather than mass servitude. Indeed the greatest difficulty for liberal-scientific readers reading Locke today is that his vast influence makes his philosophy seem as natural as fish find water.

As the founder of both empiricism and political liberalism, Locke produced a philosophy of science and of government that helped inspire the coming revolutions in America and France. Those epochal transformations were more immediately incited, however, by the journalist and amateur scientist Thomas Paine, who ranked high among the instigators of the American Revolution, the champions of the French Revolution, and the authors of the worldwide democratic revolution that followed—so much so that John Adams proposed that what is today called the Age of Reason, a borrowing from Paine's book of the same title, should instead be called "the Age of Paine." James Monroe spoke of him "as not only having rendered important service in our own Revolution, but as being, on a more extensive scale, the friend of human rights, and a distinguished and able advocate in favor of public liberty." The Marquis de Lafayette, George Washington's favorite officer, felt that "free America without her Thomas Paine is unthinkable." The poet William Blake described Paine as a "worker of miracles," comparing him favorably to Jesus: "Is it a greater miracle to feed five thousand men with five loaves than to overthrow all the armies of Europe with a small pamphlet?"

Paine's first book, *Common Sense*, altered the political landscape almost overnight. Written in an accessible, unadorned style and bristling with insightful arguments for American independence, it was published on January 10, 1776, and sold over a hundred thousand copies before the year was out—this in a colonial America with a population of under three million. "Its effects were sudden and extensive upon the American mind," wrote Dr. Benjamin Rush, a signer of the Declaration of Independence. "It was read by public men, repeated in clubs, spouted in Schools, and in one instance, delivered from the pulpit instead of a sermon by a clergyman in Connecticut." "This book was the arsenal to which colonists went for their mental weapons," wrote the nineteenth-century historian Theodore Parker. "Every living man in America in 1776, who could read, read *Common Sense*. . . . If he were a Tory, he read it, at least a little, just to find out for himself how atrocious it was; and if he was a Whig, he read it all to find the reasons why he was one."

Paine followed *Common Sense* with a pamphlet titled *The American Crisis*, which he wrote in December 1776 while serving in the Continental army during its retreat from the British—"in a rage when our affairs were at their lowest ebb." The British, aiming to crush the rebellion at its inception, had dispatched a fleet of warships to New York; their masts, said one observer, made the harbor look like a forest. The fleet disgorged twenty-three thousand British regulars and ten thousand German mercenaries, who on August 27, 1776, promptly decimated Washington's ragtag army in the Battle of Brooklyn. Only through daring and luck was Washington able to convey the remainder of his Brooklyn forces across the East River by night, fleeing north through Manhattan before the British could catch up. Washington had been in retreat ever since, through New Jersey and on into Pennsylvania, and was close to despair. "I think the game is pretty near up," he wrote to his brother John Augustine on December 18. The enlistments of three-quarters of his troops were due to expire at the end of the month; many, badly demoralized, had already declared their intention to quit fighting and go home and attend to their shops and farms. Dr. Rush found Washington on Christmas Eve "much depressed" and scribbling feverishly on scraps of paper that he then thrust aside. Retrieving a sheet that had fallen to the floor, Rush read the words "Victory or Death."

Washington's near-desperate plan was to stage a surprise attack against the 1,500-man Hessian garrison holding Trenton. If it worked, he might be able to keep his army together through the winter; if not, he felt, the American Revolution surely would be lost. Searching for sources of inspiration, Washington on Christmas Day had the opening paragraphs of Paine's *American Crisis*, fresh off the press, read aloud to his troops. Wrapped in blankets that served as capes, slapping their arms across their chests to keep warm, many wearing shoes so ruined that their footprints in the snow were speckled with blood, they listened as Paine's memorable injunction echoed in the frosty air:

> These are the times that try men's souls. The summer soldier and the sunshine patriot will, in this crisis, shrink from the service of their country; but he that stands it now, deserves the love and thanks of man and woman.

Washington's troops crossed the Delaware that night. Two of his men froze to death but Washington kept his forces together and attacked at dawn, taking Trenton and capturing nearly a thousand of the Hessians without incurring a single American fatality. Paine's words were ever thereafter associated with the ultimate success of the American Revolution, and though he would be convicted of seditious libel in England, very nearly hanged and beheaded in France, and widely excoriated in America for his criticisms of organized religion, he remained the world's most influential political writer. As Jefferson reflected in 1801, Paine had "as much effect as any man living."

Yet there is more than a little mystery about Paine. Where did his insights come from? No mere rhetorician or rabble-rouser, he had a remarkably clear sense of what was wrong with Europe's monarchies and a robust conception of how the Americans and the French might do better. His judgments were sound enough to keep us nodding in agreement with him today, his political philosophy sufficiently sophisticated for his biographer Eric Foner to declare that his ideas "have never been grasped in their full complexity." But little in Paine's personal history suggests fertile soil for such discernment.

Born in 1737 in Thetford, a village seventy miles northeast of London, Paine had only a rudimentary education before being apprenticed to his father as a staymaker at age thirteen. He disliked the job and tried various others, all without success. He twice worked as an excise officer but was fired both times, taught school for a few months but never returned to teaching, applied to become a clergyman but was turned down, and ran a tobacco shop that promptly failed. Nor did he cut the figure of a bookish autodidact. Although friends spoke of his large, clear eyes and alert demeanor, it was Paine's habit to stay in bed reading newspapers till the sun was high in the sky, take a long walk, eat a substantial meal, nap for a few hours, and then start writing and drinking. Accounts differ, but he seems to have been a world-class brandy drinker who kept a snifter by his inkwell. Domestic bliss eluded him. His first wife, an orphan who worked as a maid, died delivering a stillborn child soon after their wedding. His second, the daughter of the widow who ran the tobacco shop, left him when he mismanaged the shop, presenting him with a £35 settlement in return for his renouncing all future claims to her property. (Paine would later reflect wistfully that many marriages "end in mutual hatred and contempt. Love abhors clamor and soon flies away, and happiness finds no entrance when love is gone.") He used the money to book passage to Philadelphia, bearing a letter of introduction from Ben Franklin. In it Franklin states rather guardedly that Paine, whom he had met in London, had been "very well recommended to me as an ingenious worthy young man," who might find employment in America "as a clerk, or assistant tutor in a school, or assistant surveyor."

Paine landed in Philadelphia on November 30, 1774, so ill (from typhus, evidently) that he had to be carried off the ship on a stretcher. He was thirty-seven years old and had written little for public consumption other than a petition urging Parliament to raise the salaries of excise officers—a plea that the legislators effortlessly ignored. Nevertheless, Paine in America soon began writing oddly prescient and optimistic essays, mostly for *The Pennsylvania Magazine*. One of them begins by ruminating about geology, then characterizes America as the political equivalent of a scientific experiment. Protesting "against the unkind, ungrateful, and impolitic custom of ridiculing unsuccessful experiments," Paine notes that scientific progress often arises from such failures, and predicts that Americans will thrive in the long term even if they falter at present:

> I am led to this reflection by the present domestic state of America, because it will unavoidably happen, that before we can arrive at that perfection of things which other nations have acquired, many

> hopes will fail, many whimsical attempts will become fortunate, and many reasonable ones end in air and expense. *The degree of improvement which America has already arrived at is unparalleled and astonishing,* but 'tis miniature to what she will one day boast of, if heaven continue her happiness.

The following year he published *Common Sense.*

Scholars searching for the key to Paine's metamorphosis have come up with many possible influences—from Quakerism (his father was a Quaker), to his unhappy experiences as an excise officer, his failures as a small-business man, or even his having grown up in Thetford, which saw more than its share of poverty. But many had been Quakers, seen poverty, failed in business, or suffered dismissal from government jobs, yet there was only one Tom Paine. It is more likely that Paine's political ideas arose from his exposure to science.

No firsthand account of Paine's intellectual development survives. He appears to have written at least part of such a work—he showed two volumes of memoirs to his friend Redman Yorke in Paris in 1802, and when writing to Jefferson in 1805 mentioned that he was composing a five-volume collected works with historical prefaces—but those manuscripts never saw the light of day. When Paine died, on June 8, 1809, in a house on Grove Street in New York City being rented by his friend Margaret Brazier Bonneville, he left his papers to her. She attempted to edit some of them, avidly erasing passages that offended her Catholic sensitivities, but produced nothing for publication. Following her death the papers passed into the possession of her son, General Benjamin Bonneville, who stored them in a St. Louis warehouse that later burned down. Bonneville's widow confirmed that Paine's papers were "all destroyed—at least all which the General had in his possession."

Notwithstanding this scholarly tragedy it is possible to construct a coherent, if fragmentary, narrative of Paine's scientific awakenings and their influence on his political thought. The story begins in 1756, when a teenaged Paine enlisted to serve on the privateer *Terrible* but was dissuaded from doing so by his father. The following year he tried again and this time succeeded, shipping out aboard the *King of Prussia* on January 17, 1757, and returning some six months later with over £30 in his pocket—his share of the booty earned by the privateer for having rescued a friendly ship, the *Pennsylvania*, from attack by the French. Paine headed for London, where he educated himself by reading library books and attending public lectures, discovering that "the natural bent of my mind was to science."

There he befriended two leading science popularizers, James Ferguson and Benjamin Martin, men of common origins who supplemented their incomes by lecturing and writing about the wonders of chemistry, optics, and the stars. Ferguson came from a large, poor family in the north of Scotland and had only three months of schooling, but had taught himself to read, charted the stars as a shepherd at age fourteen, and embarked on a career in which he had, as his epitaph put it, "by unwearied application (without a master), attained the sciences." He had just published a bestselling book, *Astronomy Explained Upon Sir Isaac Newton's Principles*, that was to inspire a generation of budding amateur scientists—among them the musician William Herschel, who upon reading it built himself an oboe-wood telescope and went on to discover the planet Uranus. Martin was a farmer's son and onetime schoolmaster who made telescopes and celestial globes. Like Ferguson he gave public "demonstrations" in which experiments were staged to show how scientific findings are established. (Samuel Johnson's *Dictionary of the English Language*, published in 1755, defined science as knowledge "grounded on demonstration.") Both men sold globes, orreries, magnifying glasses, microscopes, and telescopes—including the popular "jealousy glasses," small periscope-like pocket instruments employed to discreetly spy on one's fellow audience members at the opera. "As soon as I was able," Paine recalled, "I purchased a pair of globes, and attended the philosophical lectures of Martin and Ferguson." The globes most likely were sold him by Martin, whose Fleet Street shop, called The Globe and Visual Glasses, advertised that "the use of the Globes . . . is the first consideration among those Qualities requisite for forming the Scholar and the Gentleman." Displaying a pair of globes—one celestial, the other terrestrial—in your home signified that you were up on the latest scientific developments and knew about "the plurality of worlds," the theory that the earth was just one among many inhabited planets.

Paine's enthusiasm for science shows up so often in his works that were it not for his political influence, he might today be remembered as a gifted science writer. Long passages in *The Age of Reason* investigate the implications of there being many worlds. The earth, Paine writes,

> may, at first thought, appear to us to be great; but if we compare it with the immensity of space in which it is suspended, like a bubble or a balloon in the air, it is infinitely less in proportion than the smallest grain of sand is to the size of the world, or the finest particle of dew to the whole ocean; and . . . is only one of a system of worlds, of which the universal creation is composed.

Every star is a sun, and each star may well be orbited by planets of its own, "as our system of worlds does round our central Sun." Life abounds on the earth (Paine noting that when viewed through a microscope "every tree, every plant, every leaf, serves not only as a habitation but as a world to some numerous race"), so it stands to reason that other planets may harbor life, too:

> Since then no part of our earth is left unoccupied, why is it to be supposed, that the immensity of space is a naked void, lying in eternal waste. There is room for millions of worlds as large or larger than ours, and each of them millions of miles apart from each other.

Paine saw the prospect of extraterrestrial life as underlining the importance of democracy and education. "It is not to us, the inhabitants of this globe, only, that the benefits arising from a plurality of worlds are limited," he wrote. "The inhabitants of each of the worlds . . . enjoy the same opportunities of knowledge as we do. . . . The same universal school of science presents itself to all."

Paine tells us that when he was learning science from Ferguson and Martin—and later from the astronomer John Bevis—he "had no disposition for what is called politics. It presented to my mind no other idea than is contained in the word Jockeyship"—that is, a game, a jostling for position in order to win. Although he later changed his mind on this point, Paine never stopped working on scientific and technological projects. He investigated the causes of yellow fever and the development of steamboats, invented a smokeless candle, corresponded with Thomas Jefferson about the capillary action that draws water up into the branches of trees, and designed an innovative iron bridge. "I took the idea of constructing it from a spider's web, of which it resembles a section," he recalled, "and I naturally suppose that when nature enabled that insect to make a web she taught it the best method of putting it together."

Paine once conducted a scientific experiment with George Washington, who shared the passion for empirical inquiry that animated Franklin, Jefferson, and the other scientific Whigs who helped win American independence. In November 1783, he was Washington's guest at Rocky Hill, a country estate in New Jersey. The locals maintained that a creek running through the property sometimes caught fire at night. Washington theorized that the fires occurred when "bituminous matter," similar to oil slicks or coal dust, rose to the creek's surface. Paine argued that the fires instead occurred when "inflammable air"—swamp gas or foxfire, known today to be methane—bubbled up from decaying organic matter on the river bottom. The two set out one night aboard a flat-bottomed scow to decide the question. "General Washington placed himself at one end of the scow, and I at the other," Paine recalled. "Each of us had a roll of cartridge paper, which we lighted and held over the water, about two or three inches from the surface." Washington's soldiers prodded the river mud with their poles, and

> when the mud at the bottom was disturbed by the poles, the air bubbles rose fast, and I saw the fire take from General Washington's light and descend from thence to the surface of the water, in a similar manner as when a lighted candle is held so as to touch the smoke of a candle just blown out, the smoke will take fire, and the fire will descend and light up the candle. This was demonstrative evidence that what was called setting the river on fire was . . . the inflammable air that arose out of the mud.

Paine saw that science undermines authority, erodes superstition, and requires for its own advancement that people be free to meet and speak as they please. He was not the first to have made such connections. As his biographer John Keane notes, "The circles of Newtonians" that Paine encountered in London were "breeding grounds for a new radical politics. To many in those circles, Paine included, it seemed obvious that the sciences were friends of liberty." What Paine brought to the mix was an unprecedented combination of coolheaded empirical judgment and blast-furnace rhetoric. In 1776, many Americans were still loyal British subjects who took a sympathetic view of monarchs generally and of King George

III in particular. Paine would have none of that. Monarchy, he wrote, is "a silly, contemptible thing. I compare it to something kept behind a curtain, about which there is a great deal of bustle and fuss, and a wonderful air of seeming solemnity; but when, by any accident, the curtain happens to be open—and the company see what it is—they burst into laughter." The job is empirically untenable: "The state of a king shuts him from the world, yet the business of a king requires him to know it thoroughly; wherefore the different parts, unnaturally opposing and destroying each other, prove the whole character to be absurd and useless." Paine was particularly acute about the historical foundations of the monarchies: Each had originated, he noted, in the violent coercion of "a banditti of ruffians" who, "their power being thus established . . . contrived to lose the name of Robber in that of Monarch; and hence the origin of Monarchy and Kings." This royal genesis in thievery, brutality, and usurpation is so embarrassing that a monarchist

> never traces government to its source, or from its source. It is one of the shibboleths by which he may be known. A thousand years hence, those who shall live in America or France, will look back with contemplative pride on the origin of their government, and say, This was the work of our glorious ancestors! But what can a monarchical talker say? What has he to exult in? Alas he has nothing. A certain something forbids him to look back to a beginning, lest some robber, or some Robin Hood, should rise from the long obscurity of time and say, I am the origin.

To conceal the illegitimate source of their power, monarchists are obliged "to trump up some superstitious tale, conveniently timed . . . to cram hereditary right down the throats of the vulgar."

The despotic power of monarchs becomes ludicrous when perpetuated through heredity—"All hereditary government is in its nature tyranny"—and Paine ridiculed hereditary rule by imagining what would happen if it were applied to art and science:

> As the republic of letters brings forward the best literary productions, by giving to genius a fair and universal chance; so the representative system of government is calculated to produce the wisest laws, by collecting wisdom from where it can be found. I smile to myself when

> I contemplate the ridiculous insignificance into which literature and all the sciences would sink, were they made hereditary; and I carry the same idea into governments. An hereditary governor is as inconsistent as an hereditary author. I know not whether Homer or Euclid had sons: but I will venture an opinion, that if they had, and had left their works unfinished, those sons could not have completed them.

While Paine's scientific skepticism ate away at the pomp and pretense of European nobility, the findings of eighteenth-century science afforded him an impression of the universe as based on harmonious laws, which in turn implied that human society too should be lawful and harmonious. This was very different from the widely held opinion, still voiced today, that an inherently lawless and violent humanity will descend into chaos unless held in check by authoritarian rule. "Such governments," Paine argued, "consider man merely as an animal; that the exercise of intellectual faculty is not his privilege; *that he has nothing to do with the laws but to obey them;* and they politically depend more upon breaking the spirit of the people by poverty, than they fear enraging it by desperation." Free governments work the other way round, building on foundations of liberty and justice. No king is required for them to function, any more than a master cat is needed to tell all the world's cats how to catch mice, or angels to usher the planets in their orbits. (Paine, a Newtonian deist, believed that God had fashioned the universe well enough to never thereafter require any fine-tuning via divine intervention.) "No one man is capable, without the aid of society, of supplying his own wants; and those wants, acting upon every individual, impel the whole of them into society, as naturally as gravitation acts to a center," he wrote. Systems of government are to be judged by the extent to which they embody natural law: "All the great laws of society are laws of nature."

Paine foresaw that an international community of scientists would become a model of liberalism. Science, he wrote,

> has liberally opened a temple where all may meet. Her influence on the mind, like the sun on the chilled earth, has long been preparing it for higher civilization and further improvement. . . . The philosopher [i.e., natural philosopher, or scientist] of one country sees not an enemy in the philosophy of another; he takes his seat in the temple of science, and asks not who sits beside him.

As "science and commerce" made the world wealthier and more peaceful, free peoples would find that almost their only remaining opponent was "*prejudice*; for it is evidently the interest of mankind to agree and make the best of life."

When Paine returned to Europe from America in the spring of 1787, he thought that he was leaving politics behind to pursue a career as an inventor and engineer. He described his political career as "closed" and told friends that he now intended to devote himself to "the quiet field of science." He arrived in Paris bearing another letter from Franklin, describing him as "an ingenious, honest man [who] carries with him the model of a bridge of a new construction, his own invention." A soaring arch, the bridge was meant to demonstrate the beauty and efficiency of science while symbolizing the lofty spirit of liberty. Crossing it would be like ascending in a balloon—an experience that, Paine later wrote, "would have every thing in it that constitutes the idea of a miracle." He saw the bridge as exemplifying his view that science reveals wonder in even the most ordinary things. "Every thing is a miracle," he wrote, "and . . . no one thing is a greater miracle than another."

But the bridge was never built, and politics returned to the center of Paine's life in about as vivid and immediate a fashion as can be imagined. The storming of the Bastille on July 14, 1789—which found Paine in England, supervising construction of yet another bridge—soon impelled him to resume writing about political affairs. His *Rights of Man*, published in two parts in 1791 and 1792, brought a conviction for seditious libel in a British court, but by that time Paine had fled to France. There he was welcomed as a hero, granted citizenship, and elected to the National Convention. Initially optimistic, he soon became rudely acquainted with the revolution's dark side.

An early warning came on June 21, 1791, when Lafayette burst into Paine's room in Versailles to awaken him with the news that King Louis XVI and his family had gone into hiding. Paine dressed hastily and strolled the streets of Paris with his friend Thomas Christie. The two regaled in the excitement of the mingling mobs, although Paine remarked coolly, "You see the absurdity of monarchical governments. Here will be a whole nation disturbed by the folly of one man." They joined a crowd that listened respectfully, hats in hand, as a declaration was read aloud on behalf of the National Assembly to the effect that the escape of the monarchs would not impede progress toward a new constitution. The reading ended and the crowd donned their hats, adorned with tricolor ribbons denoting their revolutionary sympathies, but Paine had left his own cap back in his room. The mob turned on him, crying "Aristocrat! *A la lanterne*!" Three men seized Paine and dragged him toward a lamppost. Everyone understood what would follow: Royalists customarily were hanged by a rope around the neck from a street post and then decapitated, after which their organs were removed and displayed on pikes while their mutilated corpses were dragged through the streets. Paine narrowly escaped this fate thanks to the fact that Christie, who spoke good French, reasoned with the mob, and that an anonymous individual intervened and assured them that Paine was "un Américain." Exhibiting remarkable coolness if also a dangerous inattentiveness to the potentials of revolutionary violence, Paine wrote languidly of this episode in a letter to the Marquis de Condorcet, observing that "during the early period of a revolution mistakes are likely enough to be committed."

Beneath the studied nonchalance, however, Paine knew that his personal prospects were bleak. "I saw my life in continual danger," he wrote Samuel Adams in 1803. "My friends were falling as fast as the guillotine could cut their heads off, and as I every day expected the same fate, I resolved to begin my Work." That work was *The Age of Reason*, a book prompted on one hand by Paine's lifelong distrust of organized religion and on the other by his concern, as a deist, that the French "were running headlong into atheism." In 1776 Paine had told John Adams that he planned to write a book about religion but, anticipating that it would arouse public hostility, intended to "postpone it to the latter part of my life." Having good reason to expect that his life would indeed soon end, he now composed the book feverishly and with his customary verve. "To believe that God created a plurality of worlds, at least as numerous as what we call stars, renders the Christian system of faith at once little and ridiculous; and scatters it in the mind like feathers in the air," he wrote. "The two beliefs cannot be held together in the same mind; and he who thinks that he believes both, has thought but little of either." Were Jesus of Nazareth the son of God, then a Jesus operating in the universe revealed by science "would have nothing else to do than to travel from world to world, in an endless succession of death." And if, as the Bible recounts, Satan carried Jesus to a high peak and showed him all the kingdoms of the world, "how happened it that he did not discover America?"

In an episode that put the "dead" in deadline, Paine completed *The Age of Reason* by Christmas Eve of 1793 and was arrested shortly thereafter. Locked up in the Luxembourg prison, he passed the time debating religion with his friend Anacharsis Cloots, a militant atheist, until Cloots was guillotined on March 24. A few days later Georges-Jacques Danton, who had sought to calm Paine's concerns during the onset of the Terror by reassuring him that "revolutions cannot be made with rosewater," joined him in prison. "That which you did for the happiness and liberty of your country, I tried in vain to do for mine," Danton told Paine, as their fellow prisoners looked on. "I have been less fortunate, but not less innocent." Danton was guillotined on April 5. "Many a man whom I have passed an hour with in conversation I have seen marching to his destruction the next hour, or heard of it the next morning," Paine wrote. "For what rendered the scene more horrible was that they were generally taken away at midnight, so that every man went to bed with the apprehension of never seeing his friends or the world again."

Paine's death sentence, signed by Maximilien Robespierre, was conveyed to the prison on July 24. By that time Paine had fallen seriously ill with a fever and been moved to a larger cell where three Belgian inmates could care for him. Concerned that the emaciated and semiconscious Paine was having trouble breathing, the Belgians got permission from a guard to leave the door open so that air could circulate through the cell. At dawn a turnkey made his way down the hall, chalking numbers on the doors of those to be executed that night. Since Paine's door was open, the guard chalked his number, *Quatre*, on the inside of the door. When night fell the Belgians quietly closed the door, then listened in horror as a death squad proceeded down the corridor, dragging screaming inmates away. One of the Belgians clasped his hand over the delirious Paine's mouth to ensure that he would not moan, cry out, or otherwise draw attention to himself, and the squad passed them by. Robespierre himself was guillotined just four days later.

Paine lived for fifteen more years, mostly in America, where he worked on various inventions—bridges, cranes, gunboats, and steamboats—and wrote tracts opposing slavery, championing women's rights, and, in a letter to President Jefferson, proposing the Louisiana Purchase. He made many friends, among them Robert Fulton, inventor of the submarine and builder of the first economically viable steamboat, and the journalist Louis-Sébastien Mercier, whom he described as "a very funny, witty, old man, who had just as sound notions of liberty as he had of astronomy."

In his will Paine declared, with his customary unadorned candor, "I have lived an honest and useful life to mankind; my time has been spent in doing good, and I die in perfect composure and resignation to the will of my Creator, God." His physician, Dr. James R. Manley, asked, "Do you wish to believe that Jesus Christ is the son of God?" "I have no wish to believe on that subject," Paine replied, then slept, then died. His body, buried in New Rochelle, New York, was exhumed in 1819 by William Cobbett, an English journalist, for reburial beneath a proper monument in England. But funds for the monument were never raised, and the disposition of Paine's remains is unknown.

Paine was widely reviled, first for his revolutionary political views and later for his caustic critiques of organized religion in *The Age of Reason*. That book, too, was largely scientific in its orientation, Paine taking the deistic view that God may be known by studying his handiwork in the natural world. "The principles of science lead to this knowledge," he wrote, "for the Creator of man is the Creator of science; and it is through that medium that man can see God, as it were, face to face." This naturalistic view of God had the virtue, in Paine's opinion, of being far more steadfast and reliable than are the Scriptures:

> The Bible represents God to be a changeable, passionate, vindictive Being; making a world and then drowning it, afterwards repenting of what he had done, and promising not to do so again. Setting one nation to cut the throats of another, and stopping the course of the sun till the butchery should be done. But the works of God in the Creation preach to us another doctrine. In that vast volume we see nothing to give us the idea of a changeable, passionate, vindictive God; everything we there behold impresses us with a contrary idea—that of unchangeableness and of eternal order, harmony, and goodness.

Paine even urged that science itself be taught as a form of theology: "That which is now called natural philosophy, embracing the whole circle of science, of which astronomy occupies the chief place, is the study of the works of God and of the power and wisdom of God in his works, and is the true theology." In doing so he drew unflattering comparisons between theology as it was and as it could be:

> The study of theology, as it stands in Christian churches, is the study of nothing; it is founded on nothing; it rests on no principles; it proceeds by no authorities; it has no data; it can demonstrate nothing; and it admits of no conclusion. Not any thing can be studied as a science, without our being in possession of the principles upon which it is founded; and as this is not the case with Christian theology, it is therefore the study of nothing.

William Blake, a religious rather than theological believer, perceived Paine to be a "better Christian" than were his critics, but the theologians were enraged. Their slanders had such lasting effect that Theodore Roosevelt in 1888 could dismiss Paine as a "filthy little atheist," thus making three factual errors in three words: Paine stood five foot ten inches tall, had the presentable appearance of an internationally celebrated man of letters, and was a deist who opposed atheism.

Napoleon Bonaparte, seeking to curry Paine's favor, visited him in Paris one day and proposed that every city should erect a gold statue in his honor. And so, perhaps, they should, but none has yet done so in the nation he named (Paine invented the term "The United States of America") and helped establish. Visitors to Washington, D.C., today may view the statues of such foreign-born contributors to the American Revolution as Lafayette, Thaddeus Kosciusko, Baron Steuben, Jean de Rochambeau, and Michael Kovats de Fabricy—but none of Tom Paine.

CHAPTER FIVE

AMERICAN INDEPENDENCE

My hope [is] that we have not labored in vain, and that our experiment will still prove that men can be governed by reason.

—THOMAS JEFFERSON TO GEORGE MASON, 1791

We had no occasion to search into musty records, to hunt up royal parchments or to investigate the laws and institutions of a semi-barbarous ancestry. We appealed to those of nature, and found them engraved on our hearts.

—JEFFERSON TO JOHN CARTWRIGHT, 1824

On the Fourth of July, 1776, the day the Continental Congress adopted the Declaration of Independence, its author, Thomas Jefferson, took time on four occasions to record the local temperature. Consulting a portable thermometer, Jefferson noted that it was a comfortable 68 degrees at 6 a.m. in Philadelphia on that first Independence Day, rising to 72½ degrees at 9 a.m., 76 degrees at 1 p.m., and falling to 73½ degrees by nine in the evening. Jefferson did this as part of an ongoing research project aimed at tracking the air temperature, wind speed, barometric pressure, and other atmospheric phenomena, including displays of the aurora borealis, in as many locations as possible across the American colonies. He frequently importuned his fellow revolutionaries to collect their own data as well, so that weather maps could be drawn up for the entire eastern seaboard. (His most faithful meteorological correspondent was James Madison, later to succeed Jefferson as president, who promoted the metaphorical use of the word "barometer" to refer to social indices such as public opinion.) The idea was to advance meteorological science, so that farmers' almanacs could be refined and weather forecasts improved.

All this was characteristic of Jefferson, who throughout his long career—he wrote the Declaration at age thirty-three and would live precisely fifty years longer, dying at Monticello on July 4, 1826—was a steadfast devotee of science and technology. Jefferson's 1787 *Notes on the State of Virginia*, replete with references to Newtonian gravitation, was recognized as the best natural history treatise to have been written by an American. His work in agriculture, which he called "a science of the very first order," culminated in his inventing an improved plow board based on Newton's calculus of least resistance. To investigate the Virginia earthen barrows that he correctly hypothesized to be Indian burial mounds, Jefferson invented a method of stratigraphic excavation still employed by geologists today. (He also compiled what was at the time the world's most extensive compendium of American Indian vocabularies.) He read the classical philosophers in the original Greek and Latin but regarded Bacon, Locke, and Newton as superior to them all, commissioning a group portrait of the three and advising the artist, "I would wish to form them into a knot on the same canvas, that they may not be confounded at all with the herd of other great men." While recognizing science to be the sole source of what he called "sure knowledge," Jefferson had a scientist's caution about the errors present in every observation. Timing the solar eclipse of September 17, 1811, through his telescope at Monticello, he noted that while the timed intervals might be internally accurate with regard to one another, he had "no confidence" that his clocks were accurate enough to yield the absolute time of the eclipse. A modern computer analysis shows that Jefferson was right on both points: His overall times, hindered by his difficulty in setting his clocks against transits of the sun across the local meridian, were off by several minutes, but his internal values—such as the interval between the onset of the eclipse and the moment when the moon was directly centered on the sun—were accurate to within 10 percent.

Fascinated by technology, Jefferson traveled with a pocket telescope, fitted his carriage with an odometer that rang a bell to mark each mile, tinkered with a document copier (and complained about the poor quality of its "fetid" copy paper), invented a cipher machine still being used in the twentieth century, and served his dinner guests at Monticello from a set of dumbwaiters modeled on those of the Café Mécanique in Paris. He prescribed a national system of weights and measures, headed up the patent office, and wrote the first scientific paper (his "Report on Desalination of Seawater" of November 1791) to be published under U.S. government auspices.

Jefferson often said that he valued science over statecraft—"Science is my passion, politics my duty"—and his behavior bore him out. Elected vice president in 1797, he tarried at Monticello to study the bones of a prehistoric ground sloth sent him by the frontier scout John Stuart, journeying to his own inauguration only once political opponents began to cite the delay as evidence of haughtiness. He brought the bones along to Philadelphia, presenting them on March 3 to the American Philosophical Society, a scientific body which made him its president, an honor he pronounced to be "the most flattering incident of my life." On the following day he finally got around to taking the oath of office as vice president of the United States.

While serving in that post he reported to his daughter Martha that he was "abandoning the rich and declining their dinners and parties" in order to consort "entirely with the class of science, of whom there is a valuable society here." During the bitter presidential contest of 1800, when Jefferson and Aaron Burr remained deadlocked in the House of Representatives through thirty-five ballots over the course of eight anguished days, Jefferson stayed out of the fray, preferring to investigate whether the full moon influences terrestrial climates. (It does not.) As president it was his custom to ride into the countryside on horseback in the mornings, collecting botanical cuttings along the Potomac: A white-flowered herb, *Jeffersonia binata*, is named in his honor. He worked at his desk with a pet mockingbird perched on his shoulder—the bird would hop up the stairs after him when he went to bed—and conducted animal husbandry experiments that involved keeping a pair of grizzly bears and a flock of Merino sheep on the White House lawn. A veteran who came to the White House seeking a pension complained of being "attacked and severely wounded and bruised by your excellency's ram."

"Nature intended me for the tranquil pursuits of science, by rendering them my supreme delight," Jefferson wrote shortly after leaving office, "but the enormities of the times in which I have lived, have forced me to take part in resisting them, and to commit myself on the boisterous ocean of political passions." He said that what he valued most in public service was the opportunity to advance learning and liberty. His self-composed epitaph describes him as "Author of the Declaration of American Independence, Founder of the University of Virginia, and Author of the Virginia Statute on Religious Freedom," but makes no mention of his having been governor of Virginia and president of the United States.

The Declaration of Independence, which Jefferson drafted in Jacob Graff's rooming house on a portable desk of his own design, is steeped in the language of science and of the Enlightenment philosophers inspired by science. The reference in its first sentence to "the laws of nature and of nature's God" echoes Descartes' "laws of motion" and Newton's "laws of nature." Its second sentence asserts as "self-evident" certain "truths," among them that all men are created equal and are endowed with inherent and inalienable rights such as life, liberty, and the pursuit of happiness. The term "self-evident" apparently was inserted by Benjamin Franklin, to whom Jefferson had submitted his draft for review—Jefferson had written "sacred and undeniable"—and its effect was to shift the argument toward a grounding in science and mathematics. Franklin and Jefferson would have first encountered the term "self-evident" in the axioms of Euclid's geometry, which in those days was usually taught from a popular textbook prepared by Isaac Newton's teacher Isaac Barrow.* It showed up as well in lectures on oratory composed by the English chemist Joseph Priestley, who was a friend of both Jefferson and Franklin, and in the first alphabetical encyclopedia published in English—John Harris's *Lexicon Technicum*, a copy of which Jefferson had in his library—where science is described as being founded on "self-evident principles." John Locke composed an essay on self-evident axioms in mathematics and mathematical science. "In all sorts of Reasoning," he wrote, "every single Argument should be managed as a Mathematical Demonstration." Jefferson echoed Locke so strikingly in the Declaration that he eventually felt obliged to maintain that he had not been imitating Locke and had indeed "turned to neither book nor pamphlet while writing it."

---

*In Euclid and Newton alike, an axiom is a statement that is either self-evidently true or at least so plausible that one may base a solid argument on it, e.g., "Things which are equal to the same thing are also equal to one another" (Euclid). Axioms form the backbone of Newton's *Principia,* one of Jefferson's favorite books. Newton's most familiar assertion, "To any action there is always an opposite and equal reaction," is posited as an axiom.

The Declaration is structured as a syllogism. Its major premise asserts as axiomatic that people may "alter or abolish" a government that denies them human rights. Its minor premise is that King George has been guilty of just such conduct: "To prove this let facts be submitted to a candid world." Its conclusion is that the colonies are thereby justified in severing their ties with England. As an example of clear reasoning this would have pleased Jefferson's teacher William Small, a Scottish mathematician who taught rhetoric, logic, and natural philosophy at the College of William and Mary. Jefferson remembered Small as "a man profound in most of the useful branches of science, with a happy talent of communication . . . & an enlarged & liberal mind," adding, "From his conversation I got my first views of the expansion of science & of the system of things in which we are placed." Along with a devotion to science and liberty, Small brought with him from Scotland a cleanly composed logic book written by *his* teacher William Duncan, a self-taught journalist, classicist, and science professor so admired by his students at Marischal College, Aberdeen, that they chipped in to equip his physics laboratory. Drawing on the thoughts of "the great Mr. Locke," Duncan taught that science is the soundest method of discovery, and that arguments ought therefore to begin with perception and proceed through reasoned judgment to demonstration. (These "demonstrations" might take the form of either scientific experimentation or mathematical reasoning.) The first example of a syllogism in Duncan's *Logic* is not the old, "All men are mortal; Socrates is a man; . . . " but the assertion of a "self-evident" link between reason and human rights:

> Every creature possessed of reason and liberty is accountable for his actions.
> Man is a creature possessed of reason and liberty.
> Therefore, man is accountable for his actions.

The lasting appeal of the Declaration of Independence arose in part from Jefferson's having firmly grounded it in mathematics, science, and logic. As the Princeton University rhetoric professor Wilbur Samuel Howell observes, the ideas set forth in the Declaration "were given added persuasive power by their adherence to the best contemporary standards of mathematical and scientific demonstration and what the best contemporary thinkers expected of proof before it could claim to convince the reason."

Abraham Lincoln believed the Declaration to be "applicable to all men and all times," adding, "I have never had a feeling politically that did not spring from the sentiments embodied in the Declaration of Independence."

Granted, Jefferson was an extraordinary individual; the dark waters of his intellect run so deep that his biographers, like Isaac Newton's, have complained after years of labor that their scholarly bathyspheres never did touch bottom. Jefferson's library was sufficiently commodious—he read not only Greek and Latin ("a sublime luxury") but Old English, French, Italian, and a smattering of German—that when the Library of Congress was put to the torch by British troops in 1814, he was able to reestablish it by shipping six thousand of his own books to the Capitol. His papers and correspondence, publication of which has been going on for decades in an edition expected to run to some sixty volumes, reflect a penetrating interest in everything from astronomy, biology, horticulture, history, philosophy, and law to the military applications of submarines armed with torpedoes. A close reader of the Bible, Jefferson agreed with Thomas Paine that biblical tales of the supernatural had the effect of "degrading the Almighty into the character of a showman, playing tricks to amuse and make the people stare and wonder"; accordingly he produced what is known today as the "Jefferson Bible," a New Testament stripped of mysticism and miracles. When a student asked for advice on how best to pursue his studies, Jefferson replied that a knowledge of "astronomy, botany, chemistry, natural philosophy, natural history [and] anatomy . . . is necessary for our character as well as comfort" and that trigonometry "is most valuable to every man [as] there is scarcely a day in which he will not resort to it for some of the purposes of common life," whereas higher mathematics, such as "spherical trigonometry, algebraical operations beyond the 2d dimension, and fluxions"—Newton's term for differential calculus—are "a delicious luxury indeed" but not essential to a liberal education. Discerning a mathematical problem in the Constitution's approach to apportioning congressional districts on the basis of population, Jefferson devised a way to solve it: Known to mathematicians today as the "method of d'Hondt," it anticipated by many years its independent discovery by the Belgian mathematician and attorney Victor d'Hondt in 1878. President John F. Kennedy was not exaggerating when he toasted a gathering of forty-nine Nobel laureates by describing them as comprising "the most extraordinary collection of talent, of human knowledge, that has ever been gathered together at the White House, with the possible exception of when Thomas Jefferson dined alone."

Jefferson was not alone among the founders in his devotion to science. George Washington was no scientist—his formal education ended at about age fifteen, and by his early twenties he was fighting in the French and Indian War—but he had a sturdily empirical habit of learning from experience and a lifelong scientific curiosity that shone out from the depths of his monumentally composed personality. Having learned mathematics as a surveyor, Washington conducted agricultural experiments at Mount Vernon, designed and tested a new variety of plow, dispatched rare birds for study at Philadelphia's new natural history museum, and offered to send soldiers to help excavate a mastodon discovered near Newburgh, New York, in 1801. Washington couldn't even order a sleigh without inquiring whether, "upon philosophical, mechanical, or practical principles, is it best to have the sliders (excepting always the curve in front) a little circular, or quite straight? The longer the bearing the greater the friction; the shorter, the weight is confined to a smaller space of the sliders, and consequently the compressure greater. . . ." Ever the objective observer, Washington is said to have been taking his own pulse at the moment when he died. John Adams, America's second president, studied physics at Harvard under the astronomer John Winthrop, helped found the American Academy of Arts and Sciences—in Boston, where, he said, "I knew there was as much love of science, and as many gentlemen capable of pursuing it [as may be found in] any other city of its size"—and lamented in a letter to Jefferson that he should have spent more time in scientific pursuits: "Oh that I had devoted to Newton and his fellows that time which I fear has been wasted on Plato and Aristotle [and] twenty others upon subjects which mankind is determined never to understand, and those who do understand them are resolved never to practice." Dr. Benjamin Rush of Pennsylvania, an influential founder who served as treasurer in the Adams administration, was a chemistry professor and epidemiologist who wrote the first American textbook on psychiatry and pioneered the treatment of alcoholism as a disease. Roger Sherman, a signer for Connecticut described by Jefferson as "a man who never said a foolish thing in his life," was a sufficiently capable astronomer to have done the celestial calculations for a colonial almanac. And Benjamin Franklin, who soothed Jefferson's temper while the delegates deleted his denunciations of slavery from his draft of the Declaration, was America's foremost scientist.

In drafting the Constitution, the founders again had frequent recourse to scientific language, logic, and metaphor. John Adams consulted Newton's laws of motion for ideas on how to structure the Constitution. James Madison, its principal author, declared, "Liberty is to faction what air is to fire . . . but it could not be less folly to abolish liberty, which is essential to political life, because it nourishes faction, than it would be to wish the annihilation of air." The founders also drew on the thinking of scientifically inclined English thinkers like the seventeenth-century political philosopher James Harrington and the eighteenth-century colonial governor Thomas Pownall. Harrington had studied democratic precepts during an extended stay in the Venetian Republic; his utopian book *Oceana* compared the "political anatomy" of a bicameral legislature with the circulation of blood through the human heart discovered by William Harvey: "The Parliament is the heart, which, consisting of two ventricles, the one greater and replenished with a grosser matter, the other less and full of a purer, sucks in and spouts forth the vital blood of Oceana." Pownall, a close friend of Benjamin Franklin's, envisioned a future political science "that might become *Principia* to the knowledge of politick operations." A free-trade internationalist, Pownall argued that it was in England's economic interest to grant freedom to its American colonies, since "North-America is become a new primary planet in the system of the world," and will "shift the common center of gravity of the whole system of the European world."

The founders viewed science as an engine that could make posterity healthier and wealthier, to be sure, but also freer; and since freedom is both good in itself and a promoter of betterment, they anticipated that humanity might be improved as the light of science and liberty spread across a benighted world. As Jefferson put it, "The progress of science offers to increase the comforts, enlarge the understanding, and improve the morality of mankind." These were not just pipe dreams but hardheaded political calculations that have in many ways been borne out in the shaping of the modern world. What, exactly, did they have in mind?

The practical potential of scientific research—the process of applying science to technology and, in turn, using the new technology to improve scientific tools—was by the eighteenth century clear enough to those who cared to read its handwriting on the wall. Jethro Tull's seed drill, invented in 1701, sharply reduced the cost of food. Improved iron and steel was being produced, owing to the innovations of Abraham Darby (coke smelting, from

1709) and Benjamin Huntsman (crucible steelmaking, from 1740). Steam-driven pumps drew water out of mine shafts, enabling coal miners to dig to unprecedented depths, while John Kay's "flying shuttle" for faster weaving brought boom times to the textile industry. The decrepit state of European and American internal transportation was beginning to yield to innovations such as England's growing network of canals (hastened by the Canal Act of 1759) and Pierre Trosanquet's road-building efforts in France. It required no great discernment to anticipate the value of improved technology to international economic competition. The feedback loop of the modern world—in which nations that nourish scientific research reap material rewards that increase their wealth and power, leading to further scientific and technological progress and even greater wealth and power—had begun to stir into action.

Technological innovation put cash in the pockets of persons excluded from such traditional sources of prosperity as inherited land and titles, creating a social mobility that undermined old political structures and strictures. Applied science amplified and accelerated the pace of technological progress, offering not only a source of labor-saving devices in the home—a better loom or plow, a more helpful farmer's almanac—but a way to make money and move up in the world. Alexis de Tocqueville viewed this dynamic as central to the American character: "Once works of the intelligence became sources of power and wealth, people were obliged to look upon every scientific advance, every new discovery and idea, as a germ of power placed within the people's grasp." For those who disdained materialistic motives—arguing that knowledge should be pursued as an end in itself, rather than as a gateway to power and wealth—Tocqueville had two rebuttals. First, it's not all that clear that pure science is somehow superior to applied science, Tocqueville suspecting that the intellectuals' habit of esteeming the abstract over the concrete was anachronistic. "Permanent inequality of conditions encourages men to limit themselves to the proud and sterile search for abstract truths," he wrote, "whereas the democratic social state and institutions encourage them to look to science only for its immediate and useful applications." Second, pure science benefits from the growth of applied science regardless of whether that growth is impelled by a desire for power rather than knowledge:

> Although democracy does not encourage men to cultivate science for its own sake, it does vastly increase the number who do cultivate it. It

> is inconceivable that from such a vast multitude there should not on occasion arise a speculative genius impassioned solely by the love of truth. . . . Such a genius will strive to penetrate nature's deepest mysteries regardless of the spirit of his country and his times. There is no need to aid his development; it is enough to stay out of his way.

Of course, any talk of science and freedom in eighteenth-century America ran afoul of the fact that slavery, empirically unjustifiable and ethically reprehensible, remained entrenched in the southern states, was legal in many of the northern ones, and was explicitly protected by the Constitution. The 1790 U.S. census listed slaves as residing in nearly every state and comprising 18 percent of the young nation's 3.8 million inhabitants. The British—who within a generation of losing their American colonies would abolish slavery throughout their empire, dispatching the Imperial Navy to intercept slave ships and deliver their human cargo to freedom in Sierra Leone—were not hesitant to accuse the founding fathers of hypocrisy. "How is it," asked Samuel Johnson, "that the loudest YELPS for liberty come from the drivers of Negroes?" But while it would be pointless to try to explain away the fact that Franklin, Jefferson, and Washington owned slaves, it may be instructive to observe how the scientific spirit worked on their thinking about race.

Franklin was a sometime slave owner—he arrived in England in 1757 accompanied by two slaves, one of whom escaped—and originally a standard-issue racist who thought of Negroes as "revengeful and cruel." When he objected to slavery at all, he did so on the grounds that it made whites lazy while sullying what he took to be a more natural American population of whites and Indians: "Why increase the sons of Africa by planting them in America," he wrote when in his midforties, "where we have so fair an opportunity, by excluding all blacks and tawneys, of increasing the lovely white and red?" But Franklin was also an acute observer who reflected on what he observed, and over the years he began to realize that the shortcomings he had attributed to slaves—the tendency of some to dawdle at their labors, of others to lie or steal—were a product of their unjust servitude rather than a justification for it. Ultimately he helped found schools for black children, coming away from a visit to one of them, in Philadelphia in 1763, with "a higher opinion of the natural capacities of the black race than I had ever before entertained. Their apprehension seems as quick, their memory as strong, and their docility in every respect equal to that of white children." Ultimately he became president of the Pennsylvania Society for Promoting the Abolition of Slavery, petitioning Congress in 1790 to free those "who alone in this land of freedom are degraded into perpetual bondage."

Jefferson was brought up to believe that African-Americans were inferior to whites, but education and experience eroded this youthful prejudice, leaving him permanently conflicted on the subject. Ethically and intellectually, Jefferson opposed slavery. He wrote what John Adams described as a "vehement philippic against Negro slavery" in his draft of the Declaration of Independence, and in his *Notes on the State of Virginia* of 1787 deplored the "miserable condition" of slaves, asking:

> Can the liberties of a nation be thought secure when we have removed their only firm basis, a conviction in the minds of the people that these liberties are of the gift of God? That they are not to be violated but with his wrath? Indeed I tremble for my country when I reflect that God is just.

Four years later he commended the work of the African-American astronomer Benjamin Banneker to the Marquis de Condorcet, writing:

> I am happy to be able to inform you that we have now in the United States a Negro, the son of a black man born in Africa, and of a black woman born in the United States, who is a very respectable mathematician. I procured him to be employed under one of our chief directors in laying out the new federal city on the Potomac [Washington, D.C.], & in the intervals of his leisure, while on that work, he made an Almanac for the next year, which he sent me in his own handwriting, & which I enclose to you. . . . I shall be delighted to see these instances of moral eminence so multiplied as to prove that the want of talents observed in [Negroes] is merely the effect of their degraded condition, and not proceeding from any difference in the structure of the part on which intellect depends.

All the while, however, Jefferson remained a slave-owning Virginia planter, who like other plantation owners was concerned that any attempt to end slavery might incite a civil war: "We have the wolf by the ear, and we can neither hold him, nor safely let him go." This stance may have been hypocritical and self-serving, but it was not groundless: Forty years after Jefferson confided to John Adams his fear that abolition might set off "another Peloponnesian war" among "confederacies" of states, the defeat of the Confederate States by the Union brought an end to American slavery at a cost of over half a million dead.

Jefferson's fellow Virginian George Washington regarded slavery as not only repugnant but dishonorable—a serious consideration for a man whose honor was his core, his compass, and his most conspicuous asset. "If there are spots on his character, they are like spots on the sun, only discernible by the magnifying powers of a telescope," declared the *Pennsylvania Journal*, in a 1777 editorial calling Washington "the best man living," but Washington felt that his owning slaves eclipsed that sun. He supported abolition—writing in 1797, "I wish from my soul that the legislature of this state could see the policy of a gradual abolition of slavery"—but was hard-pressed to imagine how it could be brought about without ruining the economies of the southern states or inciting them to secession. Meanwhile he refused to buy or sell any more slaves, converted his Mount Vernon plantation from tobacco to grain cultivation to reduce its labor requirements, tried without success to lease most of it so he could free at least some slaves, and wrote wistfully of wishing "to see some plan adopted, by which slavery in this Country may be abolished by slow, sure, & imperceptible degrees." A firsthand glimpse of Washington's perplexity was recorded by the English music-hall comedian John Bernard, who came upon the elderly ex-president extracting an unconscious woman from the wreckage of a carriage that had crashed near Mount Vernon. Bernard accepted an invitation to accompany Washington home and their conversation turned to science, Washington naming Franklin, Rittenhouse, and Rush as American scientists of international caliber. He was praising England as the cradle of liberty when a slave carrying a jug of water entered the room. "This may seem a contradiction," said a blanching Washington,

> but . . . till the mind of the slave has been educated to perceive what are the obligations of a state of freedom, the gift would insure its abuse. We might as well be asked to pull down our old warehouses before trade had increased to demand enlarged new ones. Both houses and slaves were bequeathed to us by Europeans, and time

> alone can change them; an event, sir, which, you may believe me, no man desires more heartily than I do. Not only do I pray for it on the score of human dignity, but I can clearly foresee that nothing but the rooting out of slavery can perpetuate the existence of our union, by consolidating it in a common bond of principle.

Washington's will ordered that Mount Vernon's slaves be freed once his wife had died, "and I do moreover most pointedly, and most solemnly enjoin it upon my executors . . . to see that *this* clause respecting slaves, and every part thereof be religiously fulfilled . . . without evasion, neglect or delay, after the crops which may then be on the ground are harvested." Thus he left a divided legacy—a partial eclipse, if you will—as a living slaveholder and posthumous emancipator.

Many other faults may be found with the founders and their Constitution; there is a thriving academic industry that does little else. Some scholars feel that the Constitution was insufficiently egalitarian and democratic (although it outlawed nobility and made the people the ultimate source of power) or that it did not go far enough in extending voting and other human rights (although no other constitution circa 1787 did this either). Perhaps the fairest test is to ask whether the republic eventually managed to mitigate the inequities and injustices that had attended its birth—and this it did. Slavery was abolished in 1865. A century later, racial segregation was stripped of the sanction of law if not always of custom. The term "civil rights," which arose in the context of racial equality, began to be applied to other instances in which the law was said to impede human rights—from equal treatment of women (who were unable to vote until 1920, and long thereafter discriminated against in contract, employment, and matrimonial law) to homosexual marriage. By the twenty-first century Americans had reason to expect that every legitimate civil rights cause could in the long run earn majority support. Any claim that the American enterprise was poisoned at its roots by the sins of the founders must be weighed against this record of advancement.

The founders often spoke of the new nation as an "experiment." The term, borrowed from science, had several meanings. Trivially, it meant that the United States might succeed or fail, as might any new nation; that much was obvious. Procedurally, it involved deliberations about how to facilitate both liberty and order, matters about which the individual states experimented considerably during the eleven years between the Declaration of Independence and the Constitution. Empirically—and this is the way the phrase is most often employed today—it raised the question of whether a free constitutional democracy would work. As Jefferson wrote in 1804, "No experiment can be more interesting than that we are now trying, and which we trust will end in establishing the fact, that man may be governed by reason and truth."

The new government, like a scientific laboratory, was designed to accommodate an *ongoing* series of experiments, extending indefinitely into the future. Nobody could anticipate what the results might be, so the government was structured, not to guide society toward a specified goal, but to sustain the experimental process itself. It might blow up in the process—the danger referred to by Abraham Lincoln when he spoke at Gettysburg of the Civil War as "testing whether . . . any nation . . . conceived in liberty and dedicated to the proposition that all men are created equal . . . can long endure"—but the idea was to make it strong and flexible enough to keep cranking out tolerable results in changing times. As Madison observed in *The Federalist*, any system of government will be flawed, since the imperfection of human institutions "arises as well from the object itself, as from the organ by which it is contemplated." The empirical course Madison urged was to conduct "an actual trial," by letting the experiment function and dealing with its faults as they presented themselves.

The drafters of the Constitution often compared it to Newton's mechanical model of the solar system: Its mechanisms of checks and balances were like the dynamic laws governing the motions of the planets, functioning reliably in the background while human events unfolded unpredictably in the foreground. The point was to let human frailty, inadequacy, and ignorance play out within a dynamic system robust enough to absorb the many mishaps that would inevitably result. Madison, who studied Newtonian physics at Princeton and was known as a "man of science," depicted the proposed federal government as akin to the sun, its legal precedence over the states required lest the "planets . . . fly from their orbits." John Adams similarly compared the states to Newton's planets—and was opposed by the Virginia planter John Taylor, who preferred a botanical analogy that likened the divisions of government to the parts of plants.

The Federalist Papers—written by Madison, Alexander Hamilton, and John Jay, and published under the nom de plume "Publius" from 1787 to 1788 to argue for the proposed constitution—employ the word "experiment" forty-five times, and "democracy" only ten times. Democracy was important, but underlying it was the principle that citizens be free to experiment, assess the results, and conduct new experiments. As Lincoln remarked in 1861:

> Without the Constitution and the Union, we could not have attained the result; but even these, are not the primary cause of our great prosperity. There is something back of these, entwining itself more closely about the human heart. That something, is the principle of "Liberty to all"—the principle that clears the path for all—gives hope to all—and, by consequence, enterprise, and industry to all.

Many of the vexing characteristics of today's liberal democracies—their cacophony, inefficiency, and haphazard planning—become less troubling if their inherently experimental character is taken into account. Like scientific experiments, democracies tend to be untidy, patched-up affairs that seldom work out as expected. They stumble around like fugitives in a fog, slapping together ugly compromises that seem laughably inadequate to the problems at hand. Yet this unappealing system of government, so repellent to the perfectionist in each of us, has proven to be tougher, more resilient, and better able to answer to the needs of its citizens than any other. And that is because democracy, like science, it is not built on hopes of human perfection but on an acknowledgment of human fallibility. Franklin expressed this sentiment elegantly: "I confess that there are several parts of this Constitution which I do not at present approve," he told the Constitutional Convention on the final day of its deliberations,

> but I am not sure I shall ever approve them: For having lived long, I have experienced many instances of being obliged by better information, or fuller consideration, to change opinions even on important subjects, which I once thought right, but found to be otherwise. It is therefore that the older I grow, the more apt I am to doubt my own judgment, and to pay more respect to the judgment of others. . . . Thus I consent, Sir, to this Constitution because I expect no better *and because I am not sure that it is not the best.*

Modern scientists have more often comprehended the experimental nature of liberal democracy than have modern philosophers. "Science is far from a perfect instrument of knowledge," wrote the astronomer Carl

Sagan. "It is just the best we have. In this respect, as in many others, it's like democracy." "Science is a kind of open laboratory for a democracy," noted the theoretical physicist Lee Smolin.

> It's a way to experiment with the ideals of our democratic societies. For example, in science you must accept the fact that you live in a community that makes the ultimate judgment as to the worth of your work. But at the same time, everybody's judgment is his or her own. The ethics of the community require that you argue for what you believe and that you try as hard as you can to get results to test your hunches, but you have to be honest in reporting the results, whatever they are. You have the freedom and independence to do whatever you want, as long as in the end you accept the judgment of the community. Good science comes from the collision of contradictory ideas, from conflict, from people trying to do better than their teachers did, and I think here we have a model for what a democratic society is about. There's a great strength in our democratic way of life, and science is at the root of it.

The physicist Richard Feynman, lecturing at the University of Washington in 1963, observed that political "fundamentalists"—those who imagine that everything would be fine if only everyone subscribed to a putatively perfect philosophy—make the mistake of underestimating the experimental, unpredictable nature of democratic governance. He reported talking with a group of conservatives who thought that "the Constitution was right the way it was written in the first place, and all the modifications that have come in are just the mistakes . . . I tried to explain that . . . according to the Constitution there are supposed to be votes. It isn't supposed to be automatically determinable ahead of time on each one of the items what's right and what's wrong. Otherwise there wouldn't be the bother to invent the Senate to have the votes." Speaking in his customarily rough-hewn, extemporaneous way, Feynman added a penetrating insight:

> The government of the United States was developed under the idea that nobody knew how to make a government, or how to govern. The result is to invent a system to govern when you don't know how. And the way to arrange it is to permit a system, like we have, wherein new

> ideas can be developed and tried out and thrown away. The writers of the Constitution knew of the value of doubt. In the age that they lived, for instance, science had already developed far enough to show the possibilities and potentialities that are the result of having uncertainty, the value of having the openness of possibility. The fact that you are not sure means that it is possible that there is another way some day. That openness of possibility is an opportunity. Doubt and discussion are essential to progress. The United States government, in that respect, is new, it's modern, and it is scientific. It is all messed up, too. Senators sell their votes for a dam in their state and discussions get all excited and lobbying replaces the minority's chance to represent itself, and so forth. The government of the United States is not very good, but it, with the possible exception [of] the government of England, is the greatest government on the earth today, is the most satisfactory, the most modern, but not very good.

The deeds of the founding fathers accorded with their view of the Constitution as setting a noble experiment in motion. Unlike almost all history's other victorious revolutionaries, they did not struggle to consolidate personal control of the nation they had founded, but readily relinquished power in order to let the experiment proceed. George Washington set an early example by resigning his military commission and quietly returning to civilian life as soon as the Revolution had been won, though many Americans were clamoring to crown him king. (Advised of Washington's intention, King George III reputedly said, "If he does that, he will be the greatest man in the world.") After serving two terms as president, Washington again exhibited restraint: Rather than continue in office he congratulated his fellow citizens "on the success of the experiment" and went home—publishing his farewell address in a newspaper rather than delivering it in person so that there would be no second-guessing his decision. The address urged that the American experiment be judged by its empirical results: "Is there a doubt whether a common government can embrace so large a sphere? Let experience solve it. To listen to mere speculation in such a case were criminal."

Jefferson had long championed freedom of the press, going so far as to claim that he would prefer "newspapers without a government" to "a government without newspapers." This conviction was sorely tested once President Jefferson came under vituperative attack from some of the harshest and least principled editors ever to wield a pen. Two developments fueled the acid tide—the emergence of political parties, which, as the founders had feared, amplified incendiary quarrels between factions, and a boom in printing presses. Improvements in printing-press technology by scientifically inclined experimenters like the third Earl Stanhope in England and George Clymer in Philadelphia were cutting costs at a rate robust enough to support a rapid profusion of newspapers. When Jefferson took office in 1801 there were already about two hundred American newspapers; a generation later there were over a thousand. Faced with intense competition, unscrupulous editors invented or hyperbolized accusations aimed at inciting indignant howls from the wounded parties—who, ideally, would sue for libel and in doing so stir up even more publicity. Jefferson, who as chief executive was automatically the largest target on the landscape, was assailed in print as a traitorous atheist, enthralled by French philosophy and habituated to gambling and cockfighting, who had fathered children by one of his slaves, Sally Hemings. Yet Jefferson continued to defend the shining promise of a free press, if not its tarnished actuality, as essential to the democratic process. "I have lent myself willingly as the subject of a great experiment," he wrote,

> which was to prove that an administration conducting itself with integrity and common understanding cannot be battered down even by the falsehoods of a licentious press. . . . The fact being once established that the press is impotent when it abandons itself to falsehood, I leave to others to restore it to its strength by recalling it within the pale of truth. Within that it is a noble institution, equally the friend of science and of civil liberty.

Jefferson viewed the printing press as a boon to the free discourse required by science and liberal democracy. As he wrote to John Adams in 1823, "The light which has been shed on mankind by the art of printing has eminently changed the condition of the world . . . and while printing is preserved, it can no more recede than the sun return on his course." Democratic leaders had to put up with free and open criticism in print; it was a price to be paid for the health of the republic.

Many felt that the price was too high—that harsh criticism of the government would drive qualified officials from office and undermine public confidence, promoting cynicism and apathy. Countering such fears, Jefferson in his second inaugural address noted that although his administration had been viciously attacked—"The artillery of the press has been leveled against us, charged with whatsoever its licentiousness could devise or dare"—he nevertheless had managed to function effectively and had even been reelected. Hence the experiment, testing "whether freedom of discussion, unaided by power, is not sufficient for the propagation and protection of truth," was at least provisionally successful:

> The experiment has been tried; you have witnessed the scene; our fellow-citizens looked on, cool and collected; they saw the latent source from which these outrages proceeded; they gathered around their public functionaries, and when the Constitution called them to the decision by suffrage, they pronounced their verdict, honorable to those who had served them and consolatory to the friend of man who believes that he may be trusted with the control of his own affairs.

The lesson Jefferson drew from this positive result was that "the press, confined to truth, needs no other legal restraint; the public judgment will correct false reasoning and opinions on a full hearing of all parties; and no other definite line can be drawn between the inestimable liberty of the press and its demoralizing licentiousness." The success of this process was due, Jefferson said, "to the reflecting character of our citizens at large. . . . I offer to our country sincere congratulations." This was charming, but not merely charming: The future of the democratic experiment rested with the electorate.

Science was likely to spread across the world—Jefferson noting that "every man of science feels a strong and disinterested desire of promoting it in every part of the earth"—and democracy might be expected to do the same. "Science has liberated the ideas of those who read and reflect, and the American example had kindled feelings of right in the people," Jefferson wrote in 1813. "An insurrection has consequently begun, of science, talents and courage, against rank and birth, which have fallen into contempt." The following year he spoke of the need for citizens to overcome the "dread" suffered by "those whose interests or prejudices shrink from the advance of truth and science." Ultimately, Jefferson declared, "There is not a truth existing which I fear or would wish unknown to the whole world." The same outlook is evinced in the last of Jefferson's eighteen thousand letters:

> All eyes are opened, or opening, to the rights of man. The general spread of the light of science has already laid open to every view the palpable truth, that the mass of mankind has not been born with saddles on their backs, nor a favored few booted and spurred, ready to ride them legitimately, by the grace of God.

Jefferson wrote that letter to excuse himself, on grounds of fading health, from attending a celebration in Washington of the fiftieth anniversary of the signing of the Declaration of Independence. He died on that celebratory date, after asking, shortly past midnight, "Is it the Fourth?" His old friend John Adams—to whom Jefferson had written, eight years earlier, "I am sure that I really know many, many, things, and none more surely than that I love you with all my heart, and pray for the continuance of your life until you shall be tired of it yourself"—died the same day. His last words were, "Thomas Jefferson still survives." Though Jefferson had died a few hours earlier, Adams was right. But this was not yet clear to all.

CHAPTER SIX

THE TERROR

As distant as heaven is from earth, so is the true spirit of equality from that of extreme equality.

—MONTESQUIEU, 1748

The French Revolution was as much the progenitor of modern totalitarianism as of modern democracy.

—CHARLES COULSTON GILLISPIE, 2004

In the troubled years leading up to the French Revolution, King Louis XVI frequently expressed regret over his having backed the American colonies in their rebellion against England. His principal motive had been to avenge the defeat suffered by his grandfather, Louis XV, in the French and Indian War—a debacle that cost France Louisiana and nearly all its other American holdings—and thereby regain some of France's former glory. But while the outcome of the American Revolution may indeed have restored a measure of French prestige, it had also demonstrated that a working democratic republic could be established in defiance of a powerful monarchy. This lesson was not lost on the French people. They were avidly reading translations of the Declaration of Independence and Thomas Paine's *Common Sense*, sharing stirring tales of how General Lafayette and his eight thousand French troops had helped the tattered American revolutionaries prevail against regimented redcoats on the field of battle, and gossiping about Benjamin Franklin's libidinous triumphs among the ladies of Parisian high society: Franklin, science and liberty personified, a buckskin-booted democrat adorned by his glittering wit, had become the most popular man in France.

Worst of all, France's American adventure had run up debts that threatened to bankrupt the government. The beginning of the end came on July 13, 1788, when a freak summer storm pelted northern France with hailstones large enough to kill men and livestock, destroying crops just prior to harvest and driving up food prices until a loaf of bread cost a month's wages. The coldest winter in living memory ensued. Desperate peasants, foraging at dawn for roots and bark to feed their families, stumbled over the corpses of neighbors who had died while pursuing the same desperate endeavor. These misfortunes climaxed a decade marked by crippling crop failures, caused in part by the 1783 eruption of an Icelandic volcano but blamed by conspiracy theorists on sinister government plots.

The bloated French nobility was ill-suited to deal with any such emergency. A hundred thousand aristocrats, deliberately rendered superfluous by their regent, lived tax-free amid a perpetual swirl of parties and amusements in Paris and at Versailles, and were about as much use in a crisis as peacocks in a house fire. Nor was Louis XVI capable of kindling much hope among his subjects. Corpulent, somewhat slow-witted, and habitually decked out in robes so splendid as to simultaneously summon up the delights of the dinner table, the hunt, and a starry night sky, Louis seemed more a waddling embodiment of excess than an agent of reform—an impression reinforced by the self-indulgent profligacy of his wife, Marie Antoinette, known popularly as "Madame Deficit."

Louis was also, however, an honest and earnest man who understood that something had to be done. When all else failed he called a national election in order to reconstitute the Estates-General. This old parliament not having met for 175 years, historians consulted antiquated books to acquaint themselves with its rules of procedure. Once elected, representatives of the first estate (clergy, in vestments) the second (nobles, in colorful silks), and the third (bourgeoisie, artisans, and peasants, in black) marched to the opening session at Versailles on May 5, 1789, listened to dedicatory speeches, and then collapsed into confusion. Their first priority was to revise the old rules—otherwise the clergy and nobility, representing only 3 percent of the populace, could outvote the other 97 percent—but they could not agree on how to do so, and after a few weeks the king disbanded them. Representatives of the third estate, emboldened by a recent pamphlet by Abbé Sieyès titled *What Is the Third Estate?* ("It is the whole"), defiantly met in what is known to history as a tennis court (actually a handball court) and resolved to call themselves the National Assembly and to draft a constitution. The king acceded to their terms, but by then there was rioting in the streets of Paris. On July 14 a mob stormed the Bastille and seized its stores of gunpowder. Within days, peasants in the countryside were rising up to attack their masters and burn the ledgers listing their feudal dues and debts. In response, the National Assembly abolished the feudal system, passed a Declaration of the Rights of Man and of the Citizen that made all citizens equal before the law—it affirmed their right to "liberty, property, security, and resistance to oppression" and defined liberty, sensibly enough, as "freedom to do everything which injures no one else"—and enacted a constitution limiting monarchical powers. The king and his family were politely but firmly invited to quit Versailles for Paris, where they remained as virtual prisoners of the new government. The French Revolution had begun.

The Americans initially took an optimistic view. Thomas Jefferson, who as ambassador to Paris had a front-row seat to the gathering storm, wrote complaisantly to James Madison in May 1789 that "the revolution of France has gone on with the most unexampled success hitherto," and a month later informed John Jay that "this great crisis being now over, I shall not have matter interesting enough to trouble you with as often as I have done lately." He told his friend John Trumbull that "all danger of civil commotion here is at an end, and it is probable they will proceed to settle to themselves a good constitution, and meet no difficulty in doing it." The French future looked so sunny that August found Jefferson preparing to return to America, having successfully petitioned Congress for a six-month leave of absence on grounds that, as he assured Paine, "Tranquility is well established in Paris, and tolerably so thro' the whole kingdom; and I think there is no possibility now of any thing's hindering their final establishment of a good constitution, which will in its principles and merit be about a middle term between that of England and the United States."

Panglossian as such sentiments seem in retrospect, they were not unfounded. If the French reformers faced problems spared their American counterparts—lingering feudalism, grinding poverty, an inequitable and unpopular legal system, an alarmingly rigid and vertical class structure, deeply entrenched religious biases—they seemed to have got off to a good start in addressing them. Plus France was, after all, widely regarded as the most civilized and sophisticated society in Europe, with particular strengths in science and philosophy; surely, in such a nation, common sense would prevail. And perhaps it would have, had the nature of science and its implications for politics been more clearly understood in France. But this was not the case.

French science may not have lived up to the high esteem in which it held itself, but it was admirably organized, generously funded, and peopled by researchers of undeniable capacity. While scientific societies elsewhere struggled for funding, the French Academy of Sciences, sumptuously headquartered at the Louvre, nourished a galaxy of talent that other nations could only envy—men like Antoine Laurent Lavoisier, the founder of modern chemistry; Georges Cuvier, a pioneer in paleontology and comparative anatomy; Jean Baptiste Lamarck, who formulated a pre-Darwinian theory of biological evolution; and Pierre Simon de Laplace, creator of a prescient "nebular hypothesis" that envisioned how the solar system formed. Mathematicians today still make use of the Lagrangian function of Joseph Louis Lagrange, the Fourier transformation of Jean Baptiste Fourier, and the Poisson distribution of Siméon Denis Poisson; Fresnel lenses (after Augustin Jean Fresnel) are employed in modern automobile headlamps, and everyone who has ever changed an electrical fuse is familiar with the amp, named for André Ampère, the founder of electrodynamics.

Philosophically, France was the capital of the Western world. The gleaming rays of Descartes' putatively pure reasoning had been refracted into warmer tones by latter-day Cartesians like Géraud de Cordemoy and Nicolas de Malebranche, who invoked God to bridge the mind–body gap; Blaise Pascal, who invoked a god of the heart as well as the mind; and Pierre Gassendi, who portrayed reason as an aspect of "spirit." Montesquieu had set forth a separation-of-powers approach to government that influenced the American founders, while the encyclopedists Denis Diderot and Jean Le Rond d'Alembert popularized Enlightenment ideals and called for reforms in public education. The encyclopedists' purpose, Diderot wrote, was "to change the general way of thinking."

> All things must be examined, debated, investigated without exception and without regard for anyone's feelings. We must ride roughshod over all these ancient puerilities, overturn the barriers that reason never erected, give back to the arts and sciences the liberty that is so precious to them. . . . We have for quite some time needed

a reasoning age when men would no longer seek the rules in classical authors but in nature.

Fine minds adorned French culture almost as conspicuously as did the bejeweled gowns and fantastic headdresses of the noblewomen at Versailles. It seemed that, as one modern historian has summarized the prevailing view, "Philosophers were engaged in a winning battle for progress. The world was changing, nothing was doing more than 'philosophy' to promote that change, and it was change for the better."

Encouraged by all this seeming sensibleness, Thomas Paine wrote to George Washington in 1790, "I have not the least doubt of the final and complete success of the French Revolution." Observing that "the countries the most famous and the most respected of antiquity are those which distinguished themselves by promoting and patronizing science, and on the contrary those which neglected or discourage it are universally denominated rude and barbarous," Paine gazed approvingly on the "distinguished" scientific accomplishments of the French and felt assured that they would not descend into barbarity. Yet descend they did, into the bloodbaths of the Terror—their constitution suspended, their ringing declaration of the Rights of Man set aside, their press censored in the name of "the people." What went wrong?

A great many answers to that question have been advanced, from the conspiratorial (the Revolution was led astray by the Illuminati, the Freemasons, or some other secret organization) to the cultural (Anglo-American traditions were somehow better suited to the emergence of liberal democracy), the religious (it being claimed, especially in England, that Protestantism better prepared a citizenry for political liberty than did Catholicism), and, more persuasively, the pragmatic: By declaring war on Austria, Prussia, Great Britain, the Dutch Republic, and Spain, the French revolutionaries alienated their neighbors and depleted their own police and military forces, plunging the populace into paranoia and mob rule. But what emerges most clearly is that the French revolutionaries suffered two closely related misfortunes. First, they neglected the fundamental lesson of science and liberalism—that the key to success is to experiment and to abide by the results—assuming instead that the point of a revolution was to implement a particular philosophy. Second, they chose the wrong philosophy.

As the inheritors of an illustrious philosophical heritage, the French were more inclined than most to regard science as subordinate to philosophy.* Their thinkers were apt to prefer the Cartesian side of science, which involves reasoning from first principles, to the Baconian, which is inductive and experimental. To the Cartesians, science was a variety of rational philosophy, and experiments had less to do with acquiring new knowledge than with demonstrating the power of reason. To the Baconians, experiment was the font of true knowledge, and the grand constructs of the philosophers were castles built on sand.

French popularizers of both stripes, eager to acquaint the general public with the wonders of science, produced a rising tide of *haute vulgarizations* that quadrupled the annual rate of science book publication in France between 1715 and the end of the century. If the Cartesian side of science was too coldly rational to charm many readers, the science writers warmed it up by bringing God into the picture: The marvelous mathematical clockwork of the solar system reflected the glory of its creator. If Baconian science seemed too complicated, the science writers responded by humanizing the scientists themselves, with tales of astronomers searching for new worlds and brave naturalists gathering botanical samples in steamy jungles. Ambitious works like Abbé Pluche's eight-volume *Spectacle de la nature* (Newtonian science sweetened with lovely engravings) and George Buffon's thirty-eight volume *Histoire naturelle* (a common ancestry for man and apes?) enjoyed impressive success. But in reconciling science with older institutions, all such domesticated depictions underestimated the power of science to produce not only new knowledge but new ways of learning. To regard science as but a philosophy, bounded by the rules laid down by philosophers past, was to downplay both its creativity and its political implications.

Scientific knowledge is gained by conducting experiments and adjusting one's hypotheses accordingly. In the process some hypotheses emerge as more productive than others—which is to say more efficient, useful, and true in the sense that a carpenter's level is true—but none attains the status

---

*American thinkers were more skeptical of modern philosophy, and less familiar with it. Jefferson and Franklin thought of rational certainty as limited to mathematics, and consequently were chary about rationalists like Descartes—whose vortex theory of planetary rotation, Jefferson noted with satisfaction, had been "exploded" by Newton.

of certitude. Liberal democracies similarly are based on experimentation: Every election and every piece of legislation is an experiment, the outcomes of which are understood to be conditional and hence open to additional discussion and experimentation. The members of parties and factions in a liberal-democratic state may feel strongly that their views are superior to those of the others, but when defeated in an election are generally willing to let the other side give it a try, recognizing that nobody has a monopoly on the truth. From this spirit flow such liberal traditions as toleration for free speech and respect for the loyal opposition. Science applied to politics tempers reason with reasonableness, arriving at provisional conclusions, whereas philosophers can and do reason their way from self-evident propositions to unassailable results—as Descartes did, or thought he had done.

Philosophy, not science, drove the French Revolution. Revolutionary French thinkers initially drew on John Locke's scientific empiricism, David Hume's skepticism, Montesquieu's practicality, and Voltaire's wit (a sense of humor being, as George Santayana would remark, a sense of proportion), but the revolution soon devolved into the hands of radical firebrands like Maximilien Robespierre and Jean-Paul Marat, who thought of *liberté* as but a means to imposing a monolithic state of *egalité* and whose shrinking circles of philosophical zealotry came to be rimmed in red. This misplaced rationality was combined with romanticism—a seemingly odd alliance that is actually quite common, rationality calling forth romanticism in much the same way that a fervent preacher of virtue by day may succumb to vice by night. The most reprehensible goals of French rationalists were glowingly endorsed in the romantic and wildly popular new works of a literally fantastic *philosophe* who reviled the scientific values of modesty and toleration. This was Jean-Jacques Rousseau, of whom it was said that he "invented nothing, but he set everything on fire."

"Rousseau is a strange figure," notes the twentieth-century philosopher John Herman Randall Jr. "Uneducated, he wrote the most influential book ever written on education. Almost incapable of fulfilling any social duties whatsoever, he wrote the most powerful book ever written on the supreme duty of political obligation. Hating society and the company of men, he captured their imagination as few others ever have, and became the vogue in society." A favorite philosopher among those who knew few others, Rousseau all but dominated French thought from 1760 on. He more than anyone else invented Romanticism—a Manichaean world view that champions sentiment over logic, caprice over common sense, instinct over civilization, and mysticism over clarity.

Born in Geneva, Rousseau was raised by an aunt and uncle. (His mother died shortly after giving birth and his father deserted him ten years later.) At sixteen he began a lifetime of solitary wandering, a circumstance he attributed to his unique and colorful individuality—he suggests in his *Confessions* that nature "broke the mold" after making him—but which more likely reflected the fact that few could long tolerate the companionship of a man whose habitual behavior was to wreak wanton damage, excuse himself from responsibility for his actions, and then lament his resulting unpopularity as the inexplicable lot of "the most sociable and loving of men . . . cast out by all the rest . . . alone in the world, with no brother, neighbor or friend, nor any company left me but my own." While working as a servant Rousseau was confronted with evidence of his having stolen a valuable ribbon. He blamed an innocent maid, who was punished for his crime, then pardoned himself on grounds that because "she was present in my mind," he naturally "threw the blame from myself on the first object that presented itself." Over the years he profited from his affairs with wealthy women, all the while conducting an ongoing liaison with an illiterate servant girl, Thérèse Le Vasseur; the couple had five children, all of whom Rousseau immediately delivered to orphanages. ("From the first moment in which I saw her, until that wherein I write, I have never felt the least love for her," Rousseau said of Le Vasseur. "The physical wants which were satisfied with her person were to me solely those of the sex.") Diderot, one of Rousseau's many short-term friends, soon concluded that he was "deceitful, vain as Satan, ungrateful, cruel, hypocritical and full of malice." Voltaire called him "an inconsequential poor pygmy of a man, swollen with vanity." Threatened with arrest in France and Switzerland for advocating democracy and denying the divine right of kings, Rousseau lived for three years in Motiers under the protection of Frederick the Great before the villagers became so disgusted with him that they threatened his life. He fled to England with the aid of David Hume, who found him lodging and a pension—whereupon Rousseau accused Hume of conspiring against him, grandly refused the pension, and abruptly returned to France. (Adam Smith wrote consolingly to Hume, "I am thoroughly convinced that Rousseau is as great a Rascal as you and as every man here believes him to be.") Edmund Burke, who befriended Rousseau for a time but soon thought better of it, declared that he "entertained no principle, either to influence his heart, or guide his understanding, but vanity; with this vice he was possessed to a degree little short of madness." In the end Rousseau went unambiguously mad, dying in poverty in Paris at the dawn of the Revolution, a suspected suicide.

All of which played well with the sort of readers, never in short supply and amply represented in the street mobs of revolutionary Paris, who envision philosophers as lonely, misunderstood souls who wander an uncaring world thinking deep thoughts that only a few individuals are sensitive enough to appreciate. Rousseau theorized that his father's reading to him from his late mother's favorite books accounted for what he called "my odd, romantic notions of human life." Odd they certainly were, and if a romance is "a baseless, made-up story . . . full of exaggeration or fanciful invention" (the *Random House Dictionary* definition) then Rousseau's philosophy fills the bill. His political philosophy is founded on the baseless assertion that humans originally lived in a state of peaceful equality, from which happy status they eventually fell, owing to ill-advised innovations like toolmaking and property rights. In his *On the Origin of Inequality,* Rousseau embroiders this hypothesis with scraps of explorers' tales about "savages" whose traditional way of life allegedly approximated the primordial state of nature. Most of these assertions are so far off the mark as to be laugh-out-loud funny. Rousseau writes that the inhabitants of the Venezuelan jungles live "in absolute security and without the smallest inconvenience . . . expos[ing] themselves freely in the woods . . . but no one has ever heard of one of them being devoured by wild beasts." Man in the state of nature seldom gets sick and so "can have no need of remedies," is sexually promiscuous but untroubled by jealousy, has no property and needs none: "I see him satisfying his hunger at the first oak, and slaking his thirst at the first brook; finding his bed at the foot of the tree which afforded him a repast; and, with that, all his wants supplied."

Rousseau admits that his natural Man is hypothetical, abiding in "a state which no longer exists, perhaps never did exist, and probably never will exist." He is contemptuous both of facts—"Let us begin then by laying facts aside"—and the books that contain them. "I hate books," he declares. "They only teach one to talk about what one does not know." Nevertheless we must somehow "have true ideas" about humanity's primordial state, "in order to form a proper judgment of our present state." And why must we do this? In order to figure out where civilization went astray. "Man is naturally good," Rousseau asserts, "and only by institutions is he made bad." Which institutions made him bad? Private property and civil society: "The first man who, having enclosed a piece of ground, bethought himself of saying *This is mine*, and found people simple enough to believe him, was the real founder of civil society. From how many crimes, wars and murders, from how many horrors and misfortunes might not any one have saved mankind, by pulling up the stakes." Instead, humans foolishly fashioned fishhooks and arrows, donned clothing, captured fire, built huts for themselves . . . and the world went to hell.

An astonishingly large number of people still believe in Rousseau's mythical prehistory. Books continue to be published that depict early humans as peaceful, communistic idlers corrupted by innovations such as agriculture, which one modern author calls "the worst mistake in the history of the human race." A few empirical findings may serve to alleviate this delusion.

Anthropologists and other social scientists studying preagricultural peoples have adduced fairly reliable evidence of their behavior, and the results do not look like Eden. Two-thirds of the hunter-gatherer societies they have examined were found to abide in a state of almost constant warfare, producing annual fatality rates of half a percent of the population—equivalent to the murder in New York City of a hundred persons *a day.* (The actual homicide rate in New York City is under three per day.) The Tahitians, regarded by nineteenth-century Europeans as dwellers in paradise, were so warlike that they found it necessary to maintain a military mobilization level exceeding that of the Soviet Union during World War II. The !Kung bushmen of the Kalahari, portrayed by Elizabeth M. Thomas in a widely read book, *The Harmless People*, killed one another between 1920 and 1955 at quadruple the rate of Americans in the fifties and sixties. The Murngin Aborigines in the 1880s lost nearly a quarter of their males to feuds and battles. Archeologists sifting through the crushed skulls and mutilated remains of the men, women, and children cast into mass graves at Crow Creek, South Dakota, seven hundred years ago, and near Talheim, Germany, seven thousand years ago, estimate that people back then were far more likely to meet a violent death than were the twentieth-century Europeans who suffered through two world wars. Exceptions are hard to find. The "gentle Tasaday," a modern "Stone-Age tribe" that spawned a documentary film and a book, were genuinely peaceful—but they were also a fraud, perpetuated by the Philippine government's director of indigenous peoples, who allegedly absconded with the funds he had raised for the protection of this nonexistent tribe.

As indifferent to their environment as to their neighbors, the hunter-gatherers of prehistoric Europe and America evidently drove large game animals to extinction, switching to smaller game once there was nothing else left to sustain them. The hunters tended to be taller and more physically fit than their agrarian counterparts—a phenomenon that led Hollywood moviemakers to feature Apache and Comanche warriors over more settled peoples like the Pueblo—but that was because their weaker compatriots, whom agrarian societies could afford to sustain, were killed in battle or allowed to perish in infancy. As it happens, there is a political philosophy that celebrates warlike leaders who plunder agrarian settlements while happily sacrificing misfits among their own numbers: It is called fascism. It is thanks to agriculture, technology, science, and liberalism, rather than to Rousseau and the fascists, that so many puny, weak, eccentric, and otherwise encumbered individuals have contributed to civilization—people like Voltaire, Samuel Johnson, Toulouse-Lautrec, Marcel Proust, and Stephen Hawking.

Ben Franklin was willing to strum Rousseau's harp to the advantage of his diplomatic mission. His celebrity in France—which led him to write back to his sister that in Paris "my face [is] almost as well known as that of the moon"—sprang from his canny perception that he could encourage the French to support the American cause by playing on their romantic devotion to Rousseau. Franklin "seemed a living advertisement for the virtues of Rousseauistic simplicity, the product of a sylvan paradise far from the jaded artificiality of Europe," notes one historian. He was anything but. Far from exemplifying the stoicism of the noble savage, Franklin was a gluttonous *bon vivant* who advised his friend John Paul Jones that the best way to learn French was to acquire a "sleeping dictionary," by which he meant a French mistress; John Adams found Franklin's life in Paris to be "a scene of continual dissipation." No innocent, Franklin was so politically pragmatic as to be dogged by accusations of deviousness. Rather than living in a Rousseauistic state of equality with his fellow man, Franklin was a self-made media baron who had gained control of American newspapers, almanacs, and the postal service that distributed them, earning a fortune by his early forties. His famous inventions—bifocals, the Franklin stove, the lightning rod—were the work of a man committed to a materially better future, not a return to nature.

Rousseau regarded science and technology as ignoble, since they put power in human hands: Ironmongering makes men unequal; amber waves of grain are their misfortune; Europe, since it has some of the most productive mines and richest farmlands in the world, must be the world's unhappiest continent. Science corrupts, and the Enlightenment ideal of applying scientific approaches to government is inhuman. "Throwing aside, therefore, all those scientific books," Rousseau asks how society might be reformed were humankind to return to the peace and equality that he thinks it originally enjoyed. We cannot, he concedes, go home again: Abolishing science and technology would be a cure worse than the disease, and attempting to deconstruct society would run afoul of the fact that "men like me . . . can no longer subsist on plants or acorns, or live without laws or magistrates." He does, however, proffer two remedies. One is to reform education, permitting children to develop in as "natural" a state as possible, free from booklearning and the memorization of facts. The other is to wield the power of government to hammer the population into a state of equality. Such is the popular will, Rousseau asserts, and "whoever refuses to obey the general will shall be constrained to do so by the whole body; which means nothing else than that he shall be *forced to be free*."

Robespierre revered Rousseau, whom he was proud to have met during the latter's final days in Paris. "Divine man!" Robespierre enthused. "I looked upon your august features [and] understood all the griefs of a noble life devoted to the worship of truth." Like Rousseau, Robespierre was enamored of his own virtue—he enjoyed being called "The Incorruptible"—and viewed civic virtue as "the fundamental principle of the democratic or popular government." Its core, he thought, was "the love of equality," a returning of society to the egalitarianism it enjoyed in Rousseau's state of nature. A spellbinding orator, Robespierre seldom flew higher than when describing how wonderful France would be once all its inequalities had been mowed down. "What is the end of our revolution?" he asked, in February 1794.

> We wish that order of things where all the low and cruel passions are enchained. . . . We wish in our country that morality may be substituted for egotism, probity for false honor, principles for usages, duties for good manners, the empire of reason for the tyranny of fashion, a contempt of vice for a contempt of misfortune, pride for insolence, magnanimity for vanity, the love of glory for the love of

> money, good people for good company, merit for intrigue, genius for wit, truth for tinsel show, the attractions of happiness for the ennui of sensuality, the grandeur of man for the littleness of the great, a people magnanimous, powerful, happy, for a people amiable, frivolous and miserable. . . .

In short, such a people as has never trod the earth. That was the trouble with Robespierre. Like Rousseau (who understood so little of politics that he remained indifferent to the American Revolution), Robespierre lacked practical political experience of the sort possessed by any seasoned ward heeler who, whether virtuous or corrupt, would laugh aloud at the sentiments expressed in the passage above. His inexperience doesn't seem to have bothered him, politics being one of those fields (like writing, or telling a joke) that amateurs commonly imagine can be mastered effortlessly. Virtue having shown the way, all that was required was to enforce it from above: The people must be *forced* to be free.

It followed that Robespierre had little use for experimentation, debate, or the liberties that permit them. "*Truth* and *reason* alone must reign in legislative assemblies," he wrote. To disagree with his version of either was to declare allegiance to a faction. "I abhor any kind of government that includes factious men," Robespierre warned, and his fellow radicals agreed. "Factions cannot exist in a republic," declared Georges Jacques Danton. "Factions are the most terrible poison in society," said Louis Antoine de Saint-Just. Even journalists toed the line: "The duty of journalists," wrote the editor of the *Feuille de Paris* in 1793, "is to agree with one another," otherwise "the public mind . . . is never molded." There was no such thing as a loyal opposition, and no punishment was too severe for those who tried to form one. The "great purity of the French Revolution," Robespierre declared, required that those "ambitious or greedy men," those "vicious men" who refuse to place public good above their own, must be extirpated. His maxim was "to lead the people by reason and the people's enemies by terror."

> If the spring of popular government in time of peace is virtue, the springs of popular government in revolution are at once virtue and terror: virtue, without which terror is fatal; terror, without which virtue is powerless. Terror is nothing other than justice, prompt, severe, inflexible; it is therefore an emanation of virtue.

Thus was the terrifying machinery set in motion. The Revolution was "virtuous," its detractors the opposite. They were "conspirators . . . assassins . . . intriguers . . . [no] less dangerous than the tyrants whom they serve." As the goals of the Revolution were unattainable, the nation's failure to achieve them had to be blamed on ever-greater numbers of such conspirators. The worse things got, the greater was the perceived need to excise the conspirators from the body politic. To express reservations about the Terror was to risk becoming one of its victims. Urging that the king be executed without benefit of trial, Robespierre declared that Louis

> has already been judged. He has been condemned, or else the Republic is not blameless. To suggest putting Louis XVI on trial, in whatever way, is a step back towards royal and constitutional despotism; it is a counter-revolutionary idea; because it puts the Revolution itself in the dock. After all, if Louis can still be put on trial, Louis can be acquitted; he might be innocent. Or rather, he is presumed to be until he is found guilty. But if Louis is acquitted, if Louis can be presumed innocent, what becomes of the Revolution?

Louis was tried anyway, and was beheaded on January 1, 1793, Marie Antoinette following him to the guillotine on October 16. By then the blood-dimmed tide had risen high enough to drown the optimism of even such erstwhile prior supporters of the Revolution as Thomas Paine and Joseph Priestley, both of whom had been elected to the French National Convention not long before. Paine wrote to Jefferson that "there was once a good prospect of extending liberty through the greatest part of Europe, but I now relinquish that hope."

Robespierre was brought down, in the end, more by laughter than by logic. Proclaiming the Festival of the Supreme Being on June 8, 1794—an event meant to inaugurate an official state religion, as prescribed by Rousseau—Robespierre celebrated Rousseau (who "painted the charms of virtue in strokes of fire") while wearing classical robes and standing atop an artificial mountain erected in Paris for the occasion, a sight that finally exposed his thoroughgoing absurdity to the public eye. The "public accuser"—one of Robespierre's official state posts—stood accused, and he was guillotined seven weeks later. The onetime journalist Jean-Lambert Tallien, who had been an enthusiastic participant in the Terror but now sensed that the people were sick of it, described its dynamics a month later in words that remain hauntingly pertinent. Fear, he said, may be instilled in criminals by threatening to punish misconduct, but terror must be directed at everyone:

> It is not only necessary to suspend a penalty over each action, a threat over each word, a suspicion over every silence; a trap must also be placed under every step, a spy in every house, a traitor in every family. . . . Terror must be everywhere, or it must be nowhere.

Jean-Paul Marat was a journalist who literally emerged from the sewers, where he had taken refuge from the authorities and contracted a skin disease, the discomfort of which he sought to assuage by spending his waking hours in a bathtub. His subterranean residency marked the nadir of a long decline. Trained as a physician, Marat turned to scientific studies—his research on optics, heat, and electricity attracting the attention of Franklin and Goethe—but his vainglorious insistence that he be deemed a greater genius than Newton prompted the Academy of Sciences to reject his application for membership. He thereafter spiraled into a darkening rage against authorities of all sorts, thinking them guilty of persecuting and conspiring against him: "I met but outrages, sorrows and tribulations," he wrote. Obsessed, as was Robespierre, with the unique purity of his civic virtue, he declared that "I am maybe the only author beyond suspicion since J.J."—meaning Jean-Jacques Rousseau, whose literary style Marat attempted to imitate.

The Revolution provided an excellent amplifier for Marat's assertion that the lowly were low because wicked higher-ups kept them there. Declaring himself the "wrath of the people," Marat published a series of newspapers in which he advocated violent, radical action on behalf of "the people." One of his editorials is thought to have incited the massacre of September 1792, when revolutionaries broke into Paris prisons and murdered more than a thousand political prisoners. That act repulsed many Europeans who had previously supported the Revolution—in England the poet and painter William Blake, who had taken to wearing the *bonnet rouge* of the revolutionaries, thrust it aside in disgust—but it was not enough for Marat, who advocated the execution of the king, the nobility, and all "enemies of the Revolution." Soon he was calling for two hundred thousand beheadings, and had been put in charge of composing lists of those to be executed. On July 13, 1793, he was stabbed to death—fulfilling his prophetic vision of "cowardly murderers who worm their way in the dark to stab me"— by Charlotte Corday, a twenty-five-year-old member of the moderate Girondist party who had gained an audience by promising to give Marat the names of fellow moderates fit for punishment. He wrote down the names and promised, with his last words, "They shall be soon guillotined." In a fit of exclusionary perfectionism, Marat's remains were first placed into the Pantheon—his epitaph reading, predictably, "Here Rests Marat, The Friend of The People, Assassinated by the Enemies of The People"—but were later removed to the nearby Saint Genevieve cemetery, and lost track of when the cemetery was desecrated.

All across France cemeteries were being desecrated, while cathedrals—Notre Dame and Chartres among them—were turned into "temples of Reason" and religious statues replaced with busts of Marat. "Traitors" were guillotined, as were backsliders, suspects, and "egotists." A new Law of Suspects called for the arrest of those "who have not constantly manifested their attachment to the revolution." It was proposed that every military unit be equipped with a guillotine on wheels. When a German mercenary in Lyons found that "the blood of those who had been executed a few hours beforehand was still running in the street" and proposed to a group of locals "that it would be decent to clear away all this human blood," their reply was, "Why should it be cleared? It's the blood of aristocrats and rebels. The dogs should lick it up."

Ultimately the tyranny turned against science. The astronomer Jean-Sylvain Bailly was beheaded as an enemy of the state in November 1793. Lavoisier, suspected of aristocratic tendencies, was guillotined on May 8, 1794. ("It only took them an instant to cut off that head," a mathematician told the astronomer Jean Delambre, "and it is unlikely that a hundred years will suffice to produce a comparable one.") When the Marquis de Condorcet—a revolutionary firebrand notwithstanding his aristocratic background—proposed a system of universal public education based on science, the plan was rejected by Robespierre on grounds that it might produce an intellectual aristocracy; better that the people be given a Spartan education based on "virtue." Condemned by the Committee of Public Safety, Condorcet, a mathematician, committed suicide rather than face execution.

Scientists were suspect in direct proportion to their merit, which was seen as violating the egalitarian ideal. In an eerie anticipation of the Chinese Cultural Revolution of 1966–76, eminent scientists were demoted, jailed, or executed, to be replaced by mediocrities celebrated for their very mediocrity. "Let anyone be a savant who wishes; O Happy Liberty," the astronomer Jean-Dominique de Cassini sarcastically observed, after being replaced as director of the Paris Observatory on the testimony of a former student turned government informant. (The student soon joined Cassini in jail, charged with the counterrevolutionary offense of having made an erroneous measurement of the sun.) The Academy of Sciences, its work frequently interrupted by revolutionary commands to perform such tasks as testing the quality of silver vessels confiscated from churches, came under attack from Marat and the painter Louis David as insufficiently egalitarian; it was closed down in 1793, its chambers in the Louvre given over to tailors fashioning uniforms for the army. Research slowed to a trickle; French scientists would next make an appearance on the world stage as calculators of trajectories for Napoleon's artillery.

Many lessons were learned in the revolution's long aftermath, some of them useful. Rationality was put in its place, as but one aspect of science. Science rose to prominence—and so, eventually, did liberal democracy—although nothing as democratic as the prerevolutionary French elections of 1789 would be seen in Europe until well into the next century. Intellectuals were given roles in the French government, not because they were smarter than anybody else but to avoid their relapsing into the political naïveté responsible, as Tocqueville astutely observed, for their "theory of madness, the cult of blind audacity." The contrast between the American experiment, which limited government power, and the French, which did not (the French revolutionaries having imagined that checks on power were unnecessary so long as government expressed the "will of the people"), impressed itself upon many observers of the debacle—although not to a sufficient degree to prevent all future such abuses.

Rousseau's fact-free thought, however, enjoyed a lasting influence. He had created not only a new philosophy, if it may be called that, but a new and pernicious style of philosophizing—one that consists of basing real-world arguments on bald fictions, then retreating into a wounded obscurantism should anyone question the legitimacy of the enterprise. This style was resurrected in the pseudoscience of Marx, the antiscience of Hitler, and the cynicism of the postmodernists. The pragmatist philosopher John Randall observes that ever since Rousseau, "reformers have been divided into two groups, those who followed him and those who followed Locke. . . . Hitler is an outcome of Rousseau; Roosevelt and Churchill, of Locke." Randall was not being hyperbolic: The fascist ideal of an all-powerful ruler who embodies the spirit of the people came straight from Rousseau. Napoleon, visiting Rousseau's grave, exclaimed, "It was this man who brought us to the state we are in!" adding, "Time will tell if this earth would not have been a more peaceful place if Rousseau and I had never existed." Time has now delivered its verdict, and it is that yes, the world would probably have been better off without Rousseau and Napoleon.

When Napoleon took control of France in the *coup d'état* of November 1799, George Washington was living in peaceful retirement at Mount Vernon, the key to the Bastille on display in his reception room. (General Lafayette, who named his son George Washington Lafayette, had given the key to Tom Paine, who presented it to Washington; it remains at Mount Vernon today.) When Washington died, the following month, Napoleon ordered his army to observe ten days' mourning for the man who, he said, had "put his country's freedom on a sure basis." But Napoleon soon embarked on a campaign of conquest that demonstrated that although he was a better general than Washington, he was a poorer statesman and a lesser man. If, rather than crowning himself emperor of the French, Napoleon had helped establish a French democracy, stood for office, and retired at the end of his term, as Washington did, he might have spared France many years on the long road to liberty. But Napoleon had no appetite for liberalism: "I haven't been able to understand yet what good there is in an opposition," he once remarked. Napoleon seems later to have sensed that he missed the opportunity to become a founder of the new age rather than a throwback to the old, and at the end of his career, when he had lost everything, he fretted obsessively about Washington. "Posterity will talk of Washington as the founder of a great empire, when my name shall be lost in the vortex of revolution," he sighed. "People wanted me to be another Washington," he complained, objecting that this would have been impossible under the circumstances:

> If Washington had been a Frenchman, at a time when France was crumbling inside and invaded from outside, I would have dared him to be himself; or, if he had persisted in being himself, he would merely have been a fool and only would have prolonged his country's misfortunes. As for me, I could only be a crowned Washington. And I could become that only at a congress of kings, surrounded

> by sovereigns whom I had either persuaded or mastered. Then, and then only, could I have profitably displayed Washington's moderation, disinterestedness, and wisdom. In all reasonableness, I could not attain this goal except by means of world dictatorship. I tried it. Can it be held against me?

Well, yes, it could be held against him, and many did so, among them his countryman François René de Chateaubriand, who observed that while Napoleon's empire had vanished, Washington's republic flourished: "Search the unknown forests where glistened the sword of Washington," Chateaubriand wrote. "What will you find there? Graves? No, a world!" In the end, Washington's "moderation, disinterestedness, and wisdom" embodied more of the spirit of science and liberty—and more of the future of freedom—than did all Napoleon's authoritarian brilliance. "Infallibility not being the attribute of Man," Washington had said, "we ought to be cautious in censuring the opinions and conduct of one another."

CHAPTER SEVEN

POWER

The history of empires is that of men's misery. The history of the sciences is that of their grandeur and happiness.

—EDWARD GIBBON, 1761

There is more love for humanity in electricity and steam than in chastity and abstention from meat.

—ANTON CHEKHOV TO ALEXEI SUVORIN, 1894

From about 1820, the liberal-scientific nations accrued unprecedented amounts of power—over nature, and over their fellow human beings. This technological transformation freed billions from mind-deadening toil. It replaced the charm of tradition with the shock of the new. It also fostered political and military preeminence, and while it would be pleasing to report that the empowered nations acted solely to liberate, enlighten, and enrich the rest of humankind, most instead reverted to anachronistic campaigns of conquest. A handful of nations came, for a time, to rule most of the world. The era of imperialism and colonialism left behind an enduring legacy of cynicism about science and technology, especially among those who think of science less as a way of obtaining knowledge than as an engine of power.

Since science drives so much of today's technological development, many assume that it played the same role in the past—as, indeed, it sometimes did. Chemists helped iron- and steelmongers improve the smelting process, thermodynamics showed how and why steam engines worked, and the young Guglielmo Marconi could not have made his first radio transmissions without the prior research of scientists like James Clerk Maxwell and Heinrich Hertz. But nineteenth-century technological breakthroughs more often arose from the efforts of industrialists, entrepreneurs, and amateur inventors, whose products inspired science at least as much as they were inspired by science. Henry Adams, writing about public attitudes toward the steam engine during Jefferson's presidency, noted that Americans generally were "roused to feel the necessity of scientific training" by their exposure to the practical benefits of technological advancement: "Until they were satisfied that knowledge was money, they would not insist upon high education; until they saw with their own eyes stones turned into gold, and vapor into cattle and corn, they would not learn the meaning of science."

Most inventions failed, and most of the unheralded inventors stayed unheralded. The popular mass-produced Singer sewing machines of the 1860s were preceded by a century of sewing machines that didn't work very well. Primitive dishwashers were being patented for a century before improved models came into general household use in the 1950s. Fiber optics preoccupied dozens of inventors, starting with John Tyndall in 1854 and Alexander Graham Bell in 1880, but did not start carrying telephone conversations until 1977. How could these millions of often haphazard experiments, whether conducted in corporate laboratories or a local crackpot's basement, have so often surpassed the accomplishments of educated specialists and government planners? Perhaps because, as the mathematician H. B. Phillips maintained, liberalism and free enterprise promote the emergence of what he called "thought centers." "Advances will be most frequent when the number of independent thought centers is greatest, and the number of thought centers will be greatest when there is maximum individual liberty," Phillips wrote. "Thus it appears that maximum liberty is the condition most favorable to progress." Through free experimentation the steam engine, the clock, and the dynamo gained sufficient dominion over space, time, and energy to fulfill the optimistic predictions of inventors like William Strutt, who in 1823 suggested that although he knew that his forecast would "be laughed at," the day would come when "time, distance, and expense shall be almost annihilated."

Foremost among these innovations was that emblem and embodiment of the Industrial Revolution, the steam engine.

The British began mining coal in earnest after the widespread use of wood for heating, cooking, and charcoal making had driven timber prices up by a factor of ten. As the demand for coal kept rising, the miners found it necessary to sink shafts so deep that flooding became a persistent problem. Financial opportunities emerged for anyone who could develop better engines to keep the mines pumped dry. It was for this purpose that, in 1712, an itinerant ironmonger and hardware salesman named Thomas Newcomen developed the first practical steam engine. The Newcomen engine was inefficient but this was not a major problem at the mines, where the coal it burned for fuel lay close at hand. Then one day James Watt, a self-educated instrument maker and repairman at Glasgow University, was instructed to repair a Newcomen engine. He took it apart, experimented with ways to make it work better, and produced a smaller, more efficient steam engine suitable for use in transportation.

Steam engines on rails soon emerged from the mines, thanks mainly to men who grew up doing colliery work and were comfortable getting their hands dirty. Few were what you would call intellectuals. Richard Trevithick, the son of an illiterate Cornish mine captain, was a bar-brawling wrestler, baffled by books but at ease taking apart every mechanical device entrusted to his care. He built locomotives while working in the tin mines of Cornwall, and on February 21, 1804, his *Penydarren*, the first steam locomotive to run on rails, won a competition by hauling seventy men and a ten-ton load on a tramway in South Wales. George Stephenson, the son of a colliery fireman, created a series of improved locomotives and became the chief engineer for five railroad companies. His son Robert went on to construct ever-longer rail lines. The iron rails kept breaking but steel soon fixed that. An American kettle manufacturer, William Kelly—and, more successfully, the English engineer Henry Bessemer—developed the blast furnace techniques that have remained essential to steel production ever since. Steel mills, their showering sparks and glowing rivers of molten metal emblematic of industrial advance, became giant laboratories in their own right. Andrew Carnegie, whose steel mills made him the world's richest man, recalled that "years after we had taken chemistry to guide us," his competitors "said they could not afford to employ a chemist. Had they known the truth then, they would have known that they could not afford to be without one." The mighty triumvirate of coal, steel, and steam, itself a sort of reciprocal engine, hummed into action and went to work shrinking terrestrial space and time.

The world's first commercial railway opened in 1830. Operating between Liverpool and Manchester, it ran on a timetable, charged fares by the mile, offered three classes of service, and employed two sets of tracks so trains could run simultaneously in both directions. During twenty years of boom-and-bust speculation, the railroads expanded; their steel rails, cutting across European and American landscapes like giant draftsman's lines, became part of the aesthetics of local landscapes. Some lines curved gracefully around hills, in the manner of Joseph Locke, while others carved through the terrain, as favored by Robert Stephenson—two approaches described by the civil engineer Thomas Tredgold as to "clamber over or plough through"—but all approximated what Einstein would enshrine in relativity theory as *geodesics*, lines of maximal space-time efficiency. Passengers compared the experience of rail travel to dreams of riding a magic carpet—the actress Fanny Kemble reporting, after taking a 30-mph publicity ride on George Stephenson's *Rocket*, that the "sensation of flying was quite delightful and strange beyond description." As higher-pressure steam locomotives and improved tracks reduced travel times (by 1847, a London-to-Birmingham train was routinely clocking speeds of 75 mph) the lure of rapid transit grew. By 1870 the English were taking 330 million rail trips annually, up from a tenth that many in 1845. American railroads grew from 2,800 miles of track in 1840 to over 30,000 miles by 1860.

Rail travel afforded city dwellers an opportunity to take in the fresh air and beauty of the countryside, but the intrusion of thundering, moving machines disquieted many of the Victorian intellectuals who lived through it. The poet William Wordsworth, who described himself as a "sensitive being" and "*creative* soul" and had romanticized the working class during the French Revolution (although he later became something of a reactionary on this point), protested in 1844 against a rural rail line on grounds that "uneducated persons" lacked the capacity to appreciate the beauty of the English Lake District. ("It is not against railways but against the abuse of them that I am contending," he added; by "abuse" he seems to have meant the selling of train tickets to passengers less sensitive and creative than himself.) In America it was hoped that the railroads would bind the disparate states together with staves of steel, as the customary metaphor had it. John C. Calhoun spoke for many when he called on Congress to "conquer space" by underwriting the construction of railroads, highways, and canals. "Railroad iron is a magician's rod," declared Ralph Waldo Emerson. The railroads supercharged the factory system, which expanded rapidly thanks to a reliable supply of raw materials and parts, and the outward flow of inventory to fill orders, provided by rail at steadily decreasing costs—the early stirrings of today's JIT ("Just in Time") inventory management techniques.

With the railroads came the electric telegraph—the first in a series of devices, from the telephone to e-mail, that made it possible for messages to travel faster than messengers. Here too the new technology developed mainly in the hands of amateurs. Samuel F. B. Morse, who championed telegraphy and developed the binary dot-dash code that became its native tongue, was a portrait painter who plunged into telecommunications after his young wife, Lucretia, died while he was away on a business trip and was buried before he learned of her death. ("I long to hear from you," he had written her, three days after she died.) Morse failed to persuade Congress to back his invention, but struck it rich by teaming up with private investors visionary enough to construct a network—it connected New York City with Philadelphia, Boston, and points west—before putting it on the market. Telegraph lines were soon going up along railroad rights-of-way so rapidly that no one agency could keep track of them all; by 1850, a dozen American telegraph companies managed an estimated twelve thousand miles of wires. The telegraph, as one historian writes, "severed the preexisting bond between transportation and communications. . . . At one stroke the life force of science—information—was freed of its leaded feet and allowed to fly at the speed of light." "Is it not a feat sublime?" read the masthead of *The Telegrapher*, the journal of the National Telegraphic Union. "Intellect hath conquered time."

The laying of transoceanic cables—a feat spurred by the efforts of the Irish physicist William Thomson (Lord Kelvin), who amused himself by developing low-voltage communications lines after having elucidated the science of thermodynamics—spawned fresh hopes that communication might promote international peace. "What can be more likely to effect [peace] than a constant and complete intercourse between all nations and individuals in the world?" asked Edward Thornton, the British ambassador to the United States, in a toast to Morse at Delmonico's restaurant in New York in 1868. "Steam was the first olive branch offered to us by science. Then came a still more effective olive branch—this wonderful electric telegraph, which enables any man who happens to be within reach of a wire to communicate instantaneously with his fellow men all over the world." Another toast at the same dinner party envisioned the telegraph's "removing causes of misunderstanding, and promoting peace and harmony throughout the world."

Such rosy sentiments are easy to mock, but many today would agree that improved communications have, as another technological optimist put it long ago, helped people "know one another better. . . . [and] learn that they are brethren, and that it is no less their interest than their duty to cultivate goodwill and peace throughout the earth."

As railroads and telegraph lines spread into the American West, the towns that sprang up at railheads and junctions often consisted of little more than a few clapboard buildings and unpaved streets plus a set of plat maps aimed at attracting investors. "Railroads in Europe are built to connect centers of population; but in the West the railroad itself builds cities," observed Horace Greeley. "Pushing boldly out into the wilderness, along its iron track villages, towns, and cities spring into existence, and are strung together into a consistent whole by its lines of rails, as beads are upon a silken thread." It sounded grand but did not always work: Midwestern boosterism was born of concern that one's hometown had little to recommend it beyond a railroad station, a telegraph shed, and the positive outlook of its civic leaders. Those who survived risked recapitulations of the biblical saga of Cain and Abel when gunslingers, equivalent to the hunter-gatherers of old, swooped down to plunder towns protected by a single lawman or none at all: From such origins sprang an enduring tradition among rural Americans that families require firearms for their protection.

The railroad, the telegraph, and the factory transformed society's sense of time. Clocks had been around for centuries but were mainly an enthusiasm of scientists: The first clock equipped with a minute hand was commissioned by the Danish astronomer Tycho Brahe from the Swiss mathematician Jost Burgi in 1577; the pendulum clock was invented in 1656 by the Dutch astronomer Christiaan Huygens; and the first man to wear a wristwatch was the French scientist Blaise Pascal. The rise of modern factories democratized this previously elite taste for accurate timekeeping, both by placing new temporal demands on workers—who punched a time clock and whose bosses liked to say, "Time is money"—and by producing affordable clocks and watches that became the high-tech centerpiece of many a working-class home. (Hence the gold watch upon retirement, and the cinematic trope of having commandos synchronize their watches before undertaking a mission.) Prior to the advent of the railroads, each community regulated its own clocks, from solar time as customarily measured by noting when the sun crossed the local meridian. Solar time depended on your longitude, and so was inherently local: A train conductor traveling 60 mph west from Pierre, South Dakota, would have to advance his watch seven minutes per hour in order to be on time when arriving in Livingston, Montana. This was neither practical nor safe, so in 1883 the railroads went ahead and divided the continental United States into four time zones, Congress eventually mandating the system in 1918. (A similar standardization in England was called "railroad time.") Trains became symbols of time. Much of the lasting appeal of the 1952 movie *High Noon* arises from its insistence on the unity of three sounds—a ticking clock, a clicking telegraph, and the chuffing of a steam locomotive—imposed like civilization itself on a recently lawless West. Urged by a judge to save his life by getting out of town, the sheriff, played by Gary Cooper, replies, "There isn't time." Factually, his statement makes no sense—he still has ample opportunity to run away—but we take him to mean that it is no longer a time for the West, now bound together by rails and telegraph lines, to revert to the anarchy of old.

With the rise of electrical power, dynamos became central to the new technology. The dynamo generated electricity that could be carried by wires to provide lighting or, by using another dynamo in reverse, be turned back into mechanical power on demand. Its development had been difficult. Electricity had long fascinated the public—audiences were thrilled by demonstrations in which static electricity shocked ranks of hand-holding soldiers or sparked the lips of those venturesome enough to kiss an electrified woman—but nobody had been able to get much work out of it, Ben Franklin pronouncing himself "chagrined a little that we have been hitherto able to produce nothing in this way of use to mankind." Its transformation into an engine of industry was eventually inaugurated by the research of Michael Faraday.

Faraday grew up behind his father's blacksmith shop in London and at age fourteen was apprenticed to a bookbindery, where he educated himself by reading the books he bound. Intrigued by an article on electricity that he encountered in the *Encyclopaedia Britannica*, he began conducting electrical experiments and going to public science lectures, where showman-scientists staged gasp-inducing explosions and flashes of light. In 1812 Faraday attended a lecture by the famous chemist Humphry Davy—himself an autodidact who had first learned science from James Watt's son Gregory, a boarder in the Davy household. Faraday bound his notes on the lecture in leather and sent them to Davy with a letter expressing, as Faraday recalled it, "my desire to escape from trade, which I thought vicious and selfish, and to enter into the service of science, which I imagined made its pursuers amiable and liberal." He landed a job in the Royal Institution laboratory and soon became a science lecturer himself, but his real love was experimenting in his basement laboratory, where he remained doggedly determined even when repeatedly injured and once temporarily blinded. In 1849, speculating about a link between electromagnetism and gravitation—a connection that eluded him, as it would Einstein a century later—he scrawled this memorable passage in his notebook:

> ALL THIS IS A DREAM. Still, examine it by a few experiments. Nothing is too wonderful to be true, if it be consistent with the laws of nature, and in such things as these, experiment is the best test.

Faraday found that a wire carrying an electric current could be induced to circle round a magnet, and that, conversely, the magnet would circle the wire—the basis of the electric motor and the generator. The first practical generators (called "dynamic-electric" generators, or "dynamos") were employed to power arc lamps, in which an electrical spark sustained between two poles produced a dazzling, sun-white sheet of light. Arc lamps were employed in lighthouses and theaters, and as novelties in department stores, but their intense glare made them unwelcome on city streets (the elegantly dressed boulevardiers of Paris and London retreated from them in revulsion, preferring the more flattering hue of gaslight) and hopeless for lighting private residences. Fortunes clearly were to be made if a longer-lasting and less imperious electric light, suitable for the home, could be created. The inventor who proved equal to the challenge was Thomas Edison.

Edison's hardscrabble childhood was the stuff of legend. Born in Milan, Ohio, in 1847, he dropped out of school while still in the first grade, sold newspapers to passengers on the Grand Trunk Railway, and learned by reading them to write newspaper articles himself. One day in 1862 he snatched the young son of a telegraph operator from the path of an oncoming boxcar and was rewarded with a job as a telegrapher. During twelve-hour night shifts, Edison read books on technology and science and acquired a lifelong fascination with Faraday. By age 21, he had devised an improved "duplex" telegraphy system that could handle multiple messages simultaneously, and by age thirty had established a laboratory, in Menlo Park, New Jersey. From it poured a flood of innovations that contributed to the development of the telephone, the phonograph, the fax machine, and the movies, but in a sense Edison's greatest invention was the laboratory itself: Staffed by trained technicians and aimed at attaining practical, profitable results, it was America's first R&D outfit. Edison drove his team hard, constantly reminding them that "genius is ninety-nine percent perspiration and one percent inspiration." He worked for days and nights on end, napping on a stack of newspapers in a laboratory closet, unwavering in his conviction that any properly framed problem will eventually yield to persistent experimentation. When a financial backer complained that money was being wasted on useless experiments, Edison replied, "No experiments are useless."

Among his fans was Henry Ford, who wrangled a meeting with Edison in 1896 and sketched his plan for a gasoline-powered automobile. Edison saw its potential, telling Ford, "You have the thing. Keep at it!" The two later became friends, Edison joining Ford (who viewed the automobile, like the railroads, as a way for working-class Americans to widen their horizons) on summer road trips around the West with Harvey Firestone and the naturalist John Burroughs. One day in northern Michigan these "Four Vagabonds," as they styled themselves, encountered a farmer who was trying to repair an automobile. Ford helped him fix it but refused payment, saying, "I have all the money I want." "Hell," replied the farmer, "You can't have that much and drive a Ford." Edison died in 1931, after faltering at a Ford-hosted dinner in his honor at which Albert Einstein delivered a tribute by telephone. He held 1,093 patents, still the record.

Edison's interest in electric lighting was kindled by a visit, on September 8, 1876, to a brass foundry in Ansonia, Connecticut, where another inventor, Moses G. Farmer, and his partner, William Wallace, were demonstrating a steam-powered dynamo hooked up to eight glaring white arc lamps. "Edison was enraptured," wrote a reporter for the *New York Sun*. "He ran from [the dynamo] to the lights, and from the lights back to the instrument. He sprawled over a table with the *simplicity of a child*, and made all kinds of calculations." At the end of the day, Edison bluntly told the hosts, "I believe I can beat you making the electric light." Back at Menlo Park, he recalled, the "secret" to developing practical electrical lighting "suddenly came to me. . . . I am already positive it will be cheaper than gas." The idea was to find a filament material that glowed steadily when electricity passed through it, and would keep glowing for many hours without burning out. That was the inspiration; now for the perspiration. Edison experimented with thousands of substances before finding a promising candidate, the leaf of a Japanese palm, and finally settling on a filament made of carbonized cardboard. "I speak without exaggeration," he told a visitor, "when I say that I have constructed three thousand different theories in connection with the electric light, each one of them reasonable and apparently likely to be true. Yet in two cases only did my experiment prove the truth of my theory." As with most people, when Edison said, "I speak without exaggeration," he was exaggerating; his thousands of "theories" about electric-light filaments were mostly just trials of different materials. But his general point was correct—that useful, profitable inventions could be made through persistent trial and error, without waiting for theoretical science to show the way. As Joseph Priestley had put it a century earlier, "In this business . . . more is owing to . . . the observation of events, arising from unknown causes, than to any . . . preconceived theory."

Now that he had a proper filament, all that remained for Edison was "to make the dynamos, the lamps, the conductors, and attend to a thousand details the world never hears of." This he managed to do, and the Edison company was soon providing electric lighting to hundreds of firms and homes in lower Manhattan and beyond. Living resplendently in an electrically illuminated mansion on Fifth Avenue, Edison regaled journalists with hyperbolic tales about how, just ten years earlier, he'd "had to walk the streets of New York all night because I hadn't the price of a bed."

Electric lighting turned night, if not into day, at least into a new and less threatening species of night. "Sundown no longer emptied the promenade," wrote Robert Louis Stevenson, "and the day was lengthened out to every man's fancy. The city folk had stars of their own; biddable, domesticated stars." The promise of the Enlightenment had come true in the form of light itself. Hulking dynamos became the centerpieces of science and technology expositions—dwarfing visitors like Henry Adams, who came away from the Paris Exposition of 1900 calling the dynamo "a symbol of infinity" and "a moral force."

Edison promoted his direct current (DC) system as safer than the alternating current (AC) being marketed by his competitor George Westinghouse; he even made sure that the nation's first execution by electric chair was carried out using AC, in order to impress upon the public its largely illusory dangers. AC eventually won out anyway, mainly because it could be transmitted more efficiently over long-distance wires.

The inventor of the AC motor was Nikola Tesla, a man sufficiently eccentric to remain an object of fascination to this day, although he lacked Edison's genius for self-promotion and lived the last two decades of his life in obscurity, with only a pet pigeon for company. A native of Serbia, Tesla first encountered a dynamo at the polytechnic institute in Graz, Austria, where he was studying electrical engineering. The professor demonstrated that the dynamo could function as a motor when run in reverse, although it sparked badly. Tesla blurted out that he could fix the problem of converting the dynamo into a proper electric motor. The teacher rebuked him in front of the class, declaring that "Mr. Tesla may accomplish great things, but he certainly will never do this."

A sneering teacher can be as motivating as a smiling one, and Tesla obsessed over the problem for years. (Neurotic and compulsive, he was good at obsessing. Among other quirks he counted everything—every cup of coffee he drank, every step he took, every lap he swam at dawn in a public bathing house on the Seine—and claimed to be so sensitive that the sound of "a fly alighting on a table in the room would cause a dull thud in my ear.") As Tesla told the story, he was watching the sun set from a park bench in Budapest one February afternoon and reciting lines from Goethe's *Faust* to himself—lines about the rotation of the earth—when "the idea came like a flash of lightning and in an instant the truth was revealed." He told his first biographer that he immediately drew up the design, in the dirt with a stick, resolving, "I must return to work and build the motor so I can give it to the world. No more will men be slaves to hard tasks. My motor will set them free, it will do the work of the world." He moved to New York, patented his AC induction motor, and went to work first for Edison and then for Westinghouse. In later years Tesla fell victim to neuroses, hallucinations, and accidents—a fire in 1895 destroying his Houston Street laboratory and his research papers—but as he had foreseen, the AC motor lifted the burden of heavy toil from the shoulders of millions.

The scientists, not entirely left behind in this flurry of advancing technology, managed to discern many of the principles of physics operating in the new machines. Sadi Carnot in the 1820s and Kelvin in the 1850s discovered the principles governing steam engines, and from them derived the laws of thermodynamics, while James Clerk Maxwell brought mathematical rigor to Faraday's experimental results, producing an elegant portrait of electromagnetic fields. Yet none was able to accurately predict the behavior of the whirling electromagnetic fields inside the dynamos. It took an exceptionally creative scientist to recognize the importance of this seeming technicality; he showed up in the person of Albert Einstein.

Instinctively iconoclastic, Einstein regarded authority based on anything other than reason and empirical proof as self-evidently illegitimate. When his parents took him to see a military parade as a little boy, he promptly burst into tears. Left behind in a Munich high school when his parents moved to Italy, he escaped by convincing a doctor that school discipline was literally driving him crazy. He infuriated his teachers by sitting in the back row of class, a smile playing across his lips, and was so unpopular with the faculty at the Federal Polytechnic Institute in Zurich that upon graduation in 1900 he was unable to find a job in science and was reduced to working as a patent-office clerk in Bern. The work suited him, however, and although the world knew Einstein as a great theorist, able to probe the depths of the universe armed with nothing but paper and pencil, he had a lifelong involvement in engineering and technology as well. He held a dozen patents, one of them an exotic refrigerator that proved impractical for household use but was later used by NASA, and his technological facility was displayed in the special theory of relativity, his revolutionary account of light.

Einstein from an early age was intrigued by electromagnetism, the form of energy that manifests itself as light, electricity, and magnetism. One of his earliest memories was of being shown a compass, while sick in bed at age four or five, and wondering how its needle could respond to the earth's magnetic field. It seemed to him "a miracle" that this invisible field could command the behavior of the tangible needle. "I can still remember—or at least believe I can remember—that this experience made a deep and lasting impression upon me," he recalled in his *Autobiographical Notes*. "Something deeply hidden had to be behind things." A few years later Einstein's father, Hermann, and his uncle, Jakob, went into the dynamo business, building generators in a backyard shed. The business failed but Einstein kept thinking about dynamos. At the Swiss Polytechnic Institute he took classes from the physics dean, Heinrich Friedrich Weber, whose chair was funded by the dynamo manufacturer Werner von Siemens and who regularly subjected himself to the teeth-rattling jolts of electrical current it generated. Einstein knew that a mystery resided inside each dynamo, and within every other electromagnetic field that whirled or eddied or sped through space: If you tried to apply Maxwell's elegant equations to them, you got paradoxical results. Herr Weber's response to this difficulty was to ignore it; his lectures were devoid of any reference to Maxwell's equations. Einstein, however, never let it go. He worked on electrodynamic field equations during spare moments at the patent office, discussed them with friends, and had the answer by 1905—his *annus mirabilis*, the year in which, at age twenty-six, he published not only the paper that inaugurated relativity but also one that helped create quantum physics, and three others revolutionizing atomic theory and statistical mechanics.

Einstein's special relativity reconciled Maxwell's equations with the high-speed world that he could imagine by projecting the growing speed of locomotives and spinning dynamos to spaceships approaching the velocity of light—a reconciliation purchased by portraying space and time as elastic. In relativity, the velocity of light becomes the universal yardstick—all observers measure it to be the same, regardless of the speed they themselves are traveling—while the observers' mass, and the rate at which time passes for them, are flexible. A space probe leaving the earth increases in mass—and the passage of time on board slows down—relative to an identical spaceship that stays home. The effect is small for real space probes but becomes dramatic when velocities approach that of light, giving rise to the so-called "twins paradox" (which has puzzled science students for years although it isn't really a paradox at all). Let's call the twins Stella and Terry. Stella flies to a distant star at 70 percent the velocity of light, makes a brief stopover, and then returns at the same rate of speed. From Terry's point of view her trip took fourteen years, but for Stella only two years have passed. This is in no sense illusory: If Terry and Stella were both thirty years old when Stella departed, upon her return he would be forty-four years old and she only thirty-two. People took to calling it a paradox because they imagined that the situation would be reversed if you took Stella's point of view instead of Terry's—but that is not the case, because their experiences are not the same. Stella's spaceship accelerates to reach its appreciable velocity. The acceleration buys Stella space—without it she would never have left Earth—and in doing so buys time. Terry, having experienced no such change is his inertial framework, ages at a normal rate while Stella retains her youthfulness.

Einstein's discovery of a deep link between space and time—so deep that they are now regarded as two aspects of the same entity, the space-time continuum—showed how fundamentally science can upset commonsense conceptions of nature. No sooner had people adapted themselves to the conquest of space and the consistent regulation of time than they were asked to change their ways of thinking about both, the authority of precision clocks and platinum "meter bars" giving way to Salvador Dalí's limp clocks and the paradoxical architecture of M. C. Escher's imaginary castles.

This scientific revolution had its own technological roots. Einstein's original relativity paper, "On the Electrodynamics of Moving Bodies," is full of terminology that his father and uncle might have bandied about in their dynamo shop. It begins by discussing "the electromagnetic interaction between a magnet and a conductor," which is how a dynamo works. Viewed in terms of conventional physics, Einstein notes, the interaction looks very different depending on one's point of view. To resolve this asymmetry, Einstein postulates that "we first have to clarify what is to be understood here by 'time.' . . . If, for example, I say that 'the train arrives here at seven o'clock,' that means, more or less, 'the pointing of the small hand of my clock to seven and the arrival of the train are simultaneous events." Such simultaneity pertains, however, only to a system whose elements are at rest relative to one another. Set them in relative motion—a step that Einstein portrays, suitably for the technology of his day, in terms of trains, clocks, and "rigid measuring rods"—and the picture changes. Special relativity is a railway stretching to the stars, bringing back news of a stranger but less parochial physics than this world had previously known. It is also the key to nuclear power. It starkly illuminated the chasm that had opened up between the societies that possessed science, technology, and liberty, and the many more that did not—the inequity that had already fostered imperialism and colonialism.

Traditionally, nations possessing superior power have used it to subjugate their neighbors. As Thucydides commented in the fifth century BC, directly addressing his future readers:

> Of the gods we believe, and of men we know, that by a necessary law of their nature they rule whenever they can. And it is not as if we were the first to make this law, or act upon it when made; we found it existing before us, and shall leave it to exist for ever after us; all we do is to make use of it, knowing that you and everybody else, having the same power as we have, would do the same as we do.

The advent of modern technological power produced much the same result. The nations that gained it found themselves freshly empowered to navigate across the seas and to explore and exploit distant lands, aloof and imperious as "The Flying Dutchman" in Edwin Arlington Robinson's poem:

> Lord of himself at last, and all by Science,
> He seeks the Vanished Land. . . .
> He steers himself away from what is haunted
> By the old ghost of what has been before,—
> Abandoning, as always, and undaunted,
> One fog-walled island more.

The European explorers and colonists promulgated not only their novel doctrines but their affection for novelty itself. As the Stanford University political scientist David B. Abernethy suggests, "The persistent effort of Europeans to undermine and reshape the modes of production, social institutions, cultural patterns, and value systems of indigenous peoples . . . was the outward projection of tumultuous changes in the way Europeans themselves lived." Geographically, Europe is comprised of nations with rather strong and contrasting identities located in close proximity to one another; in this situation, the rise of one nation presents immediate challenges to its many neighbors; hence European imperialism itself promoted imperialism. Germany built a navy primarily out of fear and envy of the British navy. England fretted that Russia might come to dominate Mesopotamia (modern Iraq) if England didn't get there first: Thomas Love Peacock of the East India Company, asked by a parliamentary committee, "Is it your opinion that the establishment of steam[boats] along the Euphrates would serve in any respect to counteract Russia?" replied, "I think so, by giving us a vested interest *and a right to interfere*." An inherently expansionist dynamic arose from the imperialist enterprise, in that the existence of one colonial outpost promoted the creation of others in order to defend the first. The British statesman and classics scholar Evelyn Baring, the first Earl of Cromer, wryly observed in 1910 that the expansion of the British empire, like that of the ancient Roman empire, was

> accompanied by misgivings, and was often taken with a reluctance which was by no means feigned. . . . [Rome] was impelled onwards

> by the imperious and irresistible necessity of acquiring defensible frontiers. . . . Public opinion of the world scoffed two thousand years ago, as it does now, at the alleged necessity; and . . . each onward move was attributed to an insatiable lust for an extended dominion.

Setbacks in one part of the world spurred military leaders to advance elsewhere. The British Parliament's India Act of 1784 declared that any expansion of control of India was "repugnant to the wish, the honor and the policy of this nation," but General Charles Corwallis, smarting from his having been obliged to surrender to the American revolutionaries at Yorktown in 1781, was soon embroiled in the prolonged and unpopular Mysore wars and was otherwise at work extending British control over India, until by 1815 the East India Company and the British governor-general ruled forty million Indians. Just eight European states—Britain, Portugal, Spain, France, the Netherlands, Belgium, Germany, and Italy, comprising only 1.6 percent of the earth's territory—controlled 35 percent of the world's land area by 1800, 67 percent by 1878, and over 84 percent by 1914; in 1909 the British Empire alone dominated 444 million people across a quarter of the world's land surface. The costs of colonialism seemed remote and abstract but its benefits palpable to those in London and Paris who sipped Indian tea sweetened with sugar from Caribbean slave plantations—which was why British abolitionists would use only honey in their tea. Eventually the costs became unsustainable, but meanwhile European power fostered delusions about their superior "race" bringing civilization to the inhabitants of supposedly backward regions. As the British historian Norman Davies observes:

> Raw power appeared to be made a virtue in itself, whether in popular views of evolution which preached "the survival of the fittest," in the philosophy of historical materialism, which preached the triumph of the strongest class, in the cult of the Superman, or in the theory and practice of imperialism. Europeans, in fact, were made to feel not only powerful but superior.

The main motive for imperialism was that there was money in it. Substantial profits accrued to those who could import nutmeg from Pulo Run in the Banda Islands of the Indonesian archipelago, cinnamon from

Ceylon, pepper from Sumatra and India's Malabar coast, or cloves from the volcanic islands of the Moluccan triangle. This the Europeans could do efficiently, thanks to the technical advantages they enjoyed in ship design, navigational instruments, maps, and guns. It soon emerged, however, that maintaining free trade meant protecting the trade routes against piracy, then disciplining the port cities that succored the pirates, then dealing with real or imagined threats to the ports coming from farther inland. And so an enterprise that had begun in freedom devolved into freedom's opposite.

The British and Dutch colonization of the Indies (a vague term covering India and nearby islands) was illustrative of the slippery slide to colonialism. It began with the establishment of enterprises in the spice and pepper trade. Initially independent—eight separate Dutch firms were operating in the Indies by 1599—these endeavors turned into government-sanctioned monopolies as investors, stung by the vagaries of piracy and politics, lobbied for protection. In this manner the English East India Company was established by royal charter in 1600 and the Dutch East India Company set up as a monopoly in 1602 (returning a monopolistic 18 percent a year to its investors over its two-hundred-year history). Government participation put national pride on the line, turning trade into tyranny. "How," asked a modern historian, "did a people who thought themselves free end up subjugating so much of the world?" That was how—at least at first.

Many eighteenth- and nineteenth-century imperialists, men like John Barrow and Thomas Stamford Raffles, were scientifically inclined explorers out to learn about the natural world and bring the benefits of education, political liberty, and health care to what they took to be the benighted inhabitants of uncivilized lands. Barrow served forty-one years as second secretary to the Admiralty. Fascinated by astronomy, he became a navigator and geographer (helping to found the Royal Geographical Society in 1830), and an energetic explorer who studied whales and icebergs off Greenland, married a Bohr woman in South Africa, and learned Cantonese in China. He detested slavery, championed personal freedom, and always insisted—albeit while lobbying for more British naval bases, each new one intended to protect the last—that England's goal was not to seize land or rule people but to promote trade. Raffles was born at sea, off the coast of Jamaica. He clerked for the East India Company from age fourteen, teaching himself geography, botany, zoology, and a variety of Asian languages that served him well in Malaysia and Indonesia. He founded the free port of Singapore; established a college to teach science, law, and democracy to the sons of indigenous rulers; and conducted botanical and zoological studies in the course of which he identified thirty-four new species of birds and thirteen species of mammals, chiefly in Sumatra. Back home he wrote a two-volume *History of Java* and launched the London Zoological Society. Virtually immune to racial and ethnic prejudice, Raffles denounced slavery, expressed disgust at the "cold-blooded, illiberal" practices of competing imperial powers, and always maintained that his object was "not territory, but trade" in order to foster "a better state of society." Investigating a claim by his anthropologist friend William Marsden that the Battas of Sumatra were cannibals, Raffles found that they not only feasted on their prisoners of war but ate them alive, slicing off strips of their flesh and dipping them in a savory lime-and-chili sauce. Such conduct failed to disrupt his equanimity. "The Battas are not a bad people," Raffles wrote the Duchess of Somerset. "They write and read, and think as much or more than those brought up in our National Schools."

Yet the inherent illegitimacy of empire soon spawned monsters. Militarism ascended, as justifiable rebellions came to be lumped in with the earlier predations of pirates and warlords. Mediocre martinets, wielding an authority in the colonies that they could not have had at home, fueled local rebellions the suppression of which led to calls for even more military power. Racism reared its head, as an imagined justification for the anomaly of a white minority ruling multitudes of black, yellow, and brown peoples. American imperialists invoked manifest destiny—the notion that, for some unknown reason, power was historically destined to flow westward from the Middle East to Greece, Rome, Europe, and on across the Americas. "Westward the course of empires takes its way," declared George Berkeley, coining an expansionist mantra.

It all led back to profits. "The continent lay before them, like an uncovered ore-bed," wrote Henry Adams, contemplating American attitudes toward westward expansion circa 1820.

> They could see, and they could even calculate with reasonable accuracy, the wealth it could be made to yield. With almost the certainty of a mathematical formula, knowing the rate of increase of population and of wealth, they could read in advance their economical history for at least a hundred years.

The subtle interrelationship between humans and technology, in which it is often difficult to determine whether people are using tools or the tools using them, was highlighted on the imperialist stage by the advent of the steam-powered gunboat. Imperialism *sans* gunboats might have remained largely a matter of subjugating the ports and coastlines where seagoing frigates could project military power. Imperialism *plus* gunboats extended its tentacles deep upriver, tipping the balance toward the very colonialism that many early imperialists had eschewed. Here too the road was paved with good intentions. The Scottish steamboat builder Macgregor Laird, celebrating "the immortal Watt" ("By his invention every river is laid open to us, time and distance are shortened"), predicted that

> if his spirit is allowed to witness the success of his invention here on earth, I can conceive no application of it that would receive his approbation more than seeing the mighty streams of the Mississippi and the Amazon, the Niger and the Nile, the Indus and the Ganges, stemmed by hundreds of steam-vessels, carrying the glad tidings of "peace and good will toward men" into the dark places of the earth which are now filled with cruelty.

The phrase "gunboat diplomacy" entered the language as steam-powered gunboats subdued inland China during the Opium War, decimated the swift-sailing *praus* of the Burmese, and probed the heart of Africa. Quinine, the breech-loading Prussian needle-gun, the Maxim gun, and a host of other innovations contributed to the carnage. Ten thousand Dervish cavalrymen were mowed down by Maxim guns at Omdurman in the Sudan in 1898 by a British force that incurred only forty-one casualties, in what a young Winston Churchill, covering the battle as a journalist, called "the most signal triumph ever gained by the arms of science over barbarians." Thousands of Zulu warriors armed with spears and leather shields were Gatling-gunned in the battle of Isandlwana in 1879 (in which the Zulus nonetheless prevailed). Millions of Congolese died under the sadistic rule of the notorious King Leopold II of Belgium, whose African campaigns incited Joseph Conrad, in his *Heart of Darkness*, to muse about "the fascination of the abomination." Leopold's army of African conscripts, the Force Publique, ordered to return with a severed right hand for every bullet they fired, filled the quota by amputating the hands of the living as well as the dead. The bloodstained list is long.

Yet this final spasm of colonialism was anachronistic, the work of modern Europeans who, unlike the Genghis Khans and Tamerlanes who preceded them, regarded themselves as being in some sense *liberal*. To comport their liberality with the stark fact that they ruled unfree multitudes, they were obliged to imagine that by exploiting non-Europeans they were somehow exporting the very gifts—of democracy, science, and economic freedom—that their conduct defied. For a long time they could at least take comfort in feeling that all Europe was united in the colonial enterprise—that, as King Leopold put it, each power was but imitating its neighbors. But the dream of European unity was shattered by the two world wars, in the aftermath of which the exhausted and all but bankrupt colonial powers found that they could no longer afford their colonies—which in the long run had proven unprofitable anyway, *un*free trade being no bargain. Britain by 1921 was paying more to maintain just the state of Iraq than to meet all its national health care needs, while France was drained financially and emotionally by rebellions in Vietnam in the 1950s and Algeria in the 1960s. At the same time their financial resources were diminishing; Britain's share of world manufacturing exports fell from 25 percent in 1950 to under 10 percent in 1973. The empires folded with spectacular speed. After being occupied for over a century on average, and in some instances over two centuries, more than eighty colonies gained independence in four decades, from 1940 to 1980—among them Syria and Lebanon in 1945–46; India, the largest nation ever ruled by a foreign power, the following year; Burma, Ceylon, and Indonesia in 1948; Iraq in 1958; Indonesia by 1960, then a slew of sub-Saharan African states, including Cameroon, Chad, Ivory Coast, Mali, Mauritania, Nigeria, Senegal, and Togo. By the late 1980s no substantial colonies remained anywhere in the world. The seemingly utopian visions of Joseph Priestley, who in 1791 had predicted that one day "the very idea of distant possessions will be even ridiculed [and] no part of America, Africa, or Asia, will be held in subjection to any part of Europe" had been realized.

The end of empire may have seemed little more than a footnote for the Europeans, struggling to rebuild capitals that were soon targeted on Soviet nuclear strike lists, but it was a watershed for the inhabitants of the former colonies. Whole chapters had been torn from what would otherwise have been the ongoing narratives of their own histories. Each new nation, writes the historian John Darwin, "needed a history that placed its own progress at the heart of the story."

> Each had its own heroes whose national struggle had been waged in the face of Europe's cultural arrogance. New "nationalist" histories portrayed European rule (or influence) as unjust and repressive. Far from bringing progress to stationary parts of the world, European interference had blocked the social and cultural advances that were already in train. . . . "Decolonized history" encouraged many different social, ethnic, religious, or cultural groups to emerge from the shadows. The old colonial narratives in which Europeans stood out against the dark local backcloth now seemed like cartoons; crude and incomplete sketches of a crowded reality.

The emergence of these new narratives—and, importantly, their study by social scientists—was largely responsible for the rise of the notion that there are ways of knowing preferable to that of a "Western" science now declared to have been discredited by its association with Western imperialism and colonialism. These campaigns spawned many of the antiscientific and illiberal attitudes that have bedeviled academic discourse over the past half century. Among their targets was the idea of progress, now depicted as a parochial illusion, as blinkered as racism. Was there any progress, really?

CHAPTER EIGHT

PROGRESS

Science is progressive.

—THOMAS JEFFERSON TO JOHN ADAMS, 1813

All that is human must retrograde if it does not advance.

—EDWARD GIBBON, 1776

To speak of the world as having made progress has gone out of fashion. "Writers nowadays who value their reputation among the more sophisticated hardly dare to mention progress without including the word in quotation marks," observed the economist Friedrich Hayek in 1960. The situation had got worse by 1998, when, as one journalist noted, "No one talks seriously about 'progress' any more." Yet like sex in Victorian England, progress is constantly present even if shunned in polite conversation. By almost any empirical measurement, the centuries since the advent of science and liberalism have seen the greatest improvements in human health, wealth, and well-being in all history. So why do so many sophisticated thinkers disdain the very concept of progress as inappropriate or naïve?

Some feel that the terrible violence of the twentieth century invalidated the promise of science and liberty. Others argue that material gains were purchased at the price of environmental degradation. Many object less to the general prospect of progress than to the utopian assertion that progress is *inevitable*. They critique Enlightenment thinkers who imagined that the future would produce a "new man," whose rational judgment would render war and injustice obsolete; as no such new man has appeared, the scholars dismiss the concept of progress itself.

It certainly is the case that several otherwise sensible Enlightenment thinkers pushed the prospects of progress to extremes. One such was the Marquis de Condorcet. His "Sketch for a Historical Picture of the Progress of the Human Mind" was prescient in many regards. It portrayed a future in which scientific, liberal-democratic government has become the norm, food is abundant and cheap, and the human population has grown significantly; many lethal diseases having been eradicated, the average man's lifespan extends to "the time when naturally, without illness or accident, he finds life a burden." In this happy future, prejudices are in retreat—notably "the prejudices that have established inequality between the sexes, fatal even to the sex it favors"—while universal public education works to "speed up the advances of [the] sciences." So far, so good: By and large, these predictions came true. But Condorcet did not stop there. In his enthusiasm for rationality—"The time will therefore come when the sun will shine only on free men who know no other master but their reason"—he lapsed into advocating what he called "the perfectibility of man." This notion has cropped up in thinkers from Plato to the behaviorist B. F. Skinner, misleading so many of them that the Australian philosopher John Passmore was moved to devote an entire book, *The Perfectibility of Man*, to debunking it. Human beings "are capable of more than they have ever so far achieved," Passmore writes, but their achievements will always "be a consequence of their remaining anxious, passionate, discontented human beings. To attempt, in the quest for perfection, to raise men above that level is to court disaster; there is no level above it, there is only a level below it."

Setting aside as imaginary the perfectibility of human nature, two steps may be taken to clarify the question of progress.

First, the present is compared to the past, rather than to an imagined future, our concern being empirical facts about what has actually happened rather than the rise and fall of utopian ideas about what might have happened.

Second, progress is defined as the creating of more options, for more people, over time. To eat rather than to starve is an option. To see a doctor rather than suffering or dying from a curable disease is an option. To leave your place of birth and seek education and employment, and to enjoy these opportunities notwithstanding your race, creed, or gender, is an option. To speak your mind, vote, and send money home to your aging parents are options. Such a definition may fail to satisfy those who would prefer to see the people as toiling together toward a common goal, but when people are free to do as they please they pursue many goals, some of which look contradictory. In the liberal-democratic world today people watch television, surf the Internet, *and* buy more books than in the past; play video games, *and* score higher on standardized IQ tests than their parents and grandparents did; drive cars, *and* run marathons; squander money on frivolities, *and* graduate from college in unprecedented numbers. Some eat junk food, others organic produce; some fear that science debunks religion, while others contentedly integrate science and religion into their lives. Such societies may look haphazard, but their diversity and freedom is their source of strength, and challenges the notion that progress must be planned. It may well be, as Oliver Cromwell used to say, that a man never mounts so high as when he knows not where he is going.

The greatest eighteenth-century exemplar of a commonsensical, nontranscendent view of progress was Benjamin Franklin. Unwaveringly cheerful and companionable, Franklin drew much of his optimism from his firsthand experience in conducting scientific investigations and coming up with inventions. He pioneered the science of weather forecasting (his fascination with electrical storms leading him to conceive of his most famous experiment, in which lightning captured by flying a kite is stored in a battery), made the first studies of the Gulf Stream (his maps comporting well with today's infrared satellite images), and discovered that a thin coat of oil will calm troubled waters. (To impress the ladies, always a personal priority, Franklin wielded a hollow cane loaded with oil with which he could as if by magic quiet roiled ponds at his command.) His lightning rods saved homes and lives despite the protestations of certain divines that they interfered with the fiery instruments of God's righteous wrath. Believing as he did in practical progress, Franklin championed universal education, freedom of (and from) religion, freedom of speech and of the press, free-market competition, self-improvement, and the other middle-class virtues of what he liked to call "we the middling people." Initiates into the Junto, a club of young workingmen that Franklin founded in Philadelphia in 1727, were asked to affirm that they respected the current members, felt affection for people regardless of their religion, profession, or religious beliefs, and would love and pursue the truth for its own sake. His *Poor Richard's Almanack* of 1733–58 imparted scientific and technological knowledge in a spirit of personal improvement (it was sometimes titled *Poor Richard Improved*), seasoning the mix with a sage realism: "Love your neighbor; yet don't pull down your hedge"; "Three may keep a secret, if two of them are dead."

Franklin's career, like Voltaire's, fell into three acts—first the pursuit of wealth and fame, then of scientific knowledge, and finally of liberal political reform—and he generally kept his priorities straight. When he signed the Declaration of Independence he was a wealthy, powerful, seventy-year-old celebrity who appreciated the risks he and his fellow delegates were taking—telling them, "We must, indeed, all hang together, or most assuredly we shall all hang separately"—but he anticipated the future with cheerful benignity. "The rapid Progress *true* science now makes, occasions my regretting sometimes that I was born so soon," he told Joseph Priestley, predicting that as science advanced, "Agriculture may diminish its labor and double its produce; all diseases may by sure means be prevented or cured, not excepting even that of old age, and our lives lengthened at pleasure." He suggested to Sir Joseph Banks, the president of the Royal Society in London, that, "Furnished as all Europe now is with academies of science, with nice instruments and the spirit of experiment, the progress of human knowledge will be rapid and discoveries made of which we have at present no conception." He literally bet on the future, providing in his will for the disposition of £2,000, plus accrued interest, *two hundred years* after his death.

Franklin's cheerful practicality sat poorly with Romantics like the poet John Keats, who dismissed him as "not sublime" (predictably so, since Keats thought that "the humanity of the United States can never reach the sublime") and Keats's friend Leigh Hunt, who sniffed at Franklin's having been a printer, saying that "something of the pettiness and materiality of his first occupation always stuck to him." Franklin's technological optimism was ridiculed by Henry David Thoreau, the American Rousseau, who camped out a mile from his boyhood home and wrote a book about it, subtitled *Life in the Woods*, full of lamentations about the alleged spiritual poverty of civilization. "Our inventions are wont to be pretty toys," wrote Thoreau, "which distract our attention from serious things. . . . We are in great haste to construct a magnetic telegraph from Maine to Texas; but Maine and Texas, it may be, have nothing important to communicate." Railroads, which would have delighted Franklin, aroused Thoreau's particular contempt: "A few are riding, but the rest are run over. . . . No doubt they can ride at last who shall have earned their fare, that is, if they survive so long, but they will probably have lost their elasticity and desire to travel by that time." Franklin's bourgeois values were satirized in the works of Sinclair Lewis, who denounced "the easy-going descendants of the wisecracking Benjamin Franklin" (*It Can't Happen Here*, 1935) and invited his readers to sneer at the fictional Babbitt, who, if asked about his religion, "would have answered in sonorous Boosters' Club rhetoric, 'My religion is to serve my fellow men, to honor my brother as myself, and to do my bit to make life happier for one and all' "—a flawless sentiment that Lewis discounted as self-evidently unsublime.

Yet people somehow did find uses for the telegraph, the telephone, and the railroads, and persisted in believing that they could do worse than to toil on behalf of themselves and their brethren in hopes of making the world a somewhat better place. And what did they achieve, in terms of health, wealth, and happiness?

Regarding health: In 1800, a decade after Franklin's death, half of the world's nearly one billion people died by age thirty. Two centuries later, the global population had swelled to six billion but life expectancy had more than doubled, to over sixty-five years of age. (In developing nations it was sixty-three; in the developed nations, seventy-nine. The World Health Organization estimates that average global life expectancy, which was forty-eight years for those born in 1955, will be seventy-three years for those born in 2025.) Thanks to advances in agriculture, medical research, and social hygiene, the rate of premature deaths at the end of the twentieth century had dropped to under 6 percent worldwide, and in the developed nations to under 1 percent, while the number of people aged sixty-five and higher rose to 15 percent of the general population. (In 1800 it had been under 5 percent.) These gains were attained despite such disastrous twentieth-century setbacks as the influenza epidemic of 1918–19, which killed over forty million people, the two World Wars (over seventy million dead), the communist Chinese famine of 1959–61 (fourteen to thirty million), and the HIV/AIDS epidemic (twenty-five million and counting).

People not only lived longer but became taller, more robust, and generally healthier. Stature—the maximum lifetime height attained by the average individual—increased throughout the nineteenth and twentieth centuries, due in part to significant reductions in vitamin-deficiency diseases such as scurvy, pellagra, rickets, and goiter (which are prevented by vitamin C, niacin, vitamin D, and iodine respectively). It is estimated that almost half of the labor involved in building the canals, railroads, highways, bridges, and buildings associated with the Industrial Revolution came not from machines but from the increased strength of bigger and stronger laborers, whose personal energy levels increased by roughly 50 percent from 1790 to 1980. Such progress continues today: World food production grew by 52 percent per person from 1961 to 2001, and malnutrition in the developing world dropped from 45 percent in 1945 to 18 percent in 2001. While much remains to be done in reducing such scourges as diarrheal diseases, which killed almost 3 million infants and children worldwide in 1990; measles (2 million deaths annually, almost all of them children); and tuberculosis (over 2.5 million deaths, mostly adults under age fifty), humankind has already come a long way toward realizing what the historian James C. Riley calls "the hope that all people will enjoy a long and healthy life."

Regarding wealth: In 1800, the world's per capita GDP was about seven hundred dollars per year, and was growing at a fraction of 1 percent annually. With the Industrial Revolution the global economy took off, reaching a per capita GDP of $6,460 and a growth rate of 3 percent for all humanity by 2008. Although this unprecedented increase in wealth was concentrated in the liberal-scientific nations, the poor were not entirely left behind. In 2007 the developing world enjoyed growth rates of over 7 percent, three times that of the developed world, and the peoples of emerging nations such as Mexico were earning more money than British subjects did in 1900, when England was near its peak of power. The efficacy of science and liberalism in realizing such results is underscored by the tragic example of sub-Saharan Africa, which has persistent extreme poverty, and where neither science nor democracy has yet made many inroads. By contrast, in East Asia, where science and democracy have been on the increase, the number of people living in extreme poverty has dropped from eight hundred million to under three hundred million in only three decades.

Happiness is harder to measure but a good place to start is with education, the root of which is literacy. In a very real sense, those who cannot read are unable to make much sense of the wider world. "We are all illiterate here and so we don't know what is happening," a Chinese farmer told Patrick E. Tyler of the *New York Times* in 1996. "If something is harmful, we don't know it. If something is good, we don't know it." A neighbor added, "Because I have never been to school, I am like a blind person, even though I have eyes." So it is cheering to observe that world illiteracy has long been in retreat. Europe went from a medieval adult literacy rate of under 10 percent (estimated by checking old contracts to see how many signed their names rather than making a mark) to nearly half the adult population by

1750 and 99 percent today. Japan's 1872 Fundamental Code of Education, aimed at ensuring that there would be "no community with an illiterate family, nor a family with an illiterate person," was so effective that by 1913 more books were being published in Japan than in Britain. More than 80 percent of all the adults in the world today are literate, up from 73 percent in 1985, and the literacy rate continues to grow by about half of 1 percent annually. In 1970, 37 percent of all people over age fifteen were illiterate; today, fewer than 18 percent are. Lingering illiteracy is almost entirely a matter of supply; there is little lack of demand. When several African nations eliminated school fees in 2004, they found their classrooms crammed with new students almost overnight. Uganda's public-school enrollment doubled in one year, while in Kenya, a million pupils enrolled within a matter of days. (When one of them, Joseph Lolo, age sixteen, was urged by his impoverished father to return to work to help support his family, he replied, "What will save me if I don't go to school?") Except in sub-Saharan Africa, where almost 40 percent of the children still have no access to education, nearly all the world's children are enrolled in primary schools; 70 percent of those of high-school age are enrolled, and one in four goes on to some form of higher education. In 1953, only 7 percent of adult Americans had a college degree; today nearly a third do.

Wider-reaching indicators of happiness include the United Nations Human Development Index (which combines literacy, income, and life expectancy) and various surveys of "subjective well-being." Such studies indicate that the residents of wealthy nations like France and Japan are on the whole happier with their lot than are those of poor nations like Oman and Albania, but there are some interesting wrinkles. A sense of happiness and contentment grows with increasing income up to about $15,000 per year—but above that point, more money evidently makes little difference. Intriguingly, those living in nations that distribute their income the most equitably report themselves happiest: The inhabitants of relatively socialistic nations like Iceland, Holland, Finland, and Sweden show up as happier than Americans, although the Americans make more money. (The Danes, who fork over nearly half their income to the tax collector, were named the world's happiest people in 2008.) Saudi Arabia has a per capita GDP of nearly $9,000 but distributes its wealth so vertically—Saudi princes living like princes while many ordinary Saudis cannot even go to a decent public school—that it ranks lower in perceived quality of life than do poorer nations like Poland and Costa Rica. The picture is complex—overall, the wealthiest people in virtually every nation are happier than their poorer compatriots—but it seems safe to say that the general rise in global income has at least made most people today feel happier than their grandparents were. If any deep spiritual gratification is to be realized by being poor and unhealthy, nothing of it shows up in the social-science polls.

All these gains were made following the ascent of science and liberal governance. Global spending on science and technology R&D now surpasses a trillion dollars a year, while the number of liberal-democratic nations has increased from about a dozen in 1900 to thirty-six in 1960 and sixty-one in 1990. (In 1977, only 28 percent of the world population was classified as "free"; by 2007 the ranks of the free had grown to 47 percent.) But while the connection between science and improving health seems clear enough, to what extent do science and liberty deserve credit for the world's increasing wealth?

Science generates profits, of course—most of America's annual GDP growth comes from science, technology, and invention, while nearly half of the world's one hundred largest corporations make their money chiefly from science and technology—but liberalism also plays a role. Worldwide, the growth of wealth correlates with "economic freedom"—which is not the same thing as political freedom (see China) but is characteristic of the liberal democracies. The Economic Freedom of the World index defines economic freedom as "personal choice, voluntary exchange, freedom to compete, and protection of person and property." Its 2007 survey of more than 120 countries found that the least-free fifth (or "quintile") of nations, which included Syria, Myanmar, and Zimbabwe, had a per capita GDP of only $3,305 and were growing economically by less than half of 1 percent annually, while the most-free quintile, including Hong Kong, Switzerland, the UK, and the U.S., averaged a GDP of over $26,000 and a 2.3 percent growth rate. So *economic* liberalism, at least, correlates with wealth.

Worldwide, per capita income rose *eightfold* in constant dollars from 1820 to 2000, a remarkable achicvement. But this would mean little if only a few were enriched while the majority stayed poor. (If Warren Buffett and Bill Gates walk into the restaurant where you're having dinner, statistically each diner is suddenly worth hundreds of millions of dollars, but that's no help in paying your check.) In the United States all five quintiles enjoyed rising incomes from 1980 through 2005 but the rate of rise for the top quintile outpaced the rest, making the nation more economically vertical than it had been since the Roaring Twenties. In the long term, though, the American financial picture has not been one of titans ruthlessly exploiting workers. The number of hours put in by the average American laborer declined steadily from 1870, when it stood at just under 3,000 hours per year, to less than 1,600 hours in 1998, yet per capita GDP increased nearly tenfold during the same period, rising from $6,683 to $55,618. In the late fifties, when John Kenneth Galbraith characterized the United States as an "affluent society," the average American made under $9,000 a year, compared to $20,000 in 2005, by which time the median American family's net worth (meaning assets minus liabilities) exceeded $100,000.

What about the world as a whole? Are the rich nations getting richer at the expense of the poor nations? Many have claimed that this is the case, and that globalization—meaning international free trade—is to blame: "Globalization has dramatically increased inequality between and within nations"; "global income inequalities are widening"; "inequality is soaring through the globalization period"; and even, "The attacks on the World Trade Center and the Pentagon stemmed from tensions created by the widening gulf between rich and poor nations." Alarming if true, but is it true?

A conspicuous gap opened up between rich and poor in about 1820, when the Industrial Revolution reached critical mass, and it continued to widen until about the early 1970s. Globalization got going around 1980, so the relevant question is whether economic inequality has increased since then. It may stand to reason that a booming global economy benefits the rich and leaves the poor behind, but applied reason is only as good as the data on which it is based. An objective analysis of the best available data (which, admittedly, are far from flawless) indicates that while larger inequalities did crop up here and there, global inequality overall ceased growing from 1980 to 2000 and in many respects began to shrink. More to the point, income inequalities tend to be largest within the nations least touched by globalization. The poorest 10 percent of those living in the rich, free, "globalized" nations earn 2.5 percent of the total national income, while their counterparts in the least-free, least globalized nations get only 2.2 percent of the pie. (The disparity is even worse in absolute terms: The bottom 10 percent in the least economically free nations make under $1,000 a year, while in the freest nations they earn over $7,000.) So there is little evidence that the rich got that way by exploiting the poor.

One source of confusion on this point arises from the statistical matter of how economists bin their data. Even if economic inequality were increasing within every nation (which it is not) and between all nations (which it also is not), the gap between rich and poor for all humanity could still be decreasing, if for instance large numbers of poor people in the most populous nations were making more money. And this is just what is happening, notably in India and China, the two biggest countries in the world. China in 2004 had a per capita GDP of under $4,000, compared to over $35,000 for the United States, but it also had four and half times the population of the United States and was growing economically at triple the U.S. growth rate. Nor was this just a brief spurt: From 1980 to 2000 China's economy grew an average of 8.3 percent annually, and India's 3.7 percent, while the highest-income nations grew by under 2 percent. So a massive elevation of poor people in a large nation can reduce global inequality, even if that nation grows more economically vertical within its borders.

The index customarily employed to measure economic inequality is called the Gini ratio. It ranges from zero to one, where zero represents perfect equality (everybody has exactly the same income or worth) and one indicates absolute inequality (one person or nation has all the money and the rest have none at all): The lower the Gini ratio, the less the inequality. A few studies suggest a modest rise in the Gini since 1980, but most indicate that it has been falling—meaning that nations are becoming economically more equal, not less so. This tendency holds up even if we set aside 80 percent of the world's population and compare only the top and bottom 10 percent. Growing inequality is found mainly by restricting our attention to just 1 percent at each end of the economic spectrum—the very richest of the rich, and poorest of the poor, with about 60 million persons in each category. There, inequality is indeed on the rise. But since the extremely poor, many of whom are in sub-Saharan Africa, are the least touched by globalization, the disparity argues for more, not less, global trade and economic liberalism.

These findings are consistent with what Nathan Rosenberg and L. E. Birdzell Jr., in their book *How the West Grew Rich*, call "an oddity of Western economic growth that, while it made some individuals extremely rich, it benefited the lifestyle of the very rich much less than it benefited the lifestyle of the less well-off [because] the most lucrative new products were those with a market among the many, rather than among the few." These include high-tech products once regarded as the province of the well-to-do but now widely available: Worldwide, one in every five persons circa 2008 had access to a computer, and more than half had mobile phones.

In retrospect, it is clear that the Industrial Revolution generated global economic inequalities and that colonialism may have exacerbated the situation. But inequality is to some extent the unavoidable consequence of any sudden increase in wealth. Imagine a nation in which everybody is equally poor: If the population is divided into fifths by income, each quintile has about 20 percent of the income. The citizens of this imaginary nation now set about training themselves in, say, computer science and technical support—as has happened in Ireland and India—and are rewarded with an economic boom so powerful that soon, 80 percent of them have doubled their incomes. Since nearly all this fresh money was earned by the top four-fifths of the population, the table now shows a much more unequal picture. This may strike some as undesirable or even unjust—one hopes that the bottom fifth will come to share the wealth—but for a formerly poor nation to have doubled the earnings of four-fifths of its population is a major achievement, one that cannot be dismissed as a statistical artifact of the "Warren Buffett and Bill Gates walk into a restaurant" variety.

Liberalism fosters scientific and technological progress by accommodating change without pretending to know where changes will lead. Unlike the political right, it denies that the answers to contemporary problems are necessarily to be found in the past; unlike the left, it refuses to sacrifice extant freedoms in order to attain progressive goals. "All institutions of freedom are adaptations to [the] fundamental fact of ignorance," notes Friedrich Hayek. "Nowhere is freedom more important than where our ignorance is greatest—at the boundaries of knowledge, in other words, where nobody can predict what lies a step ahead." Scientific and technological creativity stands at those frontiers, which by virtue of their inherent unpredictability are ill-suited to hierarchical organization and planning. "The Western scientific community was more, rather than less, efficiently organized by reason of its lack of a hierarchy," write Rosenberg and Birdzell. Top-down organization may spur development in the early stages of a scientific or technical endeavor, when the questions are still relatively simple and the variables few, but thereafter it will tend to hinder progress unless it gets out of the way. Joseph Needham, whose magisterial *Science and Civilization in China* helped awaken Europeans and Americans to the technological ingenuity of the East, blamed government bureaucracy for the fact that China failed to expand its scientific research after about the fifteenth century: " 'Bureaucratic feudalism' at first favored the growth of natural knowledge and its application to technology for human benefit," he writes, "while later on it inhibited the rise of modern capitalism and of modern science"; progress "was constantly inhibited by the scholar-bureaucrats." Notwithstanding a few exceptions such as the Manhattan Project's development of nuclear weapons, neither government nor industry can expect to get results by assembling a lot of scientists and ordering them to produce a wished-for result—as was confirmed by President Richard Nixon's futile War on Cancer.

Planners seek to direct human effort toward useful endeavors, but in a changing world even the most fruitful projects can look pointless at first. When an associate commiserated with Thomas Edison over his having conducted nine thousand unsuccessful experiments in trying to devise a new type of battery, saying, "Isn't it a shame that with the tremendous amount of work you have done you haven't been able to get any results?" Edison, grinning, replied: "Results! Why, man, I have gotten a lot of results! I know several thousand things that won't work!" The imperious motto of the 1933 Chicago Century of Progress exposition, "Science Finds—Industry Applies—Man Conforms," was off the mark on all three points. The most fertile facts found by science are seldom of immediate technological value, although they may become so eventually. Faraday's research laid the basis of modern electronics, but it took a century for the electronics industry to come into its own; Einstein's relativity showed the path to nuclear energy, but the path was so obscure that Einstein himself originally thought it impassable. Nor does industry merely apply scientific knowledge; it often makes its own discoveries, for its own reasons. Food preservation was invented in 1810 by the Paris confectioner Nicholas Appert, who won a cash prize, put up by Napoleon Bonaparte, by establishing that foods would not spoil if they were sealed in boiling-hot bottles with no room for air; all this happened sixty-three years before Louis Pasteur established *why* it worked. Superconductivity—the ability of an electrical current to flow at zero resistance through certain materials—was discovered decades before scientists began to comprehend its physics. Velcro was the brainchild of a Swiss electrical engineer who knew little about materials properties, and while computers were pioneered by scientists, the first PC was assembled by a couple of youthful hobbyists working in a suburban garage. Nor have the products created through free scientific and technological inquiry tended to make people conform; rather, they have presented them with new choices. People say that they hate their mobile phones but cannot live without them. If they are conforming to anything when they use their cell phones—which, among their other virtues, have proven effective in reducing poverty—it is to their own preferences rather than to the dictates of industry, which had mobile-phone technology in the 1940s but could find no market for it.

This is not to argue for anarchy. Like liberalism, scientific and technological progress requires law, an underlying structure that works to keep things fair and honest and to protect the intellectual property rights of inventors; hence the importance placed on the U.S. Patent Office by the American founders. Scientific experimentation takes place within a structured framework of publication, peer review, grant applications, and free—often scathing—debate, the general effect of which is to weed out inferior results. Free-market capitalism requires legal and ethical restrictions too, not only to protect ordinary folks from fat-cat predators but also to assure the normal functioning of markets. It is no coincidence that the British, who borrowed the old Roman phrase "laws of nature" and applied it to scientific findings, are widely respected for their legal system and the quality of their scientific research alike.

One of the most hopeful, if least appreciated, signs of world progress is growing urbanization. Improved agricultural productivity is now having much the same effect in the developing world as it previously had in Europe and America: Freed from the necessity of remaining on farms, billions of people are moving to cities. The number of urban dwellers quadrupled between 2000 and 2005; today, for the first time in history, the majority of people live in cities. Although the political left likes to portray urbanization as driven by ruthless capitalism or environmental problems, in the main it appears to be the result of free individual choice. Pastoral romances to the contrary, over the course of history the vast majority of those residing in the countryside did so because they had no real alternative. They lived as their ancestors had lived, farming or fishing at subsistence levels, and if they happened to visit a city—which typically required a costly, arduous, and often dangerous journey—they seldom did more than briefly conduct their business, gawk at the sights, and express astonishment at the prices before hastening home. This historic logjam broke when scientific agriculture diminished the demand for farm labor. No longer needed at home, inspired by the prospect of better jobs and schools in the cities, and able to get there by bus on decent roads rather than by trudging through dark forests infested by highwaymen, people—young, enterprising people especially—moved to town.

Their vast migration, involving roughly two billion people in just a half century, has put the brakes on population growth, which in the developing world slowed from 2.06 percent annually in 1955 to 1.48 percent by 2005. (Rural families tend to have more children, since they can help with the chores, than do urbanites, for whom children mean more mouths to feed in cramped quarters.) As cities now account for most of the world's population growth, the global growth rate is expected to continue to decline. The civil engineer Robert Ridgway had a point when he said, back in 1925, that "perhaps the most notable effect of the application of those laws of nature which have been brought to light by the patient investigations of the scientist during the past century and a half is evidenced in the wonderful growth of cities."

The new arrivals are seldom greeted with open arms. Poor, undereducated, and unable to compete for any but menial jobs, they have congregated in the vast shantytowns that now ring megacities like Rio de Janeiro, Mexico City, Nairobi, Lagos, Jakarta, Karachi, Shanghai, Delhi, and Bombay. There they contend with inadequate water supplies, sewage disposal, electricity, and health care. Even the most hard-hearted capitalists should like to see their living conditions improved, if only for the self-interested reason that slums are potential breeding grounds for epidemics that could break out and sweep across a jet-girdled world.

But while this new rise of slums is unprecedented in scale, the same phenomenon has occurred many times before, whenever agricultural improvements freed the rural poor to urbanize themselves. In Europe, urbanites rose from 10 percent of the population in 1800 to nearly 30 percent by 1890. The pace of urban growth was particularly swift in industrial regions such as the northern British Isles; the population of Manchester and Leeds tripled between 1810 and 1841, and Glasgow grew 40 percent in a single decade, 1831–41. Little new housing was built to accommodate the influx, and the resultant overcrowding and squalor fueled epidemics of typhoid, typhus, smallpox, cholera, and tuberculosis. Life expectancy at birth in Manchester in 1841 was 26.6 years, half that of Bangladesh and Haiti today. The grim spectacle of the working poor crammed into pestilent slums and working in dank, dangerous factories created an image of industrial progress as having been purchased at an exorbitant cost in human lives, environmental desolation, and the rise of a hard-hearted, class-conflict society in which every human interaction was reduced to cash on the barrelhead.

London, beset by successive waves of immigration from the countryside beginning in the late sixteenth century, was a conspicuous laboratory of this rapid, unplanned and widely unwanted new urbanization. Barred from entering by the restrictive employment practices of the urban guilds, immigrants from the farms squatted outside the city walls in shantytowns called rookeries (or *suburbs*, meaning "under the city"). Alarmed to find growing hordes of rustics at their doorstep—many of whom had brought their farm animals with them, giving rise to the term "pigsty" to mean a filthy dwelling—Londoners from 1662 to 1697 passed a series of Poor Laws that required the poor to return home if they wanted to get on the relief rolls or obtain certificates of urban settlement. These harsh measures had little effect, and the slums had become even more overcrowded by the nineteenth century. Charles Dickens's *Oliver Twist*, serialized in 1837–39, described "unwashed, unshaven, squalid, and dirty figures . . . thieves, idlers, and vagabonds of every low grade . . . constantly running to and fro" in "nearly ankle-deep . . . filth and mire." The London *Times* protested, in an editorial of October 12, 1843, that "within the most courtly precincts of the richest city of GOD's earth, there may be found . . . FAMINE, FILTH AND DISEASE." Fyodor Dostoevsky, visiting London in 1863, called it "a biblical sight, some prophecy out of the Apocalypse being fulfilled before your very eyes."

Reformers were quick to blame the British economic system: "Rich London is the creature of slum-London, of poor London," wrote William Morris in 1881,

> and though I do not say that the London slums are worse than those of other big cities, yet they together with the rich quarters make up the monstrosity we call London, which is at once the centre and the token of the slavery of commercialism which has taken the place of the slaveries of the past. . . . The sickening hideousness of London, the metropolis of the nation, which has worked out the sum of commercialism most completely, seems to me a mark of disgrace branded on our wire-drawn refinement to show that it is based on the worst kind of theft—legal stealing from the poor.

Yet London today has few remaining slums—none, by Dickensian standards—and its air is the cleanest it has been since 1585. The trouble had not been capitalism or industry, but a population influx so enormous that it almost overwhelmed the city's infrastructure: Dickens was describing a city that had ingested nearly three million people, quadrupling its population, within a single generation. The poor were not the problem; their poverty was. They may have brought with them more than their share of "rogues, vagabonds, and thieves," as Parliament began hearing as early as the 1500s, but they also brought creativity, energy, and innovation. Indeed a major reason that the guilds sought to exclude them was because they represented a dangerous potential for change. In 1588 Lord Cecil, reporting to Queen Elizabeth I on the residents of one shantytown, observed that "they excel . . . in policy and industry [and demonstrate] a natural ardency of new inventions annexed to an unyielding industry." London prospered because it eventually managed to incorporate its new citizens into legal, commercial, and social systems liberal enough to benefit from their intelligence and hard work.

The new megacities have no real alternative—and no brighter prospect—than to do the same, by absorbing their shantytown compatriots rather than locking them outside the (figurative) city gates. How might this be done without recapitulating generations of Dickensian suffering along the way?

This question has been ingeniously investigated by the Peruvian economist Hernando de Soto. Born in Lima, de Soto was educated in Switzerland before returning to Peru in 1980. There he studied the relationship between government, the law, and urban poverty, interviewing the poor and studying their economic situation with everything from satellite imagery to the barking of dogs—which roughly demarks the property lines of squatters' undeeded homes. De Soto found that the newly arrived poor typically want to work, pay taxes, and own or lease their dwellings, but are discouraged from doing so by dense thickets of laws and regulations that function rather like the stone walls confronted by the suburbanites of seventeenth-century Europe. When he and his colleagues tried to obtain the necessary permits to legalize a small Lima garment-making shop containing only two sewing machines, they discovered that even though with the help of a lawyer, a luxury few squatters could afford, the process took 289 days and cost thirty-one times the monthly minimum wage. For a family to obtain legal title to the shack where they lived or worked required more than two hundred bureaucratic steps taking twenty-one years of effort. The effect of all this red tape, de Soto argues, is to erect a "legal apartheid between those who can create capital and those who cannot." Peru "had become two nations: one where the legal system bestowed privileges on a select few, and another where the majority of the Peruvian people lived and worked outside the law, according to their own local arrangements." A "paper wall," de Soto concluded, "stops the poor from being able to develop private legal enterprise."

> These obstacles many times don't exist for the wealthier parts of the population, because they're continually plugged into lawmakers. Something goes wrong, you talk to your friend the minister, you're well organized in the chamber of commerce and the local manufacturers association, some kind of a guild. The poor don't have a voice, at least not a business voice.

Borrowing a phrase from the French historian Fernand Braudel, de Soto portrayed the established urban elites as living inside a "bell jar." "The bell jar makes capitalism a private club, open only to a privileged few, and enrages the billions standing outside looking in," de Soto wrote in *The Mystery of Capital.* Those inside the bell jar—in twentieth-century Lima as in seventeenth-century London—might claim that the poor don't want to work, but when de Soto's team set up storefront offices in the slums of Peru to register previously extralegal businesses, over 276,000 squatters took advantage of it. During the first four years of the experiment, the taxes paid by these newly enfranchised and self-nominated businessmen and women, who previously had paid no taxes at all, totaled $1.2 billion.

De Soto's research indicates that in many developing nations, needless government regulation has driven *the majority* of economic activity off the radar screens and the tax rolls. He estimates that 90 percent of Peru's small industrial enterprises and 60 percent of its fishing fleet and food stores operate outside of the law, that most of the new jobs being created in Latin America and Russia are extralegal, that 92 percent of all the construction and 88 percent of all business in Egypt are extralegal, and that "almost eighty percent of the [Mexican] population is either living or working in the extralegal economy," producing a third of Mexico's GDP.

> In every country we have examined, the entrepreneurial ingenuity of the poor has created wealth on a vast scale—wealth that also constitutes by far the largest source of potential capital for development. These assets not only far exceed the holdings of the government, the local stock exchanges, and foreign direct investment; they are many times greater than all the aid from advanced nations and all the loans extended by the World Bank.

But the extralegal poor cannot leverage their assets. Lacking title to the land they occupy and official permission to engage in the businesses that support them, they are unable to obtain loans, credit, or mortgages. Nor can they build up a business much past family-scale, for fear of attracting official scrutiny. Their assets are what de Soto calls "dead capital," money unavailable for investment. The potential value of small-scale "live capital" often escapes those of us in the developed world who tend to think of forming a company as a high-finance proposition, but the average U.S. start-up has one employee, under $25,000 in assets, and is funded by a home mortgage.

De Soto urges that legal systems adapt to conditions as they are (a million squatters at the gates) rather than as a ruling elite might prefer them to be. He likes to quote Oliver Wendell Holmes's dictum, "The life of the law has not been logic; it has been experience." The essential issue is squatting, and while squatters often have a hard time of it, history is on their side. Ethan Allen's hordes of squatters helped him win statehood for Vermont, and an elderly George Washington, the most esteemed man in America, was unable to evict squatters found on his Virginia farmland. (His attorney advised that even if Washington succeeded, the evictees might burn his barns in revenge.) The Homestead Act of 1862, giving 160 acres free to any settler who stayed on and developed the land for five years, was a case of the government playing catch-up ball: Most of the settlement had already taken placc. Freeing up land in England was so painful a process that as late as 1880 Benjamin Disraeli identified the two greatest problems facing the nation as being the governance of Ireland and "the principles upon which the landed property of this country should continue to be established."

In Peru, de Soto helped fashion a land-reform policy under which more than a million families gained legal title to their lands and some 380,000 previously black-market firms were legally registered in just seven years, from 1988 to 1995. He warns the leaders of other developing nations that to exclude the majority of their own citizens from full participation in a capitalist economy threatens capitalism itself, since the poor may otherwise succumb to the siren call of socialism or communism. (The communists returned the compliment; in the eighties de Soto's offices in Peru were bombed and his car machine-gunned by Maoist "Shining Path" terrorists.) So while success is not assured, "Capitalism is in trouble if it doesn't adapt."

Civilization is the product of cities (both words come from the same root, *civitat*) and the mass migration of the world's poor to cities has great potential benefits. "Einstein taught us," de Soto writes, that "a single brick can be made to release a huge amount of energy in the form of an atomic explosion. By analogy, capital is the result of discovering and unleashing potential energy from the trillions of bricks that the poor have accumulated in their buildings." His experimental approach has drawn criticism—from conservatives preaching self-reliance, and from progressives suspicious of capitalism—but the way for humankind to continue progressing is through science, not sermons. Liberalism creates the political environments that best facilitate scientific inquiry, and of the social sciences, none is more quantitative and successful—and none has been so roundly denounced—as economics.

CHAPTER NINE

THE SCIENCE OF WEALTH

Western civilization was being transformed by two great movements: the economic changes which we sum up as the rise of capitalism, and the changes in knowledge which we sum up as the scientific movement.

—GEORGE NORMAN CLARK, 1949

The band of commerce was designed
To associate all the branches of mankind;
And if a boundless plenty be the robe,
Trade is the golden girdle of the globe.

—WILLIAM COWPER, 1782

For an allegedly dismal science, economics generates lots of jokes. Economists, it is said, are frequently wrong (they "forecast twelve of the past three recessions"), tend to be noncommittal ("If you laid all the world's economists end-to-end, they would not reach a conclusion") and like to hedge their bets ("I asked an economist for her phone number and she gave me an estimate"). Harry Truman wanted a one-armed economics advisor who wouldn't keep saying, "On the other hand." Ronald Reagan speculated that if economists had invented Trivial Pursuit, the game would have a hundred questions and three thousand answers. As early as 1855, Walter Bagehot of the *Economist* was claiming that "no real English gentleman, in his secret soul, was ever sorry for the death of a political economist."

Part of this is guilt by association. As a social science, economics tends to get lumped in with "soft" disciplines like cultural anthropology, much of which reads like fairy tales translated into Esperanto, and the psychoanalysis of Sigmund Freud, who discovered nothing and cured nobody. But the social sciences labor under two hobbling limitations—a constrained ability to conduct experiments (they cannot very well force human subjects to run mazes) and the lack of a dependable metric by which to quantify indistinct matters like human happiness, hope, anxiety, or fear. Economics suffers far less from these limitations. The financial world provides a ready-made metric—in the form of money, which generates a constant flood of hard, quantifiable data—and innumerable laboratories in the form of markets, which by their very nature involve constant experimentation. This has allowed economists to make important discoveries about profit and loss, prosperity and poverty, and the financial effects of governmental policies.

When the historian Thomas Carlyle characterized economics as "dismal," he was responding to the admirably specific predictions of Thomas Malthus, who in his 1798 *Essay on the Principle of Population* argued that human population growth was unsustainable. Malthus's thesis was that a well-nourished population would grow geometrically (1, 2, 4, 8, 16 . . .) while the cultivation of crops to feed all those new mouths, limited as it was to available land, could only grow linearly (1, 2, 3, 4 . . .). A "gigantic inevitable famine" thus stalked the human future, concluded Malthus, who conceded that his prediction had "a melancholy hue."

Dismal though it may have been, the Malthusian hypothesis qualified as genuinely scientific: It was logical, quantitative—and it could be tested, by waiting to see what happened. As time passed, the number of mouths to feed did indeed increase. During the nineteenth century the population of England quadrupled, that of Europe more than doubled, and the world population climbed from about 900 million to 1.6 billion. Yet mass starvation did not result. Instead, farmers became much more productive than Malthus could have foreseen. Thanks to technological innovations and an increasing penchant for agronomic research and experimentation, English farmers by 1800 were feeding twice as many people as they had a century earlier, with half as much labor. Food production continued to soar thereafter while the number of agricultural laborers in England shrank from three-quarters of the workforce in 1690 to one-quarter by 1840. Progress in the United States was even more spectacular. From 1800 to 2000, the American agricultural workforce diminished from 70 percent to under 2 percent of the population, during which time American farms became *thirty-five times* more productive. Famine still occasionally raised its ugly head—its last European appearance being the Irish potato blight of 1845–1852, which resulted in a million deaths—but overall, Europe and the rest of the developed world managed to sustain ever larger populations with an ever smaller portion of its workforce. Malthus could still be proven right in the long run, of course, should the global population eventually outrun science and technology, but as the world becomes more urbanized and its population growth rate slows, it now appears that the human population will stabilize at something under twelve billion by about the middle of this century. So Malthus was probably wrong, and economics is not necessarily dismal.

Needless to say, our planet still faces many economic problems. Nearly 900 million people struggle to get by on less than a dollar a day, while Europe and the United States spend more on pet food than it would cost to feed them all. The world's three hundred richest individuals have more money than the entire bottom half of humankind, while the per capita GDP of wealthy countries like Luxembourg, Switzerland, and the United States runs to fifty times that of poor countries like Sierra Leone, Tanzania, and Congo. But there have also been remarkable gains. In the past quarter century, the number of people in the developing world living on less than a dollar a day has been cut almost in half—down from nearly 1.5 billion to 900 million between 1981 and 2008, despite a simultaneous increase in world population—and the proportion of the world's population living on under *two* dollars a day was halved between 1983 to 2003, although by 2009 a global slump had begun eating into some of those gains.

Confronted with this dynamic, it is important to understand not only what went wrong but what is going right. To continue to reduce hunger we need to identify its causes (they include civil war, bad government, and bad roads) rather than just feeling guilty about it or blaming the rich. (Soaking the rich to aid the poor wouldn't work. If the entire wealth of all the world's millionaires were confiscated and doled out to everybody else, the one-time giveaway would yield each person less than a year's wages, while the global economy sank into depression for want of investment.) Until recently, science was of only marginal help in understanding such processes, involving as they do billions of people living in many different cultures. But the much-maligned science of economics has now begun to have a substantial impact. Indeed, economics has done so much to analyze problems and identify remedies that it may one day be ranked with the agricultural and medical sciences when it comes to saving lives and improving their quality.

To appreciate the dimensions of the change, consider how our predecessors thought about wealth and poverty before economics came along.

For nearly a thousand years after the fall of Rome, Europe was poor and *stayed* poor. It was a time, in the words of the historian William Manchester, of "incessant warfare, corruption, lawlessness, obsession with strange myths, and an almost impenetrable mindlessness," a time when societies typically were "anarchic, formless, and appallingly unjust." Per capita GDP long remained flat, at less than five hundred dollars per person per annum, and when it finally started to grow did so very slowly, by perhaps a tenth of 1 percent annually from around AD 1000 to 1600. (The American economist J. Bradford DeLong estimates that it took twelve thousand years for humanity worldwide to rise from the $90 annual "income" of the ancient hunter-gatherer to $180 by 1750. Thereafter it climbed to a global average of over $7,000 today.) Crop failures brought famine one year in four, and even in good years only the most prosperous farmers harvested enough grain to keep their families fed until much past Easter, whereupon many were reduced to foraging for herbs and roots. Malnutrition was so persistent that in the eighteenth century, when the Brothers Grimm set about collecting old fairy tales, they encountered hauntingly recurrent themes of people going hungry and happy endings in which the hero or heroine for once has enough to eat. The median life expectancy of Europeans overall was thirty years of age; that of females, who were apt to die in childbirth, only twenty-four. Ignorance was ubiquitous: Few could read, write, do sums, or recount the history of the hamlet in which they were born and were likely to die. Innovation was almost unheard of. Waterwheels, introduced in the 800s, and windmills, which appeared in the 1100s, were the only two medieval inventions of any consequence. As Manchester describes the prevailing mindset, "The Church was indivisible, the afterlife a certainty; all knowledge was already known. *And nothing would ever change.*" Europeans were not only unaware of any prospect of improvement, "They were convinced that such a phenomenon could not exist."

Unsurprisingly, given Europe's fiscal and intellectual paralysis, the economy was regarded as a zero-sum game: There was a fixed amount of wealth in the world, so for one person to do better, another—or many others—had to do worse. As it was with individuals, so it was thought to be for states. A nation's prosperity was measured by the amount of wealth it possessed—gold, usually—and the object of foreign trade was to see that your nation wound up with more gold than the others, at their expense. (Even Francis Bacon, who strived to imagine the scientifically advanced world of the future, failed to envision a non-zero-sum economy: "The increase of any estate," he wrote, "must be upon the foreigner.") Should the citizens of another country produce better, cheaper goods than yours did, the remedy was to wall yourself off from the perceived threat by erecting protectionist trade barriers. To borrow money, a step without which only a fortunate few could build a business or launch a trade expedition, was frowned upon; to lend money at interest was a sin, punishable by excommunication. (A popular twelfth-century folk tale had it that a usurer, upon arriving at church to be married, was crushed by the toppling statue of another usurer—irrefutable evidence of God's opposition to moneylending.) Amid rampant poverty, ignorance, and fear, the monarchs and bishops who held the pursestrings were quick to affirm that only their autocratic, top-down control of economic affairs prevented financial chaos.

This long economic ice age began to thaw—spottily, with the Italian Renaissance and the rise of the Dutch Republic, then more widely by the eighteenth century—thanks to four principal innovations: readier access to capital, once religious prohibitions against moneylending were relaxed; the growth of human rights and the rule of law; improved communications and transportation (beginning with the advent of reliable steamships in the early nineteenth century and competitive land transport a half century later); and the rise of science, which demonstrated how people could, as Descartes had dreamed, become "masters and possessors of nature."

Encouraged by Bacon and other prophets of scientific progress, a few creative agrarians began experimenting. These "improving" farmers tried crop rotation, seed selection, and new tools that made farming less labor-intensive, freeing young men and women to try their hand at trades and factory work. Technological innovations and job specialization gradually amplified the productivity of the swelling industrial labor force until a growing class of merchants, capitalists, and tradesmen began to attain real wealth. The English "gentleman" whose cultivated manner, elegantly understated dress, and tasteful country house are still being imitated today belonged not to the aristocracy but to the gentry.

These changes, salutary in hindsight, were by no means universally welcomed at the time. Thousands died in riots protesting sixteenth-century English crop enclosures—the walling-off of farms to replace medieval commons with privately owned vegetable gardens and grazing pastures—although enclosure boosted production by giving farmers a proprietary stake in the land they worked. Oliver Goldsmith expressed the sentiments of many poets, for whom economics was seldom a strength, in his book-length poem "The Deserted Village":

> Ill fares the land, to hastening ills a prey,
> Where wealth accumulates, and men decay.

When British weavers' guilds protested to the king about competition from an innovative, mass-production wool factory, the factory was shuttered and its operating techniques outlawed. Sixteen thousand people were executed—many of them hanged, others broken on the wheel—in France in 1666 for unauthorized trading in imported calicoes. Clerics complained that a lust for gold was sullying their parishioners' love of God. "This new unwanted society," writes the historian Robert L. Heilbroner, was "at every step . . . misconceived, feared, and fought. The market system with its essential components of land, labor, and capital was thus born in agony—an agony that began in the thirteenth century and had not run its course until well into the nineteenth."

Yet free-market capitalism survived and grew, as an acquired taste for profits, personal freedoms, and the material benefits of scientific and technological innovation spread from Europe to America and beyond. The capsizing of the old order called for new insights: "There is nothing," wrote Samuel Johnson, "which requires more to be illustrated by philosophy than trade does." Johnson had not expected such a work to be produced by Adam Smith—whose interests were primarily scientific, who had no personal experience in business, and whom Johnson thought "as dull a dog as he had ever met"—but it was Smith's *Wealth of Nations* that began to make sense of it all.

The posthumous son of a Scottish customs officer, Smith grew up as a sickly child, closely watched over by an anxious mother whose anxieties redoubled when the boy, at age three, was briefly kidnapped by Gypsies. He studied mathematics and science at Oxford for six years, served for thirteen years as professor of logic and philosophy at the University of Glasgow—"by far the most useful, and therefore as by far the happiest and most honorable period of my life"—then traveled in France, where he was impressed by the vibrancy of experimental democratic states like Toulouse, which had its own *parlement* and institutes of art and science. Back in Scotland, Smith completed the *Wealth of Nations* and then moved to London, where he was feted by the likes of Gibbon and Burke—and by Benjamin Franklin, from whom he learned enough about the colonies to predict that America "seems very likely to become one of the greatest and most formidable [nations] that ever was in the world." He died in 1790 at age 67, having enjoyed the uncommon experience of seeing his ideas put into political practice with beneficial effect.

Smith was a deeply preoccupied character, whose many peculiarities have tempted biographers to portray him as a figure of Newtonian eccentricity. Given to shy silence in social settings, he mumbled to himself in public; walked with a strange, rolling gait; and dictated drafts of *Wealth* while rubbing his head against a particular spot on the wall of his study, symptoms which suggest to doctors today that Smith suffered from Tourette's syndrome or Parkinson's disease. He lectured without notes, in a tangled syntax that made him hard to follow, but his students adored him, James Boswell went out of his way to hear him speak, and his colleagues at Glasgow praised "his happy talent of throwing light upon the most abstract subjects [and] his assiduity in communicating useful knowledge," calling him "a source of enjoyment as well as of sound instruction." Nor did his oddities prevent his forming lasting friendships, notably with the philosopher David Hume—himself a critic of the prevailing economic philosophy who had pointed out that hoarding gold only inflated its price. The diffident Smith and the coolly acerbic Hume talked, corresponded, and critiqued each other's manuscripts for more than thirty years. Upon reading the manuscript of Smith's *Wealth*, Hume predicted that it would "fix the attention of public opinion."

That it certainly did. Smith's *Wealth of Nations* may be the one book between Newton's *Principia* and Darwin's *Origin of Species* that actually, substantially, and almost immediately started improving the quality of human life and thought. Despite its considerable length—380,000 words, more than twice that of the New Testament—*Wealth* was widely consulted, going through five editions in fifteen years. The prime minister, Lord North, took Smith's ideas to heart when drafting the British budgets of 1777 and 1778. William Pitt, one of the most powerful figures in the government, is said to have told Smith, "We are all your scholars."

Smith's influence has continued to grow across the centuries, but precisely because the liberal-democratic, scientific nations of the twenty-first century evince so much of his program, detailed accounts of it tends to strike us as largely self-evident—as is the case for Newton and Darwin. Nevertheless it may be useful to briefly recount his major findings.

Smith's approach to economics, if insufficiently quantitative to pass muster today, was scientific in at least three respects. First, Smith was more interested in understanding human affairs as they are than in urging how they ideally ought to be; he sought to establish what *is* before prescribing what *should be*. This set him apart from the moralizing and sentimental bent of many prior philosophers, among whom he counted even the rationalistic Descartes. (Smith remarked that the Cartesian philosophy "does not perhaps contain a word of truth," dismissing it as "one of the most entertaining romances that have ever been wrote.") He was equally critical of ancient philosophers like Cicero and Seneca, whom he assailed for treating the works of Greek mathematicians and astronomers with "supercilious and ignorant contempt." This opinion was not unprecedented—Bernard Mandeville too attacked what he called the "sentimental moralists"—but it was sufficiently original that Smith to this day is criticized as amoral or even immoral for refusing to imagine that he could improve people by preaching to them in print. Second, Smith stressed quantitative analysis whenever possible—seeking, as William Petty had put it in his *Political Arithmetic* of 1690, to "express myself in terms of number, weight, or measure; to use only arguments of sense, and to consider only such causes as have visible foundations in nature" rather than resorting to "superlative words and intellectual arguments." Third, Smith's approach was empirical. Rather than reasoning from first principles, as Descartes did, or from pleasing fantasies à la Rousseau, he based his arguments on dispassionate observations of the real world. *Wealth* is crammed with examples of how bakers, spinners, weavers, rice planters, coal miners, shipbuilders, herring fishermen, masons, bricklayers, clockmakers, road builders, goldsmiths, and landlords actually conduct their affairs.

Out of this empirical, somewhat quantitative, and thoroughly unsentimental analysis, Smith produced a revolutionary account of the creation of wealth and the functioning of markets. The wealth of a nation, he asserted, is properly to be measured not by its stores of gold but by the quantity, quality, and variety of goods its citizens consume. For Smith, the consumer is king—a point that in later years was often lost sight of, when his name became associated with the laissez-faire economics that critics of capitalism blamed for the perceived excesses of "rapacious industrialists" and "robber barons." The world's gold supply is limited (even today, there's not enough of it in circulation to fill the Washington Monument) but the production of consumer goods can be increased indefinitely if individuals are free to invest and to innovate.

The eighteenth century being a period of bustling inventiveness, Smith was able to find many examples of how innovations improve the quantity and quality of consumer goods. To make his case as simply as possible he opened *Wealth* with a very basic example, the division of labor in the manufacture of pins. Smith relates that he visited a small pin manufactory and found that it took "about eighteen distinct operations" to make a pin: "One man draws out the wire, another straights it, a third cuts it, a fourth points it, a fifth grinds it at the top," etc. Thanks to this division of labor, ten workers could make almost five thousand pins a day; without it, "they certainly could not each of them have made twenty, perhaps not one pin in a day." Another source of more efficient production was, of course, machinery—"beautiful machines," Smith exclaims, "which facilitate and abridge labor, and enable one man to do the work of many"—none but the most rudimentary of which could be fashioned without divisions of labor. Writing, money, and financial ledgers were other examples of inventions that increase wealth. Inventions could arise almost anywhere, Smith noting that even children had invented useful devices. "It was probably a farmer who made the original plough," he muses, and "some miserable slave . . . grinding corn between two stones, probably first found out the method of supporting the upper stone with a spindle. A millwright perhaps found out the way of turning the spindle with the hand; but he who contrived that the outer wheel should go by water was a philosopher"—that is, a scientist—"whose business it is to do nothing, but observe everything." There was almost no limit to the potential economic growth of a nation whose people were free to exercise their creativity and to profit from it.

Smith's analysis of markets has the excited air of a scientist making a great discovery—which in a sense it was, comparable to Newton's dynamics or the discovery of binary computing. Free markets link prices to production. If prices get too high—Smith disparaged "exorbitant" prices, an appropriately Newtonian word meaning "out of their orbits"—suppliers will increase production and prices will soon come down again. If prices get too low, profit margins will shrink, supplies diminish, and prices rise accordingly. In this manner, Smith noted, again in Newtonian language, prices are "continually gravitating, if one may say so, towards the natural price." Regulations and other legal restrictions are appropriate only insofar as they are required to keep the market fair and free. The beauty of a free market is that it benefits society as a whole without requiring its participants to act out of any loftier motive than the self-interested pursuit of personal gain: "It is not from the benevolence of the butcher, the brewer, or the baker that we expect our dinner, but from their regard to their own interest." In what has become the most famous passage of *Wealth*, Smith depicts the benevolent and self-regulating dynamics of marketplace as akin to the guidance of an "invisible hand."

> Every individual necessarily labors to render the annual revenue of the society as great as he can. He generally, indeed, neither intends to promote the public interest, nor knows how much he is promoting it. By preferring the support of domestic to that of foreign industry, he intends only his own security [but] by pursuing his own interest he frequently promotes that of the society more effectually than when he really intends to promote it.

As with domestic markets, so too with foreign trade:

> If a foreign country can supply us with a commodity cheaper than we ourselves can make it, better buy it of them. . . . The general industry of the country . . . is certainly not employed to the greatest advantage when it is thus directed towards an object which it can buy cheaper than it can make.

If a nation erects trade barriers on, say, wool, to prevent cheap foreign wool from cutting into the domestic trade and costing jobs, the effect is to create a monopoly that obliges people to pay an inflated price for wool, leaving them with less to spend on other goods and so tying up domestic labor and capital. Interference with free trade reduces overall domestic productivity.

Smith was neither the first to propose that production is the engine of wealth (that came from François Quesnay, a French economist he met in

Paris) nor the first to study markets. His contribution was to have combined these and other elements, along with his own experience and analysis, into a penetrating account of how the free market—what he called a "simple system of natural liberty [in which] every man, as long as he does not violate the laws of justice, is left perfectly free to pursue his own interest his own way"—facilitates both personal gain and social welfare. "Opulence and freedom," he wrote, are "the two greatest blessings men can possess."

He was at pains to show to how paltry an extent free trade was being permitted to function in the real world. Instead of encouraging individual initiative and the growth of industry, governments and guilds passed laws and rules constraining innovation and improvement. International trade was everywhere impeded by protectionist tariffs. Smith called the prevailing system *mercantilism*, meaning that it served only the shortsighted desires of merchants, about whom Smith harbored few delusions: "People of the same trade seldom meet together, even for merriment and diversion, but the conversation ends in a conspiracy against the public, or in some contrivance to raise prices." The solution resides not in government intervention—"It is impossible indeed to prevent such meetings by any law which either could be executed, or would be consistent with liberty and justice"—but in permitting free markets to function on a level playing field.

Smith anticipated the importance of information to free markets, a connection much less evident in the eighteenth century than in today's knowledge-based economies. He noted the superiority of an ordinary farmers' market, where you can see much of what's going on, to real estate transactions, which are conducted less openly and involve goods more difficult to compare with one another. Lack of information, he observed, hits unorganized working people the hardest:

> Though the interest of the laborer is strictly connected with that of the society, he is incapable either of comprehending that interest or of understanding its connection with his own. His condition leaves him no time to receive the necessary information, and his education and habits are commonly such as to render him unfit to judge even though he was fully informed. In the public deliberations, therefore, his voice is little heard and less regarded, except

upon some particular occasions, when his clamor is animated, set on and supported by his employers, not for his, but their own particular purposes.

Merchants are better informed, and hence better suited for doing business, than are gentry dabbling in trade: "Their superiority over the country gentleman is not so much in their knowledge of the public interest, as in their having a better knowledge of their own interest than he has of his." Professional traders, being constantly updated on the state of trade, are superior to regulators in setting prices:

> The inland corn dealer . . . is likely to sell all his corn for the highest price, and with the greatest profit; and his knowledge of the state of the crop, and of his daily, weekly, and monthly sales, enable him to judge, with more or less accuracy, how far they really are supplied in this manner. The interest of the corn merchant makes him study to do this as exactly as he can: and as no other person can have either the same interest, or the same knowledge, or the same abilities to do it so exactly as he, this most important operation of commerce ought to be trusted entirely to him; or, in other words, [so] the corn trade . . . ought to be left perfectly free.

Smith foresaw that free trade and a free flow of information could improve the lot of the poor and oppressed. The "dreadful misfortunes" that have befallen the inhabitants of the East and West Indies, he observed, resulted from their having encountered Europeans "at the particular time when . . . the superiority of force happened to be so great on the side of the Europeans that they were enabled to commit with impunity every sort of injustice in those remote countries."

> Hereafter, perhaps, the natives of those countries may grow stronger . . . and the inhabitants of all the different quarters of the world may arrive at that equality of courage and force which, by inspiring mutual fear, can alone overawe the injustice of independent nations into some sort of respect for the rights of one another. But nothing seems more likely to establish this equality of force than that mutual communication of knowledge and of all sorts of improvements

> which an extensive commerce from all countries to all countries naturally, or rather necessarily, carries along with it.

In other words, maximizing the free flow of information is not just admirable in itself, but can blaze a path to prosperity.

Smith had hoped that by applying what he called the "Newtonian method" to economics, it might be possible to identify natural laws of finance comparable to Newton's laws of physics. Newton himself had expressed a similar expectation. "If natural philosophy in all its parts, by pursuing this method, shall at length be perfected," he wrote of his own physics, "the bounds of moral philosophy will also be enlarged." In terms of cross-pollination it is intriguing to note that Newton wrote a paper on economics and Smith a history of astronomy—which, Smith informed his literary executor, was in his opinion the only unpublished manuscript in his archives worth preserving. Smith sought to identify Newtonian action-and-reaction forces at work in commerce, leading to states of equilibrium comparable to the steady course of planets in their orbits. The extent to which he accomplished this task is not always appreciated. Although Smith's invisible hand is sometimes compared to Newton's gravitational force—inasmuch as both agencies cannot be perceived yet are constantly and consequentially at work—Smith's resemblance to Newton goes deeper than that. Like Newton he discovered "an immense chain of the most important and sublime truths, all closely connected together." The important thing about such a chain is not its invisibility (if perceptible, it would have been discovered long before) but its universality, its ability to predict a wide variety of activities in terms of a few basic precepts. No great scientific mind was required to discern that the sun makes crops grow, or that goods may be sold for profit in a market. But to quantify how gravitation holds the moon in its orbit, or to demonstrate how free markets establish fair prices and promote the efficient production of goods, required the insights of Newton and Smith.

It is often remarked that British economics after Smith proceeded in an almost unbroken chain down to the present day. Smith's friend David Hume was close to Daniel Malthus, an eccentric botanizer and Rousseau devotee whose utopian reveries his son, Thomas Malthus, sought to deflate by arguing that social improvements were doomed to fail should a growing population breed famine. Thomas Malthus was himself a bridge: He obtained population data from Ben Franklin to support his thesis, which in turn kindled Charles Darwin's first musings about evolution through natural selection. In a curious pairing, Malthus, an underpaid academician with a cleft palate, became best friends with David Ricardo, a glittering socialite and brilliant businessman who amassed a fortune by age forty. Both had intellectual tastes that ran counter to the interests of their respective social classes: Ricardo, whose accumulations of wealth demonstrated his command of practical finance, wrote mostly theory and opposed the interests of wealthy landlords, while the academic Malthus supported the landlords and evinced an almost journalistic fascination with real-world events. Ricardo always expressed his "admiration" for Malthus's research, and Malthus said of Ricardo, who died suddenly at age fifty-one, "I never loved anybody out of my own family so much."

Ricardo had been encouraged to get into economics by his friend James Mill, an opinionated historian who applied Locke's blank-slate theory of the human mind to the education of his eldest son, John. The boy obediently studied Greek from age three and Latin from age eight. At thirteen he had mastered differential calculus, written three works of history, and learned just about everything there was to know about economics. These labors left no time for friendship, play, or leisure: "I was never a boy," recalled John Stuart Mill. He suffered a breakdown, then recovered by delving into the Romantics, falling in love, and eventually getting married, a sentimental education that took none of the edge off his clearheaded philosophizing. Mill's *System of Logic* stressed the fecundity of scientific induction and helped liberate logic from its preoccupation with syllogistic deduction, while his *On Liberty*—published in 1859, the same year as Darwin's *Origin of Species*—reinforced the foundations of political liberalism. In his *Principles of Political Economy*, Mill argued that economic productivity is based on fixed, impersonal laws of nature, so that interference with free markets, "unless required by some great good, is a certain evil." He regarded government involvement in economic matters with the same mixed feelings as economists do today. "There are some things with which government ought not to meddle, and other things with which they ought," he wrote, while conceding that the question of which is which "does not . . . admit of any universal solution." This rare ability to distinguish between questions he could answer and those better left for future generations, plus his comprehension that economics could become an objective science, made Mill's work a bulwark of the liberal-scientific tradition. One of Mill's closest readers was Alfred Marshall, who brought an unprecedented mathematical sophistication to money matters and helped create the "neoclassical synthesis" from which much of modern economics is derived. The intersecting lines of Marshall's supply-and-demand graphs are as familiar to economists as Feynman diagrams of particle interactions are to physicists.

Economics has since spread across the intellectual landscape like a river delta. Some of its theories are so exotic as to have no evident connection with the real world, while others are practical enough to prove useful when asking for a raise or mortgaging a home; overall, their benefits may ultimately rank with those of electronics and the agricultural "green revolution." Their success story has been overshadowed, however, by the spectacle of ongoing controversies, which are often cited as evidence that economists are indecisive and poor at making predictions. To glimpse the reality behind the myth, consider two of the best-known competing schools of economic theory, both of which are concerned with the role of government in financial affairs.

One school, associated with the work of Marshall's student John Maynard Keynes, values free markets but favors active government intervention to dampen the shocks of economic cycles—such as raising interest rates to slow inflation, or spending public funds to kick-start a sagging economy. Keynes saw the alleged equilibrium of the free market portrayed by Marshall and other neoclassical economists—its ability to settle on pricing points as naturally as the pulsations of a variable star—as merely a "special case." Markets, like stars, can flare up and explode, and it "is misleading and disastrous" to imagine that national leaders should ride out an economic downturn with folded arms, waiting for the market to right itself, if doing so means that millions cannot meanwhile provide for their families. Free-market forces may calm the tempest in the long run, but, as Keynes famously remarked, "In the long run we are all dead." Keynes' immediate influence was substantial—the 1944 British White Paper on Employment Policy and the U.S. Employment Act of 1946 both stated that it was government's "responsibility" to assure high and lasting employment levels—and most liberal-democratic governments today routinely take Keynesian steps to fight unemployment, inflation, and recession. Even President Richard Nixon, a Republican, declared, "We are all Keynesians now."

On the other hand (as economists allegedly like to say) there is the Chicago school. Associated with Milton Friedman and his colleagues at the University of Chicago, it stresses classical free-market values combined with the use of empirical tools to analyze how markets work. Its adherents start from the standpoint of individual liberty, regarding free markets as both an embodiment of liberty and a source of financial strength. As Friedman observed:

> I know of no example in time or place of a society that has been marked by a large measure of political freedom, and that has not also used something comparable to a free market to organize the bulk of economic activity. . . . Underlying most arguments against the free market is a lack of belief in freedom itself.

Friedman declared himself "suspicious of assigning to government any functions that can be performed through the market, both because this substitutes coercion for voluntary cooperation in the area in question and because, by giving government an increased role, it threatens freedom in other areas."

The Chicago school too has enjoyed great influence—notably on the international scale, where its championing of open markets and low taxes helped spur the free-trade revolution known rather unattractively as globalization. A striking instance of Chicago-style experimentation came in Poland in 1990, when Solidarity leaders who had recently thrown off the yoke of Soviet domination attempted to convert their nation to a market economy in a matter of months. The Polish finance minister, an economist named Leszek Balcerowicz, introduced free-market reforms and waited anxiously as food prices soared. Advised to monitor a single commodity, Balcerowicz each morning checked the price of eggs at a local market. Eggs got steeply more expensive for weeks—during which time Balcerowicz was pilloried by a populace that had known only the depressed food prices and the depressed wages of communist rule—but then farmers and merchants, attracted by the higher prices, began making their way in from the countryside to the markets. Prices leveled off by the end of the month and in some areas soon declined. Poland's agonizing period of 17,000-percent hyperinflation ended, and a lasting transition from command-and-control to a market economy was under way.

In many ways the Keynesians and the Chicago school appear to be opposites. The Keynesians are seen as coming from the left and as blind to the dangers of big government, while the Chicagoans are depicted as rightists championing personal greed over social responsibility. Confusingly, members of both camps describe themselves as liberal. Even more confusingly, both are justified in doing so.

On one level, this is simply another instance of the distinction between concern with equality of opportunity (the Chicago school) and of outcome (the Keynesians). It is for instance clearly the case that excessive regulation can retard scientific and technological creativity. Friedrich Hayek of the Chicago school observed that prior to the Industrial Revolution

> the beliefs of the great majority on what was right and proper were allowed to bar the way of the individual innovator. Only since industrial freedom opened the path to the free use of new knowledge, only since everything could be tried . . . has science made the great strides which in the last hundred and fifty years have changed the face of the world.

But corporate concentrations of power, too, can inhibit creativity and freedom, while illiberal practices such as discrimination against women and minorities in the workplace have historically been alleviated through social action and government intervention, with social safety nets such as the British Reform Acts, universal health care in Germany, and Social Security in the United States helping to protect free-market capitalism from the predations of Marxist and other socialistic illiberalisms.

In practice, the liberal democracies have elected to experiment with aspects of both schools, turning the dial between equalities of opportunity and of outcome as they see fit. England and the United States both moved far toward socialism in the 1950s—when Britain nationalized major industries and the American income tax stretched all the way up to a confiscatory 90 percent—but then backed off considerably. The liberal-democratic world today is a patchwork of experimentation. An economically liberalized England boasts Europe's largest economy, yet Ireland, where the government bet heavily on a combination of free markets and government-sponsored, information-age education, outperformed England on a per capita basis and became Europe's fastest-growing economy. A relatively socialistic France suffered from high unemployment (government regulations make it hard for employers to fire anybody, which in turn makes them reluctant to hire anybody) but the even more socialistic Danes and

Finns to the north ranked among the ten most economically competitive nations in the world.

It appears, as Adam Smith anticipated, that there is no one right solution for all peoples when it comes to the proper economic role of governments. Nor are the data terribly conclusive, even within a given nation. In the United States, the Republican party cut taxes and preached the Chicago line, yet the American economy over the past half century has performed better under Democratic than Republican administrations. In the years 1948–2007, per capita GDP grew 2.8 percent under the Democrats and 1.6 percent under the Republicans, while family income growth from 1948 to 2005 was substantially higher at all levels, from 2.6 versus 0.4 percent for the bottom 20 percent to 2.1 versus 1.9 percent for households in the top 5 percent. The situation is similar when viewed from an investment perspective: Ten thousand dollars, invested in stock market index securities solely during the forty years that Democrats occupied the White House from 1929–2008, would have yielded over three hundred thousand dollars; the same money, invested solely during Republican administrations, would have returned only fifty-one thousand dollars. This might seem to score points against the Chicago school, except that the Republican presidents Ronald Reagan and George W. Bush swelled the size of government and ran up the national debt, while Bill Clinton, a Democrat, shrank the government slightly, curbed welfare spending, and balanced the budget. All economic models are imperfect, and ideologies, because they are the least empirical, are the least perfect systems of all. George W. Bush's administration spent eight years fostering the minimally regulated economic strategies their ideology favored, yet in 2008 presided over a virtual nationalization of investment banks and insurance firms whose assets exceeded a trillion dollars. The best way to reduce the incidence of such surprises is by increasing the quality of economic theory and the quantity of sound empirical data.

In a deeper sense all economics is liberal anyway, inasmuch as economics is a science. Economists are liberal, notwithstanding their political differences, to the extent that their work promotes fact-finding over ideology. An image that comes to mind is that of Hayek and Keynes during World War II. The two disagreed with and distrusted each other—Hayek was an aristocratic Viennese war veteran, Keynes a Bloomsbury bisexual—but when Hayek fled London during the Blitz, Keynes put him up in his rooms at Cambridge and the two stood rooftop air-raid watches together, scanning the night skies for German warplanes while debating the relative merits of government intervention versus unfettered free markets. What makes such scientific collegiality possible is not just a shared interest in a particular discipline but a common commitment to enlarging the circle of scientific knowledge. So it should come as no surprise that economists of all stripes have campaigned for improving the lot of the poor. Adam Smith's case for free markets is full of admonitions about reducing poverty—the existence of which he never blamed, as his conservative contemporaries were apt to do, on the poor themselves. Malthus felt that science exposed the basic equality of all human beings. "The constancy of the laws of nature," he observed

> is the foundation of the industry and foresight of the husbandman; the indefatigable ingenuity of the artificer; the skilful researches of the physician, and anatomist; and the watchful observation, and patient investigation, of the natural philosopher. To this constancy we owe all the greatest, and noblest efforts of intellect. To this constancy we owe the immortal mind of a Newton.

Mill advocated women's rights, the abolition of slavery, and free public education for the working poor. Marshall, a bank teller's son whose mathematical brilliance was discovered while he was studying at Cambridge, used to spend his vacations visiting "the poorest quarters of several cities [and] looking at the faces of the poorest people." He majored in economics rather than physics because he thought it the best way a mathematician like himself might help reduce poverty.

The tradition of economists using their knowledge of wealth production to combat poverty continues today. Jeffrey D. Sachs, an economist who has devoted his career to alleviating global poverty, credits Keynesian economics and scientific innovation with virtually eliminating extreme poverty in the developed world and thus providing a model for how the same feat can be accomplished elsewhere. The economist Paul Collier seeks to improve the lot of "the bottom billion" by drawing on sciences ranging from sociology (the poorest nations suffer from "brain drain" as the educated emigrate) to geography (they tend to be landlocked nations with "bad neighbors" whose problems with war and infrastructure prevent their participating in meaningful levels of trade). Collier is particularly illuminating on the privations that discovery of a single profitable resource can have on a poor nation. He notes for instance that the rebel leader Laurent Kabila, "marching across Zaire with his troops to seize the state, told a journalist that in Zaire, rebellion was easy: All you needed was $10,000 and a satellite phone." The $10,000 would hire an army; the cell phone was for making deals with the oil companies; Kabila reportedly negotiated oil contracts worth $500 million on his way to seizing Kinshasa. In the midst of such privations, Collier advises, the suffering population may "see the society as intrinsically flawed" and fall victim to "the quack remedy . . . of populism." Ideologues blame globalization for the world's ills, but economists like Collier know better: "We need stronger and fairer globalization," he argues, "not less of it."

It has by now become clear that economics has helped bring about the greatest increases in wealth and reductions in poverty in all history. It has done this both by creating theories the implementation of which has improved the economic performance of nations, and by building a body of empirical information through which the theories continue to evolve.

One such finding is that freedom works. Free markets, provided that they are kept aboveboard, are more efficient, and grow economies faster, than do markets that are excessively regulated, controlled, or shielded from competition. "Market capitalism is the best economic system ever invented for the creation of wealth," writes the financier Felix Rohatyn, "but it must be fair, it must be regulated, and it must be ethical." Centralized, government-controlled economics are less efficient, because it is both theoretically and practically impossible for any central body to manage a large, complex economy with anything approaching the efficiency attained by leaving it in the hands of the many. Socialism is dead, if socialism means state command of the majority of a nation's resources, but right-wing conservatism is dead too, insofar as it favors protectionism, opposes immigration, and would have us attempt to deal with modern economic challenges by applying only the precepts of the past. What remains is liberalism. Cant and quackery still raise their heads in the making of economic policy, but the ongoing ascendance of science and liberalism make it increasingly unlikely that the world will be plunged into another Great Depression, or that billions will again fall victim to the illiberal and unscientific siren songs of communism, fascism, populism, and other creeds that have been tested and found wanting. Freedom means human rights and not just free markets, of course, but there *are* no other human rights without the right to possess property—not just land and money, but intellectual property. You cannot participate in a free market unless you own the things and ideas you wish to trade, and you cannot speak and write freely if the authorities own your phone, your computer, and your Web site. On the other hand, your ownership of your ideas—or, say, Walt Disney's ownership of Mickey Mouse's silhouette—cannot be extended forever without hampering the free flow of information, ideas, and images throughout society. Thomas Jefferson understood this inherent conflict of interests quite well, which is why he ranked his service as the first head of the U.S. Patent Office as comparable with his having written the Declaration of Independence and been twice elected president.

Economics need not be divorced from morality. Preachers may call money the source of all evil, but the data of economic inquiry suggest otherwise. The virtuous circle of science increasing productivity—which generates wealth, which in turn provides the capital required to underwrite further scientific and technological progress—may well extend to actual virtue. Slavery was abolished, human rights expanded, and education and health care provided to millions by liberal-democratic, scientifically proficient nations, while the opposite has been the case in most of those nations which remain illiberal, undemocratic, and unscientific. As the economist Benjamin M. Friedman puts it, "Economic growth makes a society more open, tolerant, and democratic, [and] such societies are in turn better able to encourage enterprise and creativity and hence to achieve ever greater economic prosperity."

Links between wealth, liberty, and virtue had been glimpsed before. Montesquieu, writing before Adam Smith, argued that commerce nourishes such virtues as "economy, moderation, work, wisdom, tranquility, order, and rule." His contemporary Auguste Comte maintained that "all human progress, political, moral, or intellectual, is inseparable from material progression." Alfred Marshall, having long studied the poor, saw poverty being reduced by economic growth during his lifetime, just as he had wished would be the case. "The hope that poverty and ignorance may gradually be extinguished," he wrote

> derives much support from the steady progress of the working classes during the nineteenth century. The steam-engine has relieved them of much exhausting and degrading toil; wages have risen; education has improved and become more general . . . while the growing

> demand for intelligent work has caused the artisan classes to increase so rapidly that they now outnumber those whose labor is entirely unskilled. . . . A great part of the artisans have ceased to belong to the "lower classes" in the sense in which the term was originally used; and some of them already lead a more refined and noble life than did the majority of the upper classes even a century ago.

William Cowper, lines from whose poem "Charity" appear at the top of this chapter, championed both free trade and human rights. His abolitionist poem of 1788, "The Negro's Complaint," ends with a slave's stinging riposte to the claim that human bondage could be justified on grounds of the superiority of European to African societies:

> Deem our nation brutes no longer,
> Till some reason ye shall find
> Worthier of regard and stronger
> Than the color of our kind.
> Slaves of gold, whose sordid dealings
> Tarnish all your boasted powers,
> Prove that you have human feelings,
> Ere you proudly question ours!

Good intentions count for little unless the scientific tools are sound, and economics models remain crude in many respects. To cite one notorious example, economics traditionally has treated consumers and investors as perfectly rational entities: *Homo economicus*, as this creature was called, always acted to maximize his economic position. However, *Homo economicus* is but a mathematical fiction. He is assumed to know all the relevant information, but even skilled investors can miss important clues. He is totally selfish, but nearly half of all Americans do unpaid volunteer work and nearly three-quarters donate to charity. *Homo economicus* changes jobs to maximize his income, but we've all encountered people who pursue careers for reasons other than financial gain. Nor do people always act rationally in the jobs they do have. Cab drivers, rather than working long hours on rainy days when they have more fares, tend instead to quit early and hence limit their earnings. New "behavioral" models are now beginning to emerge, based on laboratory research that more accurately predicts how real people think and act. Social scientists find, for instance, that most people overestimate their prospects of success in a given enterprise (the majority of students taking a college course predict that their final grade will be above average) and to inflate price estimates based on first impressions (retailers have long appreciated that people are more apt to pay $19.99 for a dicer if they've been told that it lists at $29.95). As real human behavior intrudes increasingly into the equations, economics may become not only less dismal but better at making forecasts.

CHAPTER TEN

TOTALITARIAN ANTISCIENCE

Science and peace will triumph over ignorance and war.

—LOUIS PASTEUR, 1892

Science will flourish only in a society that cherishes . . . the reason, openness, tolerance, and respect for the autonomy of the individual that distinguish the social process of science.

—GERALD PIEL, 1986

Although the twentieth century produced unprecedented improvements in health, wealth, and welfare, it also saw the rise of totalitarian regimes that killed a hundred million people, threatened the survival of the liberal democracies—and ravaged ideas as well as lives, fomenting persistent misapprehensions about liberalism and science that persist to this day. Liberalism began to look quaintly old-fashioned, an eighteenth-century indulgence in a world whose future was widely presumed to belong to fascism, communism, or some other form of socialism. Science, previously esteemed, was blamed for the machine guns that mowed down young men on the battlefields, the napalm that incinerated cities by night—and, of course, for the nuclear bombings of Hiroshima and Nagasaki. Even many scientists adopted such views. "The physicists have known sin," declared J. Robert Oppenheimer, chief scientist of the Manhattan Project, adding that in making "an evil thing," they had "raised again the question of whether science is good for man." His colleague Philip Morrison expressed concern about "a latent but growing feeling that science is somehow turning evil or blind. The people have the right to ask why must we do research if the outcome is the ruin of Hiroshima and its hundred thousand blackened corpses." The mathematician and philosopher Michael Polanyi spoke of "the destructive potentialities of the scientific outlook" being realized in wars "which shattered our belief in liberal progress." It began to be said, and not just on the fringe, that scientists ought to be held accountable for the consequences of their research, and that governments should intervene to channel research toward socially responsible goals. Such an approach was thought to be working in the totalitarian states, which were perceived as more efficient than the democracies at promoting scientific and technical advancement—as witnessed by the Nazi deployment of wartime wonders like the V-1 and V-2 rockets, and the Soviet Union's acquisition of thermonuclear weapons and its launch of the Sputnik earth satellite.

If, indeed, totalitarianism nourished science and technology more efficiently than liberalism did, then the future of liberalism looked dark. But was this the case? How *did* science fare in Nazi Germany, the Soviet Union, and communist China?

Germany prior to the rise of the Nazis went through fitful excursions into liberalism while enjoying considerable scientific success. Following the European revolutions of 1848—when an economic depression touched off rebellions in the German Confederation, France, and Italy—the vested interests sought to preserve their traditions of nationalism, militarism, and monarchism by enacting just enough reform to stave off further unrest. The man who best managed this balancing act was Otto von Bismarck, appointed prime minister by King William in 1862. Bismarck was personally conservative but regarded all passionate political convictions as impediments to effective statecraft, preferring to absorb—and almost to embody—many points of view: "It was not that Bismarck lied [but] that he was always sincere" wrote Henry Kissinger, one of his many admirers. Crown Princess Victoria of Germany judged Bismarck to be "mediaeval altogether and the true theories of liberty and of government are Hebrew to him, though he adopts and admits a democratic idea or measure now and then when he thinks it will serve his purpose." Neurotic and insecure—he was an insomniac, a hysteric, and a morphine addict who, said a contemporary, "eats too much, drinks too much and works too much"—Bismarck projected an image of unbending self-confidence. "I want to make music in my own way," he said, "or not at all."

Bismarck's way of maintaining the monarchy was to build the power of the state while playing off the liberals and progressives against one another and otherwise maneuvering as necessary to forestall any real threat of Germany's becoming a genuine democracy. He drove a wedge through the liberal party, then the nation's largest, by forming an alliance with its left wing (which supported labor unions and big government) in order to weaken its moderate center (home to the liberals, who favored small government and free enterprise). The progressives got state-sponsored health insurance, workplace safety measures, and an eight-hour workday. Liberals got women's rights, a freer press, freer trade, and freer elections; Germany for a time had the only effective secret ballot in Europe. Conservatives got to retain the monarchy, the aristocracy, and the real power. "In exchange for lavish trinkets from an all-powerful state," writes the conservative commentator Jonah Goldberg, "Bismarck bought off the forces of democratic revolution. Reform without democracy empowered the bureaucratic state while keeping the public satisfied." This cynical recipe would prove chillingly effective in Nazi Germany and Soviet Russia.

It was, however, inherently unstable. Europe in general was unstable—as was demonstrated when the assassination of Archduke Franz Ferdinand of Austria-Hungary by a Serb nationalist on June 28, 1914, resulted, to everyone's surprise, in the Great War. Instabilities continued to bedevil the Germany of the Weimar Republic (1919–1933), when innovations like expressionist art contended with the conservatism of monarchists, militarists, university professors, and government bureaucrats. Much of our ongoing fascination with Weimar libertinism—nudity and drugs in the nightclubs, Marlene Dietrich in *The Blue Angel*—derives from the fact that these romps were played out by young adults who, like drunken teenagers skinny-dipping in the swimming pool just before the parents come home, had too little political power to accept the responsibilities that went along with their newly acquired freedoms.

During its time of limited liberal and progressive reform, Germany emerged as a center of scientific research and development. Germans could boast of scientific accomplishments like Wilhelm Konrad Röntgen's discovery of X-rays in 1895, Max Planck's founding of quantum physics in 1900, and Einstein's 1905 and 1915 theories of relativity. Their technological accomplishments included the invention of aspirin and heroin (advertised in tandem in the late 1800s, as relieving headaches and coughs respectively); Gottlieb Daimler's early automobiles, from

1887; Otto Lilienthal's gliders (he died testing one, in 1896, saying on his deathbed, *Opfer müssen gebracht werden!*—"Sacrifices must be made!"); Count Ferdinand von Zeppelin's dirigibles, from 1900; the first electric typewriter; the electric locomotive; the Geiger counter; and the first machine gun synchronized to fire between the blades of a spinning aircraft propeller. It was said that the dream of every German mother was to have a son who was an engineer. When Hitler came to power in the midst of the Great Depression—having won 34.3 percent of the vote in 1930, and his party 44 percent in 1933, in elections sadly consonant with Benjamin Franklin's grim warning that "those who would give up essential liberty to purchase a little temporary safety deserve neither liberty nor safety"—he took over a technological ship of state that made headway on sheer inertia even as he went to work dismantling its engines.

Hitler wanted to harness the *power* of science, of course. Like any dictator he wanted all the power he could get his hands on, demanding the "total mobilization" of science toward one or another hellish goal. But he understood little of what science is or how it works, relying for news of promising scientific developments on conversations with his barber. Comprehending nothing *but* power, Hitler assumed that science would quicken to the sharpened spurs of ruthless rule. "The triumphant progress of technical science in Germany and the marvelous development of German industries and commerce," he asserted, "led us to forget that a powerful State had been the necessary prerequisite of that success." He imagined that science was useful principally "as an instrument for the advancement of national pride." Schools should teach "will-power" instead. "Instruction in the sciences," Hitler decreed, "must be considered last in importance."

Personally, Hitler was entangled in an obscuring web of pseudoscientific enthusiasms ranging from fad diets—he drank a toxic gun-cleaning fluid as a *digestif*—to avoidance of harmful "earth-rays," the emanations of which the physician Gustav Freiherr von Pohl mapped out for him with a dowsing rod. Cosmologically, Hitler favored the "glacial cosmogony" theory concocted by an amateur astronomer, Philipp Fauth, and an Austrian engineer, Hanns Hörbiger, according to which the stars were balls of ice. He was blind to the prospects of technological progress other than those useful for killing people (Hannah Arendt: "The totalitarian belief that everything is possible seems to have proved only that everything can be destroyed"). Even there he displayed little foresight, dismissing the combat potential of rockets and jet aircraft while disdaining the possibility of nuclear weapons as a fantasy promoted by "Jewish physics."

To the notion that science could be flattened under the boots of power and still deliver the goods—a fallacy described by Jacob Bronowski as "trying to buy the corpse of science"—the Nazis added leaden layers of superstition and pseudoscience. Looming large among these was a racial doctrine that proclaimed the superiority of an Aryan Nordic race, destined to rule a world cleansed of genetic impurities. Although decked out in costumes of mythological antiquity, these ideas were actually crackpot novelties: "Aryan," a Sanskrit word for Persian nobility, had only recently been imported into European discourse, and "Nordic" was coined in 1898. Neither term described a race, much less a "pure" race, whatever such a thing might be. Nor would genetic purity confer an advantage on any population: The strength of a species resides in its genetic *diversity*, which improves its likelihood of surviving environmental changes. As Joseph Needham said of Nazi racial doctrines, "A more shameless flying in the face of established scientific fact has never been known in human history."

It would be rather astounding if a party founded on pseudoscience and maximally illiberal power politics had presided over any significant number of scientific or technological breakthroughs—and indeed, there is scant evidence to support the popular image of the Third Reich as a futuristic war machine. The German military was an imposing force to be sure, arising as it did from generations of military professionalism, lavish infusions of fresh spending by the Nazi regime, and a passion for vengeance following Germany's humiliating defeat in the Great War, but it was no monument to science. Starved for steel, the Nazis sent storm troopers to scavenge iron fences from public parks and cemeteries. Industrial chemists were ordered to develop synthetic rubber and gasoline, tasks at which they mostly failed. Nearly half of German's wartime artillery came not from German factories but from conquered neighbors, mainly France. (Such shortages were a major reason behind Hitler's disastrous decision to invade Russia.) The backbone of Germany's military transport system consisted of railways plus seven hundred thousand horses. "In weapons and technology," writes the historian Alan J. Levine, "the German forces were greatly superior only to their weakest and most backward foes, i.e., Poland, Norway and the Balkan countries. Generally speaking, Germany's victories were due to good leadership, training, and the revolutionary *use* made of tanks and tactical airpower." In cryptology, which provides a reasonable arena for comparing the scientific and technological capacities of adversaries in wartime, the Nazis' greatest achievement was the Enigma encoding device. It was so thoroughly cracked by British scientists that German submarine attacks in the Atlantic came almost to a halt, obliging British leaders to exercise restraint in acting on what they learned from deciphered Enigma dispatches for fear the German high command might otherwise realize what was up.

When people think of "Nazi science" today they usually have in mind either eugenics or the pointless and sadistic "experiments" carried out by Nazi doctors in the death camps. But such obscenities can be described as science only in the rather distant sense that, say, Theodore Kaczynski's "Unabomber Manifesto" can be called philosophy. The death-camp doctors discovered little beyond the fact that it is possible to kill a great many defenseless prisoners through the use of poison gases like the pesticide Zyklon B. Nazi eugenics research consisted of studies by physicians such as Eugen Fischer, who measured the "racial purity" of various individuals by looking for "Negro blood"; Julius Hallervorden, who studied the "feeble" brains of euthanasia victims; Robert Ritter, whose data were employed by the SS to dispatch Gypsies to Auschwitz; Ernst Rüdin, who helped draft a Nazi sterilization law aimed at preventing "genetically diseased" offspring; Otmar Von Verschuer, who campaigned for forced sterilization of the "mentally and morally subnormal"; Ernst Wentzler, who coordinated a pediatric euthanasia program that killed thousands of children; Carl Clauberg, who sterilized women at Auschwitz; and Josef Mengele, who murdered Jews and Gypsies to study their organs. It scarcely need be added that their results were of no scientific value.

Yet many thinkers continue to overestimate the quality of Third Reich science. Some are dazzled by German technological achievements, such as the development of jet engines, proximity fuses, and infrared night-vision goggles—but technological applications can lag decades behind the scientific discoveries that made them possible. One way to separate scientific research from technological applications is by tallying the citations in leading scientific journals. Such studies indicate that German research tumbled into a dying fall once Hitler came into power. Initially the impetus of prior research carried it forward, with Rudolf Schoenheimer employing natural isotopes as radioactive tracers in the human body in 1935 and Otto Hahn splitting uranium atoms in 1938, but Germany's scientific citations thereafter dwindle until the pages are almost blank. Five years into Hitler's reign, the number of German scientific papers appearing in one leading international physics review had fallen from 30 to 16 percent of the total. Membership in the nation's oldest national scientific organization, the Society of German Natural Researchers and Physicians, shrank from 6,884 in 1929 to 3,759 in 1937. There was no dramatic moment when storm troopers burst into the German research funding agency to cry "Halt!" to German science. "Instead," notes the historian Ulrich Herbert, "contrary positions and voices were simply eliminated."

The Nazis presided over a nuclear-physics brain drain of startling proportions. The computer pioneer John von Neumann departed for the United States in 1930—the same year that saw the emigration of Hans Bethe, who would help discover the nuclear processes that make the sun shine, and of Leo Szilard, who while soaking in his bath in a London hotel suddenly realized how a nuclear weapon could be made. (To keep the idea secret, Szilard patented it and assigned the patent to the British Admiralty; he then drafted a letter, which he had his friend Einstein sign, warning President Roosevelt that "it appears almost certain" that an atomic bomb could be built "in the immediate future.") The quantum-physics virtuosos Max Born and Erwin Schrödinger fled, as did James Franck, who would help build the atomic bomb and then petition to have it demonstrated to the Japanese rather than being dropped on a city.

Hitler was untroubled by the scientific exodus. When a physicist tried to alert him to the corrosive effects that Nazi anti-Semitism was having on scientific research, Hitler reportedly replied, "If the dismissal of Jewish scientists means the annihilation of contemporary German science, then we shall do without science for a few years!" To direct the Reich Ministry for Science, Education and Popular Culture—which was chartered to "unify and control all of German science by the Reich both within and outside the universities [and manage] the control and methodical shaping of all of scientific life especially at the university"—Hitler named Bernhard Rust, a former provincial schoolmaster who had been dismissed for molesting a schoolgirl but had escaped prosecution on grounds of his documented mental illness. For Rust, the whole purpose of education was to create Nazis. When Rust asked David Hilbert whether the once great mathematics center at Göttingen had suffered from the expulsion of its Jewish faculty members, Hilbert replied, "Suffered? It hasn't suffered, Minister. It doesn't exist anymore!"

A few first-rate scientists did remain in Germany throughout the war. One of them was Max Planck, the founder of quantum physics—a patriot whose eldest son died fighting in World War I and whose second son was executed by the Gestapo for attempting to assassinate Hitler with a bomb on July 20, 1944. Planck noted that intellectual midgets were being promoted to the academic posts vacated by professors who had fled the Nazi regime or perished at its hands: "If today thirty professors get up and protest against the government, by tomorrow there will be also one hundred and fifty individuals declaring their solidarity with Hitler, simply because they're after the jobs." He took comfort in the objectivity of science. "The outside world is something independent from man, something absolute," he wrote, "and the quest for the laws which apply to this absolute appeared to me as the most sublime scientific pursuit in life." Another who stayed behind was Werner Heisenberg, discoverer of the uncertainty principle in quantum physics. Something of a Romantic, Heisenberg was given to long alpine walks and to the formulation of oracular queries like, "Why is the one reflected in the many, what is the reflector and what the reflected, why did not the one remain alone?" He thought of the war as a passing storm, upon the subsiding of which intellectuals like himself would restore German culture to its proper prominence. "I must be satisfied to oversee in the small field of science the values that must become important for the future," Heisenberg wrote in 1935. "That is in this general chaos the only clear thing that is left for me to do. The world out there is really ugly, but the work is beautiful."

Jewish scientists who remained in Germany were soon dismissed from their posts and in many cases liquidated. Among those who perished at the hands of the Nazis were the mathematicians Ludwig Berwald, who died in the Lodz Ghetto; Otto Blumenthal, killed in the "model" camp at Theresienstadt; Robert Remak, who died at Auschwitz; Stanislaw Saks and Juliusz Pawel Chauder, murdered by the Gestapo; and Paul Epstein and Felix Hausdorff, who committed suicide. The chemist Wilhelm Traube was beaten to death in his apartment by Gestapo agents. The physicist Lise Meitner, a Viennese Jew who along with her two sisters had converted to Christianity, worked on the prospect of nuclear fission in Berlin with Otto Hahn and the chemist Fritz Strassmann until the summer of 1938, when she slipped away to Sweden on an expired passport. Soon thereafter Hahn and Strassmann arrived at the fission results for which Hahn would win a Nobel Prize—a finding that prompted Niels Bohr to exclaim, "Oh, what idiots we have been! Oh but this is wonderful! We could have foreseen it all! This is just as it must be!" Although by escaping from Germany Meitner may have forfeited her share of the credit for the discovery, she blamed herself for not having departed sooner. "Today it is very clear to me that it was a grave moral fault not to leave Germany in 1933, since in effect by staying there I supported Hitlerism," she wrote to Hahn, adding an unflinching indictment:

> You all worked for Nazi Germany. And you tried to offer only a passive resistance. Certainly, to buy off your conscience you helped here and there a persecuted person, but millions of innocent human beings were allowed to be murdered without any kind of protest being uttered.

Meitner refused a 1943 offer to work in the Manhattan Project, declaring, "I will have nothing to do with a bomb."

Ultimately, though, the decay of German science under the Nazis resulted not just from the brain drain and the harebrained ministrations of Hitler and his henchmen, but from fundamental differences in the way science operates under totalitarianism as opposed to liberalism. Science demands free, open discussion and publication, not only in order to circulate fresh information and ideas but to expose them to lively criticism. A totalitarian regime can afford little of either. Having seized a measure of power to which it has no legitimate claim, in order to solve real or imaginary problems that it cannot in fact solve, such a regime is highly vulnerable to criticism and so must stifle it. One way it accomplishes this, aside from jailing and murdering dissenters, is to create a cult of secrecy and power in which access to secrets is perceived as a source of power. Such vices are infectious, and the history of Nazi Germany is rife with examples of corporations and government agencies needlessly duplicating their R&D efforts by playing their cards too close to their chests.

The development of radar in Britain, essential to the Royal Air Force's defeat of a Luftwaffe that had it outgunned four to one, demonstrated some of the differences between the way the Germans and the English governments interacted with their scientists in wartime. The theory of radar was simple—a radio pulse that strikes an airplane will bounce back and so reveal the plane's location, even at night or under cloudy skies—and had been understood since the 1880s. Naturalists found acoustic analogues in dolphin and whales, which locate fish in the depths of the sea by pinging them with pulsed squeals of sound, and in bats that navigate inside ink-black caves by emitting high-pitched squeaks and mapping their echoes. The difficulties arose with implementation. Since radio waves travel at the velocity of light, a competent radar kit has to receive echoes arriving a fraction of a second after the pulse was emitted. And, since shorter wavelengths produce sharper resolution (which is why dolphin and bats use high-pitched sounds for echolocation), radar required microwave radio equipment that did not yet exist when the war began.

Overcoming these obstacles called for an openness to new ideas that was scarce in wartime Germany. The Nazis' institutionalized paranoia produced a stovepipe array of mutually suspicious public agencies and private corporations, with university scientists—such as they were—largely excluded from radar work altogether. The effort was further retarded by the low quality of Nazi appointees—men like Ernst Udet, who knew next to nothing about science but was appointed technology chief of the Luftwaffe on the basis of his fame as a World War I flying ace, and who objected that if radar systems were deployed, "Flying won't be fun anymore."

In Britain, a young and little-known Scottish electrical engineer named Robert Watson-Watt—descended from James Watt—found a ready audience in the war ministry for the idea that radar could win the air war. In the summer of 1940, when German bombs were raining down on London on a nightly basis, Prime Minister Churchill took the advice of the scientist-inventor Henry Thomas Tizard and, overruling the security concerns voiced by his cabinet, authorized the dispatch of a steel box containing radar blueprints and a new microwave transmitter prototype across the Atlantic, to see if the American allies could help speed things along. The physicist John Cockroft purchased the box in a surplus store and took it on an unescorted Canadian ship, the *Duchess of Richmond*. (Always attentive to detail, he drilled holes in the box so that it would sink should the *Duchess* be torpedoed.) The navy men aboard asked for a lecture by the famous physicist. Wanting to stick to a subject that he felt certain would have no wartime use, Cockroft spoke on atomic energy, telling the sailors that theoretically, the nuclear energy in a cup of water could blow their ship out of the water. In America, Cockroft found that radar work was being pursued by a variety of ad hoc teams involving university scientists, government engineers, and even talented amateur scientists like the financier Alfred Loomis, who tested the world's first Doppler-radar "speed gun" at his estate in Tuxedo Park, New York (one of his colleagues remarking, "Hey, don't let the cops get a hold of that"). Loomis immediately grasped the importance of the British inventions to the imminent deployment of radar in the war, and put the new transmitter into production the next day.

Churchill took a personal interest in science, numbered top scientists among his friends, and understood the importance of free communications in developing new devices and tactics. There was "no time to proceed by ordinary channels in devising expedients," he said, boasting that his military service ministers "stood on no ceremony . . . had the fullest information . . . and constant access to me," and that "anyone in this circle could always speak his mind." In his view the conflict was not

> a war of masses of men hurling masses of shells at each other. It is by devising new weapons, and above all by scientific leadership, that we shall best cope with the enemy's superior strength. . . . The multiplication of the high-class scientific personnel, as well as the training of those who will handle the new weapons and research work connected with them, should be the very spear-point of our thought and effort.

"Unless British science had proved superior to German," Churchill later wrote, "we might well have been defeated, and, being defeated, destroyed."

His seriousness on this point was evident to the young physicist Reginald Victor Jones, who had been researching the infrared spectrum of the sun when the war intervened. Appointed Britain's first scientific intelligence officer at age twenty-eight, Jones investigated the possibility that the Germans were using intersecting radio beams to signal their pilots where to release their bombs. Summoned to a meeting at Ten Downing Street in June 1940, the young Jones suspected a practical joke, but instead found himself seated at a table where Britain's top air force officers, with Churchill presiding, were discussing the German radio-beam puzzle. After listening for a while to their groping conversation, which to Jones "suggested that they had not fully grasped the situation," he was asked by Churchill to clear up a technical point. Instead he said, "Would it help, sir, if I told you the story right from the start?" Churchill was startled but replied, after a moment of hesitation, "Well, yes it would!" Jones spoke for twenty minutes, explaining his research and urging that British pilots fly along the German beams for themselves to learn how they worked. Such flights began the following day. British engineers were soon jamming the Germans' signals, a key step in ending the night-bombing raids and ultimately the Blitz—which had destroyed over a million homes, more than were consumed in the Great Fire of London in 1666. Jones recalled of Churchill that "he valued science and technology at something approaching their true worth."

This is not to say that the Allies were immune to the difficulties posed by bureaucratic opacity and conservatism—indeed British and American scientists often complained of just that—or that German scientists were never able to cut through red tape and get anything done. But on the whole, wartime scientific and technological development fared far better in the liberal democracies.

The failure of the Germans to develop nuclear weapons—a failure that surprised the Allies, who had invested in the Manhattan Project out of the quite sensible fear that the Nazis would otherwise get there first—also reflected the hobbled state of communications among the scientists and engineers involved. Historians differ sharply over whether Heisenberg, whom Hitler appointed to head up the German A-bomb project, deliberately let the project languish, but whatever his motives, his approach was a far cry from the egalitarian ethos of the Manhattan Project. Disinclined toward laboratory work, Heisenberg played Bach fugues on the chapel organ at Hechingen while his subordinates there conducted nuclear experiments with uranium, graphite, and heavy water. ("Had I never lived," he mused dreamily, "someone else would probably have formulated the principle of indeterminacy; if Beethoven had never lived, no one would have written Opus 111.") A similar insularity appears to have afflicted Walther Bothe, the eminent experimentalist whose mistaken calculation of the absorption characteristics of graphite led the German bomb project astray. Bothe, whose heart was not in the work anyway—the Nazis having hounded him over his antifascist political views so severely that he sought treatment in a sanatorium—concluded that graphite would not work as a neutron absorber in sustaining a nuclear chain reaction. Actually he was testing the wrong grade of graphite, but his results were taken at face value, leading German bomb scientists to conclude that only heavy water could do the job. This set them on a path that dead-ended once the Allies disabled the German heavy water plant at Vemork, Norway, in a series of raids conducted by Norwegian, French, and British commandos flown in on American bombers. A mistake like Bothe's would have been unlikely to persist in the atmosphere of Los Alamos, where Gen. Leslie Groves had agreed to let the scientists work with their customary informality and ordinary engineers would have felt free to question an Oppenheimer about where he'd got his graphite.*

By the end of the war it was starkly evident that the Nazi campaign of top-down totalitarian science had failed. "In the early days of the war the world was amazed by the efficiency of the Nazi war machine," wrote Needham, but "in every theater of war . . . the technology of the democracies has proved superior to that of the fascist powers." Needham concluded that "the Axis powers have carried out a great social experiment. They have tested whether science can successfully be put at the service of authoritarian tyranny. The test has shown that it can not."

Yet the notion persisted that totalitarianism was more efficient than liberal democracy. It was said that Mussolini had at least "made the trains run on time" (which as a matter of fact he did not) and that it *must* be more efficient to control science, technology, and industry via centralized planning. Adding to the confusion were the distortions created by a general reliance on a one-dimensional, left–right political spectrum. If you assume that the Nazis and the communists define the ends of the spectrum, and you've just fought a hideous war against the Nazis, it is easy to imagine that democracies will in the future evolve toward something like communism—whereas if you instead think in terms of a triangular relationship, you are more likely to see that communism is just as far removed from liberalism as the Nazis were. But this was not the mind-set of many postwar Europeans and Americans, for reasons dating back many decades.

Prior to the disaster of the two world wars, Europe was becoming politically liberal—by 1914, Russia and Turkey were the only thoroughgoing autocracies left in the region—and economically liberal enough that most

---

*When Cambridge physicists made the same error, it was quickly discovered. Their results failed to comport with those obtained by the Italian physicist Enrico Fermi, who on December 2, 1942, engineered the world's first controlled fission reaction in a makeshift laboratory set up in a doubles squash court under the University of Chicago's Stagg Field. Fermi, furiously devoted to free speech, objected indignantly when his graphite work was classified in order to prevent the news from reaching Germany.

Europeans were growing rapidly wealthier. But the wealth was not being equally shared, and with swelling masses of the working poor huddled in urban slums, the middle classes were tormented by two economic concerns—that their newfound wealth might be taken from them, and that they had no moral right to it in the first place. Some feared that the poor might vote to abrogate property rights altogether, others that unskilled workers sick of toiling in poverty would lobby to decouple merit from financial reward so that every laborer, regardless of his skill or diligence, got more or less the same wages. That all these fears should be realized—human rights overrun, property confiscated, and workers rewarded according to their needs rather than for their productivity—was precisely the communist prescription. Karl Marx and Frederick Engels knew what they were talking about when they declared, in the opening passage of the *Communist Manifesto* of 1847, "A specter is haunting Europe—the specter of communism."

Not even Marx and Engels could foresee the extent to which communism would indeed come to haunt Europe and the world. Chiefly an enthusiasm of intellectuals and political opportunists, communism failed to capture the affection of the working masses whose interests it claimed to promote, yet came to rule a third of the world's peoples. Marx and Engels talked about liberation, yet communism presided over the most craven sacrifices of liberty for the promise of material gain to have darkened human history. They described communism as a science, yet few more noxious assaults on science had ever been concocted.

Marx and Engels are easy enough to disdain. Innumerable crimes having since been committed in their names, it is possible today to take a certain grim satisfaction in reading that Marx neglected his family and that Engels campaigned against capitalism while living off the wealth that his father accrued as owner of a Manchester thread factory. But this is rather unfair. Marx was one among many radical philosophers and journalists in an age of revolutionary tumult, and his conduct was more that of an otherworldly scholar than of a scheming conspirator. And what may be said of Engels? A man born into wealth loses either way: He stands accused of complacency if he defends the existing regime, and of posturing if he objects to it. In financially supporting his friend Marx, and editing the two posthumous volumes of *Capital* following Marx's death, Engels believed he was aiding a genius whose insights would liberate the poor. Neither man thought that communism, put into practice, would become an unbridled horror. But as it did, and was taken to be scientific, questions arise as to communism's actual relationship to science and liberty.

Communist theory was inherently reactionary—a reaction against the brutalities of the early industrial era—and pseudoscientific, in that it pretended to be scientific while ignoring the need to test theories experimentally. Marx was stricken, as any right-minded individual would be, by the appalling lot of the working poor and by child labor in particular. *Capital* is rife with descriptions of children being

> dragged from their squalid beds at two, three, or four o'clock in the morning, and compelled to work for a bare subsistence until ten, eleven, or twelve at night, their limbs wearing away, their frames dwindling, their faces whitening, and their humanity absolutely sinking into a stone-like torpor, utterly horrible to contemplate.

Engels, similarly affronted, recalled walking through the slums of Manchester with "a bourgeois" to whom he complained of

> the frightful condition of the working people's quarters, and asserted that I had never seen so ill-built a city. The man listened quietly to the end, and said at the corner where we parted: "And yet there is a great deal of money made here: Good morning, sir."

Communists regarded Engels' middle-class companion as the villain of the piece, yet capitalism and liberal democracy went on to alleviate the privations of far more working people than communism ever did. Liberalism proved to be empirical enough to address the plight of the working poor, while communism—an absolutist philosophy, equal parts religion and pseudoscience—was too inflexible to respond to much of anything.

Engels thought of Marx, who admired Darwin, as a scientific thinker of Darwinian stature. "Just as Darwin discovered the law of development of organic nature, so Marx discovered the law of development of human history," he said at Marx's funeral, adding that in his opinion, Marx had been "the greatest living thinker." And indeed, Marx in his *Capital*, a book sprinkled with references to mathematics, geometry, and chemistry, claimed to have identified "natural laws" of history that move "with iron necessity toward inevitable results." According to the allegedly inexorable workings of these laws, capital in proportion to production was destined to increase, profits and wages to decrease, and capitalist states to polarize into two cliques—wealthy capitalists on top, a seething mass of impoverished proletarians below—whereupon capitalism would succumb to socialistic revolution. All this sounded Darwinian to those only vaguely familiar with Darwin, as were Marx and Engels. In fact, Darwinian evolution has no inevitable or even foreseeable direction. Evolutionary biologists can make only weak, tentative predictions about such rudimentary matters as whether a given bacterium is likely to survive in a lab rat's intestinal tract, much less about the destiny of economic and political institutions. Had Marx comprehended this he might have made more modest claims, and accompanied them with suggestions as to how communists could monitor the results of their social experiments and adjust their theories accordingly. Instead, he was long on what must be done to establish a communist society but short on how, once property had been abolished and the state had withered away, that society would actually operate: Marx's communist world was as ethereal as a Christian's heaven. Yet this very insubstantiality appealed to middlebrow thinkers seeking personal salvation through political revolution. As Michael Polanyi observed, Marxism

> predicted that historic necessity would destroy an antiquated form of society and replace it by a new one, in which the existing miseries and injustices would be eliminated. Though this prospect was put forward as a purely scientific observation, it endowed those who accepted it with a feeling of overwhelming moral superiority.

Engels imagined that science was a branch of philosophy—specifically, that the great accomplishment of modern science had been to validate the dialectical materialism of Georg Wilhelm Friedrich Hegel, whose notion that all processes can be understood in terms of a dialectic (thesis, antithesis, and synthesis) has enthralled many a muddled mind. Marx was of a similar opinion; he borrowed much of his scientific terminology from Hegel's *Science of Logic*, which he had read immediately before getting started on *Capital*. Hegel had once enjoyed an intellectual suzerainty conferred on few living philosophers. But his star had since fallen, his disciples morphing into left-wing and right-wing radicals and Arthur Schopenhauer dismissing him as "the clumsy and stupid Hegel," whose "mad sophistry" "will remain as a monument of German stupidity." None of this seems to have bothered Engels, who bought into Hegel with the blinkered enthusiasm of a vacationer boarding a burning blimp. "Nature is the proof of the [Hegelian] dialectic," he wrote, "and we must give to modern science the credit of having furnished an extraordinary wealth and daily increasing store of material towards this proof." Working from this groundless premise, Engels was quick to conclude that "a *correct* notion of the universe . . . can *only* be had by means of the dialectic method." To make such a progression—from shallow-draft philosophizing and patronizing praise to a finger-wagging dismissal of all the alternatives—might have been harmless enough in the hands of amateur scholars contributing to fact-free journals of "analysis," but it became quite a darker matter once the communists came to power and started prating Marxist-Leninist dogma across steel desks to those whose dissenting opinions would earn them a one-way ticket to the insane asylums, prison camps, and firing squads. If the world is relatively anti-intellectual today, it is because the world got a bellyful of the communists' pseudoprophetic intellectualism and turned its broad back on the lot of it. The philosopher of science Karl Popper expressed just such a change of heart in the introduction to his *The Open Society and Its Enemies*. "This book," he wrote,

> does not try to add to all these volumes filled with wisdom, to the metaphysics of history and destiny, such as are fashionable nowadays. It rather tries to show that this prophetic wisdom is harmful, that the metaphysics of history impede the application of the piecemeal methods of science to the problems of social reform. And it further tries to show that we may become the makers of our fate when we have ceased to pose as its prophets.

Although the communists spoke enthusiastically of conducting social experiments, the inherent certitude of their philosophy barred them from providing the mechanisms by which a society could respond and change course when the experiments failed. Instead it soon became a crime even to call attention to communism's failures, so that the social experiments collapsed amid a graveyard silence. To cite one example among thousands, when a 1937 national census revealed the damage done by Joseph Stalin's having uprooted more than a million "kulaks" (meaning well-off peasants, a particular object of his and Lenin's hostility), Stalin suppressed the census results, arrested the census takers, and had many of them shot. Such offenses were not unique to communism; what set them apart was the notion that something scientific was going on. It was not. Communism was characterized by the willful suspension of disbelief that is the polar opposite of scientific skepticism. The Soviet political activist Lev Kopolev recalled that when he saw people dying of hunger in Ukraine in the spring of 1932, he did not

> curse those who had sent me to take away the peasant's grain in the winter, and in the spring to persuade the barely walking skeleton-thin or sickly-swollen people to go into the fields in order to "fulfill the Bolshevik sowing plan in shock-worker style." Nor did I lose my faith. As before I believed because I wanted to believe.

The influential Hungarian Marxist Georg Lukacs spoke for many communists when he stated in 1967 that he would continue to believe in Marxism even if every empirical prediction it made were proven to be false. You can't get much further away from science than that.

Where communists ruled, real science was pitilessly pruned and uprooted to accommodate the party line. Russian astronomers who favored non-Marxist accounts of sunspots were accused of terrorism, while other scientists were persecuted simply for espousing the liberal values of science itself. The physicists Matvei Bronshtein and Lev Shubnikov were shot; their colleagues Semyon Shubin and Alexander Vitt were arrested and died in the camps; Lev Landau, who would receive the 1962 Nobel Prize in physics, nearly died in prison but was released when his colleague Pyotr Kapitsa warned Stalin that he would stop doing research unless Landau was set free. In all, more than a hundred members of the Soviet Academy of Sciences were sent to labor camps or otherwise incarcerated. Meanwhile government-sponsored pseudoscience flourished, reaching a tragicomic apotheosis in the Lysenko affair.

Trofim Denisovich Lysenko—who graduated from the Uman School of Horticulture in 1921 and ruled Russian agronomy for a quarter of a century, from 1940 to 1965, as director of the Soviet Academy of Sciences' Institute of Genetics—was a crank whose fantasies happened to dovetail with communist cant. In normal circumstances he might have puttered away his career in some backwater agricultural station, but in the USSR he became a hero.

Agriculture was one of the crowning glories of the United States and the other liberal democracies, where scientific experimentation had produced soaring crop yields, so the Soviet leaders made it a priority to outproduce Western farmers through the exercise of "socialist science." There is, however, no socialist science, no Western or Eastern science, no capitalist or communist or feminist or ethnic science. There's just science, and while one scientist may do it better or worse than the next, nobody can simply invent a different science and expect it to compete successfully with the real thing.

Stalin thought otherwise. He claimed to feel "reverence for science," but only of the communist kind—"that science whose people . . . have the boldness, the resolution to smash old traditions." Lysenko fit the bill. He presented himself as a "barefoot scientist" who had learned his trade from peasant farmers, and he promised that his earthy methods would outperform "bourgeois" science just as communist economics would outperform capitalism. As the communist theoretician and party leader Nikolai Bukharin put it in an address to the 1931 International Congress for the History of Science, Lysenkoism would help eliminate

> the rupture between intellectual and physical labor [facilitating] the entry of the masses into the arena of cultural work, and the transformation of the proletariat from an object of culture into its subject, organizer and creator. This revolution in the very foundations of cultural existence is accompanied necessarily by a revolution in the methods of science. . . .

What Lysenko actually did was to mix Lamarckian notions of acquired genetic characteristics with a romantic zeal for the health of wild plants—which, he maintained, could among other wonders morph from wheat to rye, a feat comparable to wild pigs awakening in the morning to find themselves turned into storks. (This was similar to the Nazi celebration of "natural" organisms as stronger than those allegedly weakened by cultivation.) Lysenko could never quite decide whether genes—the basic units of heredity, from which the word *genetics* is derived—existed, but he felt that mainstream genetics overemphasized competition, which smacked of capitalist market economics, whereas his own theories stressed a more socialistic biological "cooperation." Lysenko maintained that oak seedlings planted in dense clusters do not enfeeble one another by competing for sunlight and nutrients, as Darwin had found, but thrive, the weaker seedlings willingly dying out so that the stronger might prosper—a claim with unnerving parallels to communism's demand that citizens sacrifice themselves for the good of the state.

Lysenko responded to any hint of criticism with reflexive anger, political denunciation, and blanket dismissals of every scientific approach that threatened his prestige. After guaranteeing to double grain yields in a decade, he ordered that his methods be instituted on large scales without first being tested experimentally. "When we put forward a measure that is as yet founded only on theory, such as 'freshening the blood' of varieties of most important agricultural crops," he asserted, "do we have the right to lose two [or] three years in preliminary testing of this measure on little plots at several breeding stations? No! We don't have the right to lose a single year." When Lysenko's methods failed to produce the predicted results, he discarded the discouraging data and publicized only those of the few farms that seemed to have succeeded—data most often concocted by local party bosses eager to keep their jobs. These "successful" farms were then held up as "beacon-lights" for less hardworking (or more honest) farmers to emulate. "It would be hard to imagine a more effective way to make genuine science useless," notes the historian David Joravsky in his book *The Lysenko Affair.* "What need was there for a lot of experiment stations and basic research institutes, if the scientist's job was to be a publicity agent for isolated beacon-light farms?"

Under Lysenko, crop failures spread across Russian farmlands like a blight. Grain yields per acre fell 14 percent from 1930 to 1934, while 85 percent of the fruit trees planted according to Lysenko's prescriptions withered and died. The Soviet officials cheered him on. As people are apt to do once their ideology comes into conflict with empirical fact, Communist Party bosses resorted to an impenetrable bureaucratese that combined claims of great progress with exhortations to do better. A 1932 declaration by the Russian Congress for Planning Genetics and Breeding implored the Russian people to "raise more decisively the question of the reconstruction of our science itself, the rethinking of its methods of work, the introduction of the principle of the classness and partyness [*klassovost' i partiinost'*] of science on the basis of Marxist-Leninist methodology, the rethinking of trends and interrelationships with other sciences . . ." and so on. Russian wags coined a term, *boltologiia* or "jabberology," equivalent to "gobbledygook," to describe such verbiage.

Meanwhile those scientists who dared to point out the facts were persecuted, jailed, or reassigned to tertiary jobs. Prominent among such cases was that of Nikolai Vavilov, a devoted young biologist and geneticist who won the Lenin Prize and was made a member of the Soviet Academy of Sciences at age forty-two. Vavilov saw clearly that genetics research could save millions of Russians from malnutrition, if only Soviet science were released from the constraints of political dogma. The conflict came to a head one day in May 1939 when Vavilov was confronted by Lysenko, then head of the Lenin Academy of Agricultural Sciences, and a vice president of the academy named I. E. Lukyanenko. When Vavilov admitted that his Institute of Plant Botany and New Crops "bases its selection work wholly on Darwin's evolutionary teaching," his superiors pounced:

> LUKYANENKO: Why do you speak of Darwin? Why don't you choose examples from Marx and Engels? . . .
>
> LYSENKO: I understood from what you wrote that . . . evolution must be viewed as a process of simplification. Yet in chapter four of the history of the [communist] party it says evolution is [an] increase in complexity . . .
>
> VAVILOV: There is also reduction . . .
>
> LUKYANENKO: Couldn't you learn from Marx? . . .
>
> VAVILOV: I am a great lover of Marxist literature . . .
>
> LUKANENKO: *Marxism is the only science.* Darwinism is only a part; the real theory of knowledge of the world was given by Marx, Engels, and Lenin.

Vavilov ran out of patience and seized Lysenko by his suit lapels, exclaiming that Lysenko was destroying Soviet science. Lysenko shrieked like a schoolyard bully, declaring that he would tell on Vavilov. On August 6 of the following year, four Soviet agents emerged from a black sedan and arrested Vavilov while he was gathering botanical specimens in a sun-drenched field in western Ukraine. Interrogated nearly four hundred times in the ensuing eleven months, Vavilov was convicted of sabotaging Soviet agriculture and was sentenced to death. The sentence was commuted to twenty years but he soon died of malnutrition in Saratov prison, at age fifty-five. His "World Collection" of plant samples remains a resource for biologists today.

Such examples sufficed to silence Lysenko's critics, leaving him free to make extravagant promises and to blame his colleagues when they failed to come true. Ultimately the rot of mutual deception reached so deep that the Soviet Union was exporting grain while its own people starved. Stalin at one point deliberately shipped food out of districts in Ukraine, the North Caucasus, and Kazakhstan where peasants had resisted communist reforms, then deployed troops to prevent them from escaping. Six million people died in this concocted famine.

Soviet leaders were quick to credit their regime with all sorts of scientific and technological breakthroughs. Most were spurious, but two Soviet technological endeavors—rocketry and nuclear weapons—were genuinely imposing, especially when combined in the form of nuclear bombs affixed atop intercontinental ballistic missiles. A single bomb could destroy a city; an ICBM could loft the bomb from Russia to New York in a matter of minutes. The situation seemed bad enough when the Cuban missile crisis of 1962 brought the world to the brink of nuclear war, encouraging talk of whether the West might not be "better Red than dead," but at that point the USSR had only a few dozen nuclear-tipped ICBMs while the United States had a few hundred. In the accelerating nuclear arms race that followed, the number of such weapons spiraled into the thousands. Lurid war-gaming scenarios proliferated on both sides: "We" would hit "their" missile sites first, obliging them to sue for peace to save their cities; or hit their cities too, minimizing the risk of retaliation at the cost of our becoming the worst mass murderers in history, then hit them again, just to make sure, a measure known as "bouncing the rubble." We would take shelter, in foxholes or under our little desks at school, and ride it out, then rebuild, somehow, on soil poisoned by radioactive plutonium and cesium. In times of international crisis our leaders would consult over the hotline, or deliberately act crazy in order to persuade the other side that we really were deranged enough to carry out our threats. Both sides relied on the policy of Mutual Assured Destruction, or MAD. American generals tried to make the Soviets believe that they were willing to incinerate German cities to halt the progress of an invading Russian army; when that threat wore thin the Americans developed the neutron bomb, a nuclear warhead that could kill people without destroying property. The more frightening a weapons system was, the better its perceived potential for deterrence. A single "boomer"—a submarine carrying dozens of missiles, each missile tipped with multiple warheads—could wreck an entire nation overnight, yet boomers were also potent preservers of the peace. That was the trouble with MAD: To prevent a nuclear war, you had to act as if you were willing to start one and thought you could win it. Another problem with MAD was that it kept fueling the nuclear arms race. Intelligence officials habitually overestimated the adversary's nuclear strike forces—understandably so, since few cared to *under*estimate so grave a threat to their nation's security—and no chief of state could risk ignoring their warnings. Moreover, each leader wanted the other side to know that his arsenal was increasing, the better to deter them. They in turn responded by deploying still more nuclear weapons, which led to a further response, and so on.

In the end MAD worked, staving off nuclear war until the Soviet Union folded its tents on Christmas Day 1991, but it left behind an image of the USSR as having been a scientific and technological powerhouse. Now that the records are opening up, it can be seen that this was never the case.

Russia, like Germany, had many capable scientists and engineers—and unlike Nazi Germany it did what it could to hang on to them—but asking them to compete effectively against their colleagues in the free world was like fielding a shackled track team. The historian of science Loren R. Graham notes that although Russian scientists did manage to contribute to population genetics, soil science, and the design of the Tokamak nuclear reactor, "The overall record of Soviet science and technology, considering the enormous size of the Soviet science establishment, was disappointing." At the height of the Cold War the Russian biologist Zhores Medvedev made a quantitative study of Soviet versus American scientific research, but the results were so unfavorable to the USSR that he did not bother trying to publish them. The Soviet Academy of Sciences was advised in 1965 that Russia was producing only half as many research papers as the United States, despite having about the same number of scientists. The reasons for this gap were not obscure. Denied the freedom to attend conferences abroad, Soviet scientists had trouble keeping abreast of the latest developments. Hampered by the Soviet economy's inability to provide adequate laboratory equipment, many were obliged to concentrate on mathematical fields far from empirical tests; Russian astrophysicists did important work in exotic areas such as black hole theory and the origin of the universe but could not get their hands on a decent telescope. Scientists whose work was particularly important to national defense were cosseted in isolated communities where they enjoyed luxuries and liberties unavailable to their colleagues, but confinement in such gilded cages impelled the best among them to become vocal critics of the communist regime. The Russian physicist Andrei Linde recalled that physics,

> because of its importance for the development of the atomic bomb, was able to survive and develop, harmed to a much lesser degree by the dictatorship of official ideology. This circumstance gave rise to a very unusual phenomenon: The scientific culture associated with the Soviet school of physics became a culture of free political thought. As a result, many [of the] leading dissidents in the USSR, such as Andrei Sakharov and Yuri Orlov, were physicists.

Yuri Orlov helped formulate the Helsinki Accords, in which participating nations agreed to "respect human rights and fundamental freedoms, including the freedom of thought, conscience, religion or belief" and to report "interference with the free exchange of information." For this he was convicted of anti-Soviet agitation and spent nearly a decade in the Siberian Gulag, much of it in solitary confinement. In 1986 he was swapped for a Soviet spy and emigrated to the United States, where he joined the physics faculty at Cornell.

Sakharov, called the "father of the Soviet hydrogen bomb," became a passionate campaigner for democracy and human rights. He had worked on nuclear weapons from 1946 to 1968, spending eighteen years in a secret laboratory called "The Installation," which he described as a combination "between an ultra-modern scientific research institute . . . and a large labor camp." The scientists there, though showered with awards and bonuses, were virtual prisoners all the same. When Sakharov and a few colleagues wandered out toward the Installation's wooded perimeter one day, preoccupied by the conversation at hand, they were arrested by security guards and brought back to the barracks at gunpoint. Sakharov called for an end to nuclear testing and the establishment of a Russian "democratic, pluralistic society free of intolerance and dogmatism," in an article that was smuggled out of the Soviet Union, published in the *New York Times*, and reprinted in editions that sold more than eighteen million copies. He was awarded the

1975 Nobel Peace Prize, the citation calling him "the conscience of mankind," but was forbidden to travel to Norway to receive it. (In his written acceptance speech, Sakharov pointedly mentioned that his friend Sergei Kovalev, a biophysicist, had just been sentenced to seven years' hard labor.) Constantly harassed and sometimes beaten by communist agents, Sakharov was obliged to write his thousand-page autobiography three times from scratch, the original draft having been stolen by the KGB while he was in a dentist's chair and the second confiscated by agents who dragged him from his car and drugged him. Sakharov's international reputation made him an inopportune candidate for execution so the authorities sent him into six years of internal exile in Gorky, where he was forbidden to communicate by telephone or to entertain foreign visitors. He died of a heart attack on December 14, 1989, hours after addressing the Soviet Congress on behalf of political pluralism and free-market economics.

To understand how the United States and the USSR seemed during the decades of the Cold War to have had scientific and technological parity—to have resembled, as was often said, two scorpions in a bottle—requires a brief look at the history of their respective work on rocketry and nuclear weapons.

Rocketry is as old as gunpowder. It operates on the simple principle that any explosive, whether solid or liquid, expands when it burns. If you pack the explosive into a tube that is left open at one end—the Chinese originally used gunpowder tamped into bamboo—the expanding gases released by the burn will rush out the open end, pushing the tube off in the opposite direction. (Newton's third law: For every action there is an equal and opposite reaction.) Rockets clearly had military applications—the Chinese fired their bamboo rockets at invading Mongols in the year 1232—but what really mattered was that they could function in the vacuum of space. Fly a jet fighter to the edge of space and its engine will falter and shut down, since it needs oxygen from the atmosphere to burn the kerosene in its fuel tanks. But a rocket carries everything it needs on board—typically, liquid oxygen in one tank and a propellant such as hydrogen in the other—so it requires no air. Rockets are born for space.

It takes a lot of fuel and costly hardware to put even a small payload into space, and as there was no immediate practical benefit in doing so, the visionaries who laid the foundations of modern rocketry were not just engineers but futuristic dreamers as well. The three leading pioneers, all born in the 1800s, were a Russian, Konstantin Tsiolkovsky; an American, Robert H. Goddard; and a German, Hermann Oberth. Goddard was a physicist whose fascination with rockets, much ridiculed at the time, arose from a vision he had of going to Mars that came to him while he was climbing a cherry tree at age sixteen. He belonged to the American Rocket Society (originally called the American Interplanetary Society) whose founders were science-fiction enthusiasts like himself. He published a paper titled "A Method of Reaching Extreme Altitudes" in 1919, and on March 16, 1926, launched the world's first liquid-fueled rocket, from a snowy field at his Aunt Effie's farm in Auburn, Massachusetts. Tsiolkovsky, an autodidactic schoolteacher, proselytized for rocketry as the key to human spaceflight. He derived the "Tsiolkovsky equation" (demonstrating that rockets capable of sufficient thrust could loft large payloads into space) foresaw the development of multistage missiles (which increase their efficiency by shedding dead weight as they climb), envisioned space-suited astronauts assembling space stations in orbit, and studied the perils of returning to Earth amid the fireball generated by friction with the upper atmosphere. "The Earth is the cradle of humanity," he declared, "but one cannot live in a cradle forever." Oberth, a Transylvanian gymnasium teacher, read Jules Verne at about age eleven and launched his first rocket at age fourteen. Ultimately he succumbed to the UFO craze, declaring "that flying saucers are real and that they are space ships from another solar system . . . manned by intelligent observers [whose] present mission may be one of scientific investigation." When his physics doctoral dissertation on rocketry was rejected as too "utopian," Oberth published it via a vanity press as *Die Rakete zu den Planetenräumen* ("The Rocket into Planetary Space"). A copy found its way into the hands of a thirteen-year-old astronomy buff studying at a fashionable boarding school set up near Weimar in Ettersburg castle, where Goethe is said to have worked on *Faust*. "Opening it," the young man recalled, "I was aghast. Its pages were a hash of mathematical formulas. It was gibberish. I rushed to my teachers. 'How can I understand what this man is saying?' I demanded. They told me to study mathematics and physics, my two worst courses." He buckled down and went on to a celebrated career as a rocketry engineer and organizer, without ever working as a scientist. He was Wernher von Braun.

The big, powerful rockets requisite for space travel could not have been developed without the heavy investments made in them by the defense departments of Germany, Russia, and, more tardily, the United States. For von Braun, the Faustian bargain began on the day in early 1932 when a black sedan carrying three Army officers in mufti pulled up at the *Raketenflugplatz* (Rocketport) where his college rocketry club was about to conduct a test. The Nazis were not yet in power, premonitions of war were almost nonexistent, and for von Braun, the scion of a Prussian Junker family, the equation was clear. "We needed money, and the Army seemed willing to help us," he recalled. "We were interested solely in exploring outer space. It was simply a question with us of how the golden cow could be milked most successfully."

The affair began innocently enough. The rocketeers set up a government-funded launch site at Peenemünde—on the advice of von Braun's mother, who recalled that his father used to go duck hunting there—and as late as 1944 their rockets remained so inaccurate that von Braun took to observing test launchings from the intended point of impact, on the theory that "the bull's eye is the safest spot on the map." (He was nearly killed one day when, to his horror, an on-target missile loomed out of the blue sky and exploded overhead.) Before it was over, Polish and Soviet slave laborers were being worked to death in tunnels beneath Peenemünde, building the V-1 ("V" for *Vergeltungswaffe*, or "vengeance weapon") buzz bombs and V-2 missiles lobbed at London. When Berlin fell, von Braun and key members of his team surrendered en masse to the U.S. Army—their rationale being, as one put it, "We despise the French; we are mortally afraid of the Russians; we do not believe the British can afford us; so that leaves the Americans."

In the United States, von Braun popularized space exploration through books, magazine articles, and appearances in Disney films and television shows, his genial personality and deeply checkered past personifying the two faces of rocketry as both a way to explore other worlds and the possible agency of a nuclear war. (The comedian Mort Sahl suggested that an adoring Hollywood movie about von Braun, *I Aim at the Stars*, should be subtitled, "But Sometimes I Hit London.") A scientist or engineer living in a totalitarian state, von Braun told an American audience in 1957, is "coddled to some extent, and has practically all the advantages and privileges he enjoys in a free country. But there is always the looming danger that a brick may suddenly fall on his head."

Soviet rocket designers enthralled by space exploration similarly hoped that their military work would produce rockets capable of lofting humans into orbit, but the differing circumstances of the USSR and the United

States put the two nations on contrasting paths toward that decidedly subordinate goal. The problem for Russian generals at the dawn of the nuclear era was that in the event of war, American bombers could fly missions against Moscow and other Soviet cities after refueling at forward bases in Europe and Asia, whereas the USSR lacked such refueling bases and so could muster only a limited nuclear bomber threat. The answer was to develop intercontinental missiles capable of targeting American cities from within Russia's borders. But while American scientists and engineers had been able to miniaturize their thermonuclear weapons, Soviet H-bombs were still so big and heavy that hurling one over the North Pole was like using a rocket to deliver a loaded meat freezer to New York, complete with a power source for its refrigeration unit. To compensate for this technological limitation, the Russians built big boosters.

That was good news for Sergey Korolyov, the top Soviet missile designer, whose career spanned the idealistic visions and worldly torments that afflicted so many Russian scientists and engineers. As a student Korolyov had met Tsiolkovsky and absorbed his visions of space exploration, declaring, "Soviet rockets must conquer space!" His R-7 missile inaugurated a series of powerful, reliable boosters so cheap to mass-produce that while the Americans spent months testing new satellites by shaking and baking them in laboratory vacuums, the Russians often found it more cost-effective to just launch them into real space and see what happened. But along the way, Korolyov spent six years in the *sharashka* ("deceit") prisons established to punish scientists, engineers, and the *intelligentsia*. Beaten and tortured, he came home with a broken jaw, a heart condition, and a case of scurvy, and was thereafter kept under constant surveillance by Soviet agents, one of whom ominously wondered aloud, "Maybe you're making rockets for an attempt on the life of our leader?" Haunted by recurring nightmares in which guards burst into his bedroom to take him back to the camps, Korolyov habitually muttered to himself the incantation, "We will all vanish without a trace."

The Americans, meanwhile—protected by Strategic Air Command bombers, substantial ground forces in Western Europe, and a navy that roved the oceans almost at will—felt no great sense of urgency about developing ICBMs. Nor would such missiles need to be terribly powerful, since American nuclear warheads were already smaller and more sophisticated than the Russians' and were being further miniaturized—a process that would eventually produce H-bombs the size of Super Bowl trophies. This satisfied the generals, but von Braun feared that it meant the Americans would never build boosters big enough to put humans in space—as indeed would have been the case, had John F. Kennedy not made it a national goal to dispatch men to the moon. To provide an alternate vision, von Braun went public with books, TV shows, and plastic models illustrating how big, fat boosters could carry a giant space station, piece by piece, into orbit. He even went so far as to argue that the station might function as a military asset, the astronauts hurling bombs down onto an unwitting enemy.

On July 29, 1955, President Dwight D. Eisenhower announced that the United States planned to develop what everyone assumed would be the world's first earth satellite. Devoted solely to scientific research, it would contribute to the International Geophysical Year of 1957–58, a cooperative research effort involving scientists in sixty nations worldwide. To underscore its peaceful intent, the satellite would be launched by an allegedly civilian rocket (actually a rebranded Navy missile) called Vanguard. The skinny, crayon-shaped Vanguard and its grapefruit-sized scientific satellite would demonstrate both the Americans' technological superiority and their desire to prevent space from becoming a Cold War battleground.

Back in the USSR, Korolyov realized that the feel-good Vanguard project presented an opportunity to beat the Americans into space. His team already had powerful enough rockets to orbit a satellite heavier than Vanguard's little grapefruit, provided that the Soviet leaders weren't too worried about the public-relations downside of using military hardware to get the job done. He pitched the idea to Nikita Khrushchev, the first secretary of the Communist Party, to whom he had given a tour of a Soviet missile launch facility four years earlier. ("We gawked at what he showed us as if we were sheep seeing a new gate for the first time," recalled Khrushchev, who liked to portray himself as a simple farmer. "We were like peasants in a market place. We walked around the rocket, touching it, tapping it to see if it was sturdy enough—we did everything but lick it to see how it tasted.") Although Khrushchev was loathe to waste military resources, he appreciated that a satellite launch might improve the morale of his best rocket designer—who was, after all, literally a beaten man—while making it appear that the USSR had leapfrogged the United States technologically. So Khrushchev gave it the nod, and on October 4, 1957, Korolyov's team put Sputnik into orbit. "The freed and conscientious labor of the people of the new socialist society," crowed *Pravda*, "makes the most daring dreams of mankind a reality." Korolyov's childhood dreams had come true, but he had never entirely recovered from the physical and psychological abuse he suffered in the Gulag, and he died a few years later at age fifty-eight.

The American public was flabbergasted. Sputnik "created a crisis," recalled James Killian, the first White House science advisor. "Confidence in American science, technology, and education suddenly evaporated." Gliding overhead like a beeping Sword of Damocles—it did nothing *but* beep, since it contained nothing but a radio transmitter—Sputnik suggested to politicians and pundits alike that the United States, awash in a hedonistic brew of martinis, bikinis, and Cadillacs sporting tailfins larger than slabs of barbecued ribs, was losing out to the stern efficiency of totalitarian technology. "The communists have established a foothold in outer space," Lyndon Johnson warned his colleagues in Congress. "It is not very reassuring to be told that next year we will put a better satellite into the air. Perhaps it will also have chrome trim and automatic windshield wipers." The journalist Edward R. Murrow wrote, three days after the Sputnik launch:

> It is to be hoped that the explosion which flung the Russian satellite into outer space also shattered a myth. That was the belief that scientific achievement is not possible under a despotic form of government. . . . We failed to recognize that a totalitarian state can establish its priorities, define its objectives, allocate its money, deny its people automobiles, television sets and all kinds of comforting gadgets in order to achieve a national goal. The Russians have done this with the intercontinental ballistic missile, and now with the Earth satellite.

Murrow added that Americans should have learned "from the Nazis that an unfree science can be productive."

But that was not the lesson to be learned, because the Americans had not fallen behind. In many ways they were ahead—a secret that Eisenhower grimly kept while a campaigning John F. Kennedy accused the administration of having permitted a "missile gap" to loom between the United States and the USSR. There was no missile gap. Eisenhower's real reason for staking the nation's space plans on the nonmilitary Vanguard project was that he understood the intelligence potential of secret spy satellites, which were being developed by the military, but feared that their deployment would provoke an international incident should the Soviets view them as violators of Russian airspace. Meanwhile the Americans had to rely for their aerial intelligence on U-2 spy planes, which normally flew at altitudes above the reach of Russian air defenses. In 1960, the final year of Eisenhower's presidency, the Soviets managed to shoot down a U-2, capturing its pilot alive and its surveillance cameras intact. Eisenhower felt obliged to insist that the U-2 was a "weather plane," creating the embarrassing spectacle of an American president lying to his people while a Soviet premier for once told *his* people the truth. Eisenhower had hoped that the scientific Vanguard project would establish a legal precedent that orbiting satellites do not violate nations' sovereignty. Ironically enough, Sputnik laid that worry to rest.

Von Braun and his team at the American rocket center in Huntsville, Alabama, had been insisting since 1952 that they could launch a satellite on demand, using existing military rockets, and were frustrated by Eisenhower's reluctance to do so. On the day Sputnik went up, von Braun sought out Neil McElroy, recently nominated to be secretary of defense, at a cocktail party at the Huntsville officers' club and repeated this claim. "We have the hardware on the shelf," von Braun told McElroy. "For God's sake turn us loose and let us do something. We can put up the satellite in sixty days, Mr. McElroy! Just give us a green light and sixty days!" General John Bruce Medaris, head of the Army Ballistic Missile Agency, interjected a note of moderation. "No, Wernher," he said. "Ninety days."

In the end it took 119 days to launch Explorer, the first American satellite, on January 31, 1958. The rocket was stock, the satellite hastily constructed by scientists and engineers at the Jet Propulsion Laboratory in California. "We liked the difference between our satellite and Sputnik," said one of the conspirators. "Ours flew science." Indeed, Explorer made the first scientific discovery in space, when data sent back by its detectors suggested the presence of the doughnut-shaped bands of intense radiation surrounding Earth known today as the Van Allen belt—after the physicist James Van Allen, who designed the instruments involved. Meanwhile a Vanguard test on December 6, promoted by the White House to the status of an actual launch attempt and broadcast live on network television, failed in spectacular fashion: The slim rocket rose only four feet before subsiding into a blooming fireball, its nose cone toppling off and its gleaming chrome-plated satellite rolling piteously across the ground. A subsequent Vanguard launch succeeded—hurled high, as if from pent-up frustration, the satellite remains in orbit to this day—but only after the first attempt had been dubbed "Flopnik" and "Stayputnik" in the newspaper headlines, adding to an American inferiority complex that deepened in 1961 when Yuri Gagarin became the first human to orbit the earth.

Spaceflight enthusiasts had been arguing since the start of the Cold War that a space race between the two superpowers could function as a peaceful, nonmilitary arena of technological competition. Now something like that was actually happening. President Kennedy said as much in his speech of May 25, 1961, committing the nation to "the goal, before this decade is out, of landing a man on the moon and returning him safely to the earth." The idea, Kennedy told Congress, was to win the Cold War by winning the space race—"To win the battle that is now going on around the world between freedom and tyranny" so that "men everywhere" could decide "which road they should take." The Americans prevailed, landing on the moon five months ahead of Kennedy's deadline, and in the process a lot of exploration and some science got done. Even the rosy scenario of international cooperation was eventually realized, when the opening decade of the twenty-first century saw the stodgy but reliable rockets of a postcommunist Russia ferrying astronauts, cosmonauts, and supplies to an International Space Station, much as Tsiolkovsky, Goddard, Oberth, and Korolyov had dreamed.

The Soviet nuclear weapons program was launched on a river of espionage opened up by the Lend-Lease Act, under which the United States dispatched $11 billion worth of materiel to its Soviet allies—much of it flown out from Gore Field in Great Falls, Montana, under the supervision of an Army Air Force officer named George Racey Jordan. In his book *Dark Sun: The Making of the Hydrogen Bomb*, Richard Rhodes recounts Jordan's tale of how he became suspicious of all the black patent-leather suitcases, bound with sash cords and sealed with red wax, that he saw dispatched to Moscow. "The units mounted to ten, twenty and thirty and at last to standard batches of fifty, which weighed almost two tons and consumed the cargo allotment of an entire plane," Jordan recalled. "The [Soviet] officers were replaced by armed couriers, traveling in pairs, and the excuse for avoiding inspection was changed from 'personal luggage' to 'diplomatic immunity.'" When the Russians invited Jordan to a vodka-soaked dinner one night in March 1943, he became suspicious and broke away from the endless rounds of toasts to inspect a C-47 cargo plane whose crew was demanding immediate clearance to take off for Moscow. Arriving at the airstrip, he pushed aside a "burly, barrel-chested Russian" who tried to block his entrance to the plane and found "an expanse of black suitcases" filling its cavernous cargo hold. He summoned an armed GI and started cutting open the suitcases. They were full of documents. Puzzled, he wrote down the words in them that he did not understand: "Uranium 92—neutron—proton and deuteron—isotope—energy produced by fission. . . ." Jordan soon realized that he had been witnessing not only the departure of "a tremendous amount of America's technical know-how to Russia" but that the Lend-Lease planes were also being used to import Soviet agents into the United States:

> The entry of Soviet personnel into the United States was completely uncontrolled. Planes were arriving regularly from Moscow with unidentified Russians aboard. I would see them jump off planes, hop over fences, and run for taxicabs. They seemed to know in advance exactly where they were headed, and how to get there.

In addition to their own agents, the Soviet intelligence establishment was able to recruit American and European scientists and intellectuals who viewed Russia as the prime mover in driving back fascism—credit for which the Russians certainly deserved, having lost 8.6 million troops and more than 10 million civilians in the war—and thought of communism as the wave of the future. Some of these amateur spies were astoundingly naïve. David Greenglass, an engineer who spied for the Soviets while working in the nuclear facilities at Oak Ridge and Los Alamos, wrote to his wife, Ruth, in 1944:

> I have been reading a lot of books on the Soviet Union. Dear, I can see how far-sighted and intelligent those leaders are. They are really geniuses every one of them [and] I have come to a stronger and more resolute faith and belief in the principles of Socialism and Communism.

Klaus Fuchs, a theoretical physicist who joined the Communist Party in the early thirties out of opposition to Hitler and later sent nuclear secrets to the Soviets from a British laboratory at Birmingham, said blithely, "It was always my intention, when I had helped the Russians to take over everything, to get up and tell them what is wrong with their system." Harry Gold, a Manhattan Project spy, claimed that he decided "to do everything possible to strengthen the Soviet Union" after a drunken anti-Semite called him a "sheeny bastard" in the smoking room of the Philadelphia train station. Another spy, Julius Rosenberg, boasted to Greenglass in 1943 that his "powerful friends" in Moscow would set him up in a phony "screen" business after the war. "Victory shall be ours and the future is socialism's," he wrote to his wife, Ethel.

Such escapades ended badly for some of the spies—Greenglass served ten years in prison, Fuchs nine years, and Harry Gold fifteen years, while Julius and Ethel Rosenberg died in the electric chair at Sing Sing on June 19, 1953—but the espionage proved invaluable to the Soviets. By 1942, having seen enough secret files about nuclear science and engineering to conclude that something was afoot, Beria secured Stalin's agreement to try to build a bomb. Work began the following year, based on thousands of pages of captured documents: "The materials are magnificent," declared Peoples Commissar Vyacheslav Molotov in 1942, and they kept getting better. The bombing of Hiroshima on August 6, 1945, ended any lingering doubts. "Now that the Americans have invented it," said Col. Nicolai Zabotin, the head of Soviet military intelligence, "we must steal it!" An entire city, named for Beria and devoted to nuclear research, was built on a site conveniently located near four Gulags whose laborers were pressed into service clearing the land. Stalin doubled the salaries of the physicists and chemists sent to work there.

The first Russian fission device (or A-bomb) was tested in 1949. Based on espionage, it was an exact replica of the American bomb dropped on Nagasaki. A Russian A-bomb of indigenous design was exploded two years later, soon followed by the fusion or H-bomb. ("H" stands for hydrogen; H-bombs work, as stars do, by fusing hydrogen atoms to make helium and releasing energy in the process.) The Americans had hesitated about developing hydrogen bombs, owing both to the moral concerns of some scientists—there is no upper limit to the power that can be released by nuclear fusion, as a glance at the sun will verify—and by the fact that half the Los Alamos staff had departed at the end of the war. But President Truman in January 1950 directed that the work proceed apace, and the first H-bomb was detonated on Eniwetok Atoll in the Pacific on November 1, 1952. Seven hundred times more powerful than the A-bomb dropped on Hiroshima, it resembled a visiting star. When the Soviets exploded their H-bomb, less than a year later, the arms race was indisputably on. Following the collapse of the Soviet Union the United States cut its nuclear arsenal in half and funded a program that eliminated more than ten thousand Russian nuclear warheads, recycling their reactive materials for use in power plants.

By rights the end of the Soviet Union should have ended communism, which had lost the space race, the nuclear arms race, and every other major scientific and technological competition it entered; had failed to win the loyalty of the peoples it ruled or of those beyond; and had committed a multitude of hideous crimes against humanity. Yet communism still controlled the destinies of more than a billion Chinese, and so provided another unbidden test of the status of science under authoritarian rule.

The Chinese have a long tradition of technological innovation (mechanical printing, the magnetic compass, the world's first seismograph) and of such protoscientific investigations as charting the night sky. In recent years Chinese researchers working abroad have made important scientific contributions, with three Chinese physicists winning Nobel Prizes—Chen-Ning Franklin Yang and Tsung-Dao Lee, who did their major research at the University of Chicago, and Daniel C. Tsui, who earned his doctorate at Chicago and then worked at Bell Labs. China's indigenous scientists might have contributed much more to science by now, had their nation enjoyed liberal governance. But instead, China continued to labor under a Communist Party that claimed to be "guided by scientific theories," and "to ensure that decision making is scientific and democratic," but which by "scientific" meant Marxist and by "democratic" meant communist. The education of generations of Chinese scientists was blunted by mind-numbing classroom indoctrinations in communist doctrine, those who emerged with any freedom of thought intact having to choose between keeping silent or being oppressed as "dissidents." As Alfred Zee Chang of the Library of Congress noted in 1954, just four years after the Chinese Communist Party came to power:

> It is apparent that the communists have little appreciation of the talents and ability of leading scientists. In fact, these scientists, with their objective method of reasoning, represent a potential threat to communism. . . . Indigenous technology and the scientific approach on the China mainland is in a highly static condition today, and the future of science in this area presumably will be that contained within the general Soviet scientific pattern.

The Chinese communists, like the Russians, were eager to demonstrate how Marxist methods could transform agriculture. Mao Zedong's "Great Leap Forward" of the late fifties did just that, and the result was the greatest famine in recorded history: Some twenty million people perished. Survivors got by on a government ration of 1,200 calories a day (less than was provided by the Nazis to slave laborers at Auschwitz) while others resorted to cannibalism or took the government's advice and tried to live on chorella, a freshwater algae grown in urine. As in the USSR, the Chinese communist leadership continued to export food during the famine, in order to be able to boast of the remarkable productivity of their socialist farms. The State Statistical Bureau, which might otherwise have reported that millions were dying of hunger, was disbanded and replaced with "good news reporting stations." And, as in the USSR, the Chinese campaigns of agrarian reform were characterized by an overt disdain for science. Stalin during his first five-year plan waved away the warnings of scientists, who were derided as "bourgeois specialists." Lectures were preceded by the singing of an anthem to "the eternal glory of Academician Lysenko," whose campaign "protects us from being duped by Mendelist-Morganists"—a reference to Gregor Mendel and Thomas Morgan, two founders of genetics. In China in 1958, a Mao toady named Kang Sheng mounted a national lecture tour to advise the public that if schoolchildren took "action" rather than thinking or reading, they could revolutionize genetics and launch earth satellites:

> Science is simply acting daringly. There is nothing mysterious about it. . . . There is nothing special about making nuclear reactors, cyclotrons or rockets. You shouldn't be frightened by these things: As long as you act daringly you will be able to succeed very quickly.

Many knowledgeable Chinese could foresee the dire consequences of following such advice, but few objected to it. An exception was the defense minister, Marshal Peng Dehuai, who unlike Mao had grown up in poverty and had seen members of his own family starve to death. "I have experienced famine," he wrote Mao in 1959.

> I know the taste of it and it frightens me! We have fought decades of war and the people, poorly clothed and poorly fed, have spilt

> their blood and sweat to help us so that the Communist Party could win over the country and seize power. How can we let them suffer again, this time from hunger?

Mao was unperturbed. Death, he had told the May 1958 party congress, is "to be rejoiced over. . . . We believe in dialectics, and so we can't not be in favor of death. . . . There should be celebration rallies when people die." He forbade mourning the dead, since corpses "can fertilize the ground." In response to Marshal Peng's letter, Mao convened a meeting of the Politburo and gave them a speech worthy of Stalin. Progress would take time, he said: "When you eat pork, you can only consume it mouthful by mouthful, and you can't expect to get fat in a single day. Both the Commander-in-Chief [Zhu De] and I are fat, but we didn't get that way overnight." The dire warnings of the economists were to be ignored:

> Why can't our commune cadres and peasants learn something about political economy? Everybody can learn. Those who cannot read may also discuss economics, and more easily than the intellectual.

For writing his letter, Peng Dehuai was condemned as a "right opportunist" and put under house arrest. Interrogated and tortured during the Cultural Revolution, he died in prison in 1973.

The Cultural Revolution was promulgated on August 8, 1966, by the central committee of the Chinese Communist Party, which promised that it would launch "a new stage in the development of the socialist revolution in our country" and "meet head-on every challenge of the bourgeoisie." In practice this meant that thousands of scientists, authors, artists, and intellectuals were shot, beaten to death, or defenestrated by the Red Guards and other thugs, many of them students encouraged to humiliate or assault their teachers. Although no accurate census of the carnage exists, it is estimated that roughly one million people died in the Cultural Revolution and that an equal number were permanently injured. Because the party uses the term "science" to mean Marxism-Leninism, this campaign to murder scholars and destroy Chinese cultural traditions has since been blamed in China on "scientism"—a word originally coined by right-wing French reactionaries to describe the doctrine that science is superior to other systems of thought.

Meanwhile the Chinese struggled to surge ahead in real science, with Mao ordering that scientists and engineers "must be absolutely protected" from persecution, and that those educated in Europe or America were to be "neither labeled nor denounced." Top scientists were afforded the privileges enjoyed by all but the very highest government officials; during the famines, for instance, they were given cherished soybeans. But the result, as in Russia, was to turn many of them into vocal critics of the communist regime.

The physicist Fang Lizhi, often called the Chinese Sakharov, entered Peking University at sixteen and rose to become vice president of the University of Science and Technology, but when he began speaking out against socialism he was dismissed from his posts and from the Chinese Communist Party for fomenting "bourgeois liberalization." Sent off for reeducation on a communal farm, Fang spent a year of solitary confinement in a cowshed with only a single book to read, Lev Landau's *Classical Field Theory*. Studying it, he began thinking seriously about cosmology, the science concerned with the history and structure of the universe. This otherworldly pursuit brought further confrontations with the authorities. Lenin and Engels had decreed that the universe must be infinite. Otherwise, they thought, the universe would have a boundary, whereas Lenin had declared that "dialectical materialism insists . . . on the absence of absolute boundaries in nature." Actually, as the non-Euclidean geometers had already demonstrated mathematically and as Einstein would demonstrate in physics, the universe can be both finite and unbounded—just as the surface of the earth is finite and unbounded. To understand this is fundamental to scientific cosmology, but when Fang taught it to his students he was rebuked in a physics journal for the "sheer folly" of introducing alternatives to the communist world model. "We must ferret out and combat every kind of reactionary philosophical viewpoint in the domain of scientific research," the article concluded, "using Marxism to establish our position in the natural sciences."

It didn't help that Fang habitually spoke and wrote about democracy and liberalism as if he were living in a free country. "Many of us who have been to foreign countries to study or work agree that we can perform much more efficiently and productively abroad than in China," he said. "Foreigners are no more intelligent than we Chinese. Why, then, can't we produce first-rate work? The reasons for our inability to develop our potential lie within our social system." Blamed for fomenting the 1989 demonstrations in Tiananmen

Square—where students erected a Statue of Liberty and rallied to cries of "Science and Democracy!" before being mowed down by government troops—Fang was persuaded to take refuge in the U.S. embassy.* He was there when his speech accepting the Robert F. Kennedy Human Rights Award was read in absentia on American network television. "The universe has no center," it said. "Every place in the universe has, in this sense, equal rights. How can the human race, which has evolved in a universe of such fundamental equality, fail to . . . build a world in which the rights of every human from birth are respected?" Fang and his wife escaped to America the following year, Fang joining the faculty at the University of Arizona and composing several evenhanded essays about science and liberal democracy. One of them, written with the Princeton University Sinologist Perry Link, identifies five links between science and liberalism as seen from a Chinese perspective:

1. *Science begins with doubt.* "In order to make a scientific advance, one must begin by wondering about the received version of things. . . . This is also called 'dissent' [but] even an elementary course in physics takes up the problem of error. Students are taught that the only defense against error lies in the scientist's willingness always to question."

2. *Doubt leads to independence.* While studying at Beijing University, Fang realized that "the path from physics to democracy began with 'independence of thought'" and "reached the conclusion that science placed the burden of finding truth upon each individual person."

3. *Science is egalitarian.* "Objective scientific truth is something that lies beyond the variety of subjective views; statements of objective

*"Science and Democracy" was a slogan of the 1919 May Fourth student movement for Chinese modernization. In commemorating its anniversary, one of the Tiananmen Square demonstrators, Wuer Kaixi, read a statement declaring, "This student movement has but one goal, that is, to facilitate the process of modernization by raising high the banners of democracy and science . . . and by promoting freedom, human rights, and rule by law." Han Minzhu [a pseudonym], *Cries for Democracy: Writings and Speeches from the 1989 Chinese Democracy Movement* (Princeton, NJ: Princeton University Press, 1990), 136.

truth are formed only by a consensus of many observers, and are confirmed by independently repeating experimental results. No single observer is privileged; anyone may form hypotheses, and any hypothesis has to be tested by others before 'truth' emerges." As in democracy, "One person's vote, like his or her subjective view, contributes to a public consensus without determining it. . . . Just as everyone stands equal before the truth, similarly everyone should be equal before the law."

4. *Science needs a free exchange of information.* In the words of the historian of science Xu Liangying, scientific research requires "an atmosphere of freedom conducive to exploration. . . . Political democracy and academic freedom are necessary to guarantee the flourishing of science."

5. *Science is universal.* "There is no such thing as Chinese or Indian science—or, as the Nazis once seriously claimed, 'German science.' " Fang notes that although his students at the University of Arizona come from many nations, he teaches them all the same relativity and quantum mechanics, and that "what they learn holds in exactly the same way everywhere, even light-years away." As he said in 1989, "In science, we approach a situation by asking if a statement is correct or incorrect, if a new theory is an improvement over an old one. These are our criteria. We do not ask if a thing originates with our race or nationality. . . . Where it comes from is irrelevant. There are no national boundaries in scientific thought."

Like any other totalitarian ideology, communism must pretend to enjoy a total command of the truth; otherwise it could not justify its claimed authority over every aspect of people's lives. (Fang: "Authoritarianism needs authoritative statements in order to exercise its power.") That is why millions of students had to memorize Lenin's ignorant assertions about the universe, why Stalin was described in Russia as "the greatest genius of mankind," why Mao's stultifying "Little Red Book" was treated as a work of wisdom on a par with Confucius, and how North Korea's Kim Jong-il came to be described as not only "the fatherly leader" but as a poet, philosopher, historian, opera composer, film director, and the world's best golfer. When the Chinese premier Zhao Ziyang said during a visit to Italy, "Were it not for Copernicus, we still wouldn't know that the earth is round," his entirely forgivable mistake was repeated by his translator and reported without correction by the editors of the *People's Daily*, presumably out of fear of what would happen to them should they reveal that the great leader was flawed in any way. In the liberal-democratic world, everybody appreciates that politicians and scientists are ordinary mortals who sometimes say dumb things. Such mistakes seldom matter much, because the democracies, like the sciences, are broad-based and largely self-correcting; a democratic nation can limp along even when its chief of state is widely understood to be a lazy, bumbling simpleton. Fallible leadership is the only kind of leadership any nation ever has. Since totalitarians cannot afford to admit this, their domains start and end in fantasy.

It often happens that new scientific knowledge not only builds on prior knowledge but also exposes prior ignorance and error. This process poses few problems for democracies, since they incorporate fallibility as a given, but it leaves totalitarian leaders clinging to outmoded doctrines in a changing world. Hence Chinese communism became increasingly antiquated as the sum of scientific knowledge grew. The China scholar H. Lyman Miller, of Johns Hopkins University, notes in a study of scientific political dissent in post-Mao China that "Marxism-Leninism was formulated originally on the basis of nineteenth-century science, and the Chinese version of it was updated not much beyond Lenin. So the chasm between philosophical doctrine and scientific theory was wide"—and it kept getting wider. Even more alarming, from a communist point of view, was the fact that science in itself exemplifies democratic values. The scientific ethos "is inherently antiauthoritarian," writes Miller.

> Just as the scientific community operates according to antiauthoritarian norms of free debate . . . so science prospers in an external environment that similarly tolerates pluralism and dissent. . . . Scientific dissidents espoused a strong form of liberal political philosophy that grew out of the norms of their profession.

The Chinese people, as creative and energetic as any in the world, became increasingly to resemble prisoners in a decaying castle, from which they could distantly see the wider world becoming healthier, wealthier, and more knowledgeable than the folks at home.

Their isolation began to end in 1978, when the government finally started inching away from a totally planned economy by introducing limited free-market reforms. The result was an era of double-digit economic growth: China's GDP soared by an average of 10.3 percent annually during the 1980s and by 9.7 percent from 1990 through 2002. Thanks to these gains, China by 2008 had become the world's fourth most productive nation (although, being large and recently poor, it ranked only one-hundredth in per capita GDP) and the party's efforts to keep it cut off—as by throwing up a "great firewall" limiting Internet access—looked increasingly quaint.

Without denying any of the credit due to the Chinese people for the hard work that went into their economic advances, it should be said that for a nation today to liberalize its trade policies means that it is joining an international economic system constructed by the world's liberal democracies—a club devoted to economic science and economic liberalism. The benefits that it reaps are therefore made possible by global liberalism, regardless of whether the nation in question has liberalized internally, as China circa 2008 assuredly had not.

Will economic liberalism lead the Chinese to political liberalism as well? One way to gauge whether this is happening is to look at China's scientific productivity. China increased its investment in scientific research and development to an historic high in 2001, when it spent 1.10 percent of its GDP on R&D. This remained well below the R&D spending rates of wealthy liberal democracies like the United States, which spent 2.74 percent during the same period, or Japan (3.06 percent), but it was impressive for a nation whose per capita GDP was only a thirtieth of the American and Japanese. Rich nations tend to spend a higher proportion of their GDP on scientific research than do poor countries, who often lack the educational infrastructure to support world-class research and who feel that they have more pressing problems to solve first. Yet China's scientific citation rates remained far below what would be expected from a nation of its size and history of intellectual attainment, in part because the Communist Party continued to stress applied science over pure research.

Even Chinese Communist Party officials eventually began conceding that political liberalization would come, once the Chinese were "ready." This argument may have some merit, if "ready" means rich enough. During the past century, democracies established in nations with a per capita GDP of $1,500, as was China's in 2008, have on average survived for less than a decade. In any event, this book predicts that no matter how rapidly the Chinese economy grows, China will become scientifically prominent only once its people have won political—and not just economic—freedom.

The Chinese Communist Party has begun experimenting to a limited extent with grassroots democracy, but it describes even these modest efforts in language worthy of *Alice in Wonderland*. The 2005 party report on democratic reform, titled "Bid to Build Democracy Comes to Fruition," stated:

> Through painstaking exploration and hard struggle, the Chinese people finally came to realize that mechanically copying the Western bourgeois political system and applying it to China would lead them nowhere.

This is vintage rhetoric from communist functionaries, who are always talking about how hard they work and struggle but whose toil seems mainly aimed at degrading the meanings of words. By "the Chinese people" they mean the party; to "lead them nowhere" means that if free and fair elections were to be held, the communists would lose. So there won't be many such elections; this is called "building democracy." By way of historical background, the report states that "in 1921, some progressive intellectuals who had studied the ideology of democracy and science combined Marxism and Leninism with the Chinese workers' movement, and founded the CPC." In fact, the Chinese Communist Party was founded at the behest of, and with funding from, the Soviet Union. "The CPC creatively combines the general truth of Marxism-Leninism with the actual situation of the Chinese revolution, setting out such democratic concepts as 'democracy for the workers and peasants,' 'people's democracy,' and 'new democracy.'" The reference to the "*general* truth" of Marxism-Leninism means that it is false, but that the party can bring itself to admit to only a few of its many falsehoods. When the party says that it is "guided by scientific theories," it means communism. Its promise "to ensure that decision making is scientific and democratic" means that elections will be confined to candidates selected from party ranks.

In discussing communism, this book has focused on monsters like Stalin and Mao. Many other monsters could be added to the account—men like Lavrenti Beria, head of the Soviet nuclear weapons program, a sadistic rapist of Russian actresses and athletes who, when asked by Sakharov, "Why do we always lag behind the USA and other countries, why are we losing the technological race?" blandly replied, "Because we lack R&D and a manufacturing base" before proffering his "plump, slightly moist, and deathly cold" hand in farewell. Such an account has little to say about the millions of well-intentioned communists who joined the party out of a genuine desire to assuage the sufferings of the poor and to help deliver humankind from feudal ignorance and superstition into what they thought would be an age of science and social justice. But that is precisely the point: Good intentions didn't matter. Communism was inherently incapable of preventing the rise of monsters and of voting them out of office once they got there. Constitutionally unable to respond to experimental results, the communist regimes ultimately *became* failed experiments. As Boris Yeltsin, Russia's first popularly elected president, fatalistically declared:

> It was decided to carry out this Marxist experiment on us—fate pushed us precisely in this direction. In the end we proved that there is no place for this idea. It has simply pushed us off the path the world's civilized countries have taken.

An eighty-year-old communist in Hanoi put it more simply: "From the bottom of my heart, we didn't want to do any bad things. We tried to be good. But it all became such a mess."

At least three lessons may be learned from all this.

First, while science and liberalism encourage experimentation, social experiments must only be instigated with the permission of the citizens involved, must be consistent with the legal rights of all, and must remain vulnerable to repeal if they fail to attain the wished-for results. None of these provisions were fulfilled by the fascist and communist regimes, which were "experiments" only in the tragic sense employed by Boris Yeltsin. The essence of scientific experimentation is to include a feedback loop through which the researcher examines the data, draws conclusions, and then alters or ends the experiment accordingly. There is, fortunately, a system of government that fulfills these conditions. Liberal democracies and free markets conduct experiments—and respond, however imperfectly, to the results—every day. The outcomes are not ideal, but they are better than all the known alternatives.

Second, any proposed social system that fails to provide such a feedback loop—that does not, for instance, provide for free and fair elections—should be rejected out of hand. It is utterly irresponsible to hand over power to any movement not vulnerable to such review, regardless of its putative qualifications. Especially to be rejected are utopian claims that people by surrendering their freedom shall enter into a wonderful new age in which everybody will be changed for the better—as when Leon Trotsky promised that through communism a new "superman" would emerge, "incomparably stronger, wiser, more subtle" than any seen before. This is not to say that new and somehow better sorts of persons will never appear; in a sense this happens all the time. But no peoples may justifiably barter away their fundamental rights and those of their descendants in the hope of such a development, or for any other reason. This tenet, central to liberalism, is paralleled in science by the demand that new theories ought to answer to the data of past experimentation and observation; no new scientific theory is apt to find much acceptance if it claims to break with *everything* that has already been learned.

Finally, ends cannot be separated from means. If the means are illiberal, or immoral, or fly in the face of common sense, they are not to be justified by recourse to a predicted outcome. To proceed as if this were not the case is to put too much faith in our ability to predict the future. Science is better at making accurate predictions than any other system of thought, but even science makes no claim to the sort of knowledge that would be required to lead humans into darkness on the promise of eventual light. The very least that can be done to honor the memory of those who died under political repression is to act today in ways that can be justified today, without borrowing against an imagined tomorrow.

CHAPTER ELEVEN

ACADEMIC ANTISCIENCE

Stupidity is the twin sister of reason: It grows most luxuriously not on the soil of virgin ignorance, but on soil cultivated by the sweat of doctors and professors.

—WITOLD GOMBROWICZ, 1988

Science and ideology are incompatible.

—JOHN S. RIGDEN AND ROGER H. STUEWER, 2004

Once the liberal democracies had prevailed against fascism and communism—vanquishing, with the considerable help of their scientific and technological prowess, the two most dangerously illiberal forces to have arisen in modern times—you might think that academics would have investigated the relationship between science and liberalism. The subject had come up before. In 1918 the president of Stanford University suggested that "the spirit of democracy favors the advance of science." In 1938 the medical historian Henry E. Sigerist allowed that while it might be "impossible to establish a simple causal relationship between democracy and science and to state that democratic society alone can furnish the soil suited for the development of science," it could hardly "be a mere coincidence . . . that science actually has flourished in democratic periods." In 1946, the historian of science Joseph Needham argued that "there is a distinct connection between interest in the natural sciences and the democratic attitude," such that democracy could "in a sense be termed that practice of which science is the theory." Inquiring into why this might be so, the Columbia University sociologist Robert K. Merton considered, as a "provisional assumption," that scientific creativity benefits from the increased number of personal choices found in liberal-democratic nations. "Science," he wrote, "is afforded opportunity for development in a democratic order which is integrated with the ethos of science." In 1952, Talcott Parsons of Harvard argued that "only in certain types of society can science flourish, and conversely without a continuous and healthy development and application of science such a society cannot function properly." As befits scholarly inquiry, the tenor of these investigations was modest, the researchers cautioning that the subject was complex and their findings preliminary. Bernard Barber, a specialist in the ethics of science, portrayed the relationship this way:

> Examination will show that certain "liberal" societies—the United States and Great Britain, for example—are more favorable in certain respects to science than are certain "authoritarian" societies—Nazi Germany and Soviet Russia. We say that the latter countries are "less favorable"; we do not say that science is "impossible" for them. This is not a matter of black-and-white absolutes but only of degrees of favorableness among different related societies.

A good start, you might think; something to build on. But instead, academic discourse took a radical turn from which it has not yet fully recovered. Rather than investigate how science interacts with liberal and illiberal political systems, radical academics began challenging science itself, claiming that it was just "one among many truth games" and could not obtain objective knowledge because there *was* no objective reality, just a welter of cultural, ethnic, or gendered ways of *experiencing* reality. From this perspective what are called facts are but intellectual constructions, and to suggest that science and liberalism benefit each other is to indulge in "the naive and self-serving, or alternatively arrogant and conceited, belief that science flourishes best, or even can only really flourish, in a Western-style liberal democracy."

These recondite theories went by a variety of names—deconstructionism, multiculturalism, science studies, cultural studies, etc.—for which this book employs the umbrella term *postmodernism*. They became so popular that generations of educators came to believe, and continue to teach their students today, that science is culturally conditioned and politically suspect—the oppressive tool of white Western males, in one formulation.

Teachers—some of them—loved postmodernism because, having dismissed the likes of Newton and Darwin as propagandists with feet of clay, they no longer had to feel inferior to them but became their ruthless judges. Students—many of them—loved postmodernism because it freed them from the burden of actually having to learn any science. It sufficed to declare a politically acceptable thesis—say, that a given thinker was "logocentric" (a fascist epithet aimed at those who employ logic)—then lard the paper with knowing references to Marx, Martin Heidegger, Jacques Derrida, and other heroes of the French left. The process was so easy that computers could do it, and did: Online "postmodernism generators" cranked out such papers on order, complete with footnotes.

It seemed harmless enough at first. The postmodernists were viewed as standing up for women and minorities, and if they wanted the rest of us to prune our language a bit to make it politically correct, that seemed fair enough at a time when many white males were still calling women "girls" and black men "boys." Not until postmodernists began gaining control of university humanities departments and denying tenure to dissenting colleagues did the wider academic community inquire into what the movement was and where it had come from. What they found were roots in the same totalitarian impulses against which the liberal democracies had so recently contended.

An early sally in the postmodernist campaign came in 1931, when a communist physicist named Boris Hessen prepared a paper for the Second International Congress of the History of Science, in London, interpreting Newton's *Principia* as a response to the class struggles and capitalistic economic imperatives of seventeenth-century England. In it, Hessen deployed three tools that became indispensable to his radical heirs. First, the "great man" under consideration—in this case Newton—is reduced to an exemplification of the cultural and political currents in which he lived and on which he is depicted as bobbing along like a wood chip on a tide: "Newton was a child of his class." Second, any facts the great man may have discovered are declared to be not facts at all but "useful fictions" that gained status because they promoted the interests of a prevailing social class: "The ideas of the ruling class . . . are the ruling ideas, and the ruling class distinguishes its ideas by putting them forward as eternal truths." Finally, the great man's lamentable errors and hypocrisies, and those of the society to which he belonged, are brought to light through the application of Marxist-

Leninist or some other form of illiberal analysis aimed at reclaiming science for "the people": "Only in socialist society will science become the genuine possession of all mankind."

Ordinarily a clear writer, Hessen couched his London paper in obfuscatory jargon, declaring it his central thesis to demonstrate "the complete coincidence of the physical thematic of the period which arose out of the needs of economics and technique with the main contents of the *Principia*." Put into plain English, this says that the economic and technological currents of Newton's day *completely* dictated the contents of the *Principia*—that the noblemen and tradesmen of sixteenth-century England got what they wanted, which for some reason was a mathematically precise statement of the law of gravitation, written in Latin. Hence the need for obscurantist rhetoric: Had Hessen's thesis been clearly stated, it would have read like a joke.

Which, in grim reality, it appears to have been. Loren R. Graham, a historian of science at MIT and Harvard who has written widely on Soviet science, looked into the Hessen case in the nineteen eighties. He was puzzled by the fact that this dedicated communist, whose London paper was a model of Marxist analysis, was soon thereafter arrested, dying in a Soviet prison not long after his fortieth birthday. Why, Graham wondered, had this happened?

What he found was disturbing. Hessen's communist credentials were indeed impressive. He had been a soldier in the Red Army, an instructor of Red Army troops, and a student at the Institute of Red Professors in Moscow. His scientific credentials were equally solid. A gifted mathematician, he studied physics at the University of Edinburgh and then at Petrograd before becoming a professor of physics at Moscow University. And yet, Graham found, by the time Hessen "went to London in the summer of 1931 he was in deep political trouble."

The cause of the trouble was science. As a physicist, Hessen taught quantum mechanics and relativity, but neither of these hot new disciplines comported with Marxist ideology. Marxism is strictly deterministic—full of "iron laws" of this and that—whereas quantum mechanics incorporates the uncertainty principle and makes predictions based on statistical probabilities. Relativity is deterministic but was the work of Einstein, who came from a bourgeois background and, even worse, had taken to writing popular essays about religion that showed him to be something of a deist rather than a politically correct atheist. Had Hessen been a cynic he might have faked his way through the conflict, but being an honest communist he stood his ground. He implored his comrades to criticize the social context of relativity and quantum mechanics if they liked, and to reject the personal philosophies of Einstein and other scientists as they saw fit, but not to banish their science—because the validity of a scientific theory has little or nothing to do with the social or psychological circumstances from which it arises. In a way his position resembled that of Galileo, a believing Christian who sought to save his church from hitching its wagon to an obsolete cosmology. As a faithful Marxist, Hessen tried to dissuade the party from continuing to oppose relativity and quantum mechanics, since to do so would blind students to the brightest lights in modern physics.

For his trouble, Hessen was denounced by Soviet authorities as a "right deviationist," meaning a member of the bourgeoisie who fails to identify with the proletariat (his father was a banker), and as an "idealist" who had strayed from Marxist-Leninist determinism—which Marx and Lenin had learned from popular books about nineteenth-century science and then enshrined in the killing jar of their ideology. To keep an eye on him, the party had Hessen chaperoned in London by a Stalinist operative, Arnost Kooman, who just three months earlier had published an article declaring that

> Comrade Hessen is making some progress, although with great difficulty, toward correcting the enormous errors which he, together with other members of our scientific leadership, have committed. Nonetheless, he still has not been able to pose the issue in the correct fashion, in line with the Party's policy.

Hessen evidently hoped that his London paper would awaken the party to its folly by demonstrating that a social-constructivist analysis could be made *even of Newton*. Marxists accepted the validity of Newtonian physics, even though Newton lived in a capitalist England. So, therefore, should relativity and quantum mechanics be acceptable, regardless of their alleged political infelicities. Writes Graham:

> The overwhelming impression I gain from the London paper is that Hessen had decided "to do a Marxist job" on Newton in terms

relating physics to economic trends, while imbedding in the paper a separate, more subtle message about the relationship of science to ideology. He must have realized that by interpreting Newton in elementary Marxist economic terms, he could accomplish two important goals: first of all, he could demonstrate his Marxist orthodoxy, something being seriously questioned by his radical critics back in the Soviet Union; second, he could, by implication, defend science against ideological perversion by pointing to the need to separate the great merit of Newton's accomplishments in physics from both the economic order in which they arose and the philosophical and religious conclusions which Newton and many other people drew from them.

If so, Hessen's talk represented a scientist's last desperate effort to preserve his life and his Marxist faith by exposing the absurdity of the party line. It failed, and Hessen was crushed. Yet his paper lived on, through decades of heedless scholarship on the part of radical academics who neither got the joke nor much cared about the fate of its author. Thousands of academic papers were published supporting precisely the points that Hessen had lampooned—that science is socially conditioned and *ought to be* subordinated to state control. Many teachers and students today continue to believe that scientific findings are politically contaminated and that there is no objective reality against which to measure their accuracy. In the postmodernist view any text, even a physics paper, means whatever the reader thinks it means. What matters is to be politically correct. The term comes from Mao's Little Red Book.

The postmodern assault on science involved two main campaigns. One was to undermine language—to "deconstruct" texts—by claiming that what a scientist or anybody else writes is really about the author's (and the reader's) social and political context. Central figures in this effort included the German philosopher Martin Heidegger, the French philosopher Jacques Derrida, and the Belgian literary critic Paul de Man. The other campaign took on science more directly, as a culturally conditioned myth unworthy of respect. Here an influential figure was the Austrian philosopher Paul Feyerabend, who drew on the relatively liberal works of Thomas Kuhn and Karl Popper.

Derrida got the term "deconstruction" from Heidegger (who got it from a Nazi journal edited by Hermann Göring's cousin, and used it to advocate the dismantling of ontology, the study of the nature of being) but found it difficult to define: "A critique of what I do is indeed impossible," Derrida declared, thus rendering his work immune to criticism. In essence, deconstructionism demands that the knowing reader tease out the meanings of texts by discerning the hidden social currents behind their words, emerging with such revelations as that Milton was a sexist, Jefferson a slave driver, and Newton a capitalist toady.

The doctrines of Heidegger and Derrida were imported into America by de Man, who as a faculty member at Yale became, in the estimation of the literary critic Frank Kermode, "the most celebrated member of the world's most celebrated literature school." His was the appealing story of an impoverished intellectual who had fought for the Resistance during the war but was too modest to say much about it (except that he had "come from the left and from the happy days of the *Front populaire*"), was discovered by the novelist Mary McCarthy and taken up by New York intellectuals while working as a clerk in a Grand Central Station bookstore, and went on to become one of America's most celebrated professors, demolishing stale shibboleths with proletarian frankness. "In a profession full of fakeness, he was real," declared Barbara Johnson, famous for having celebrated deconstructionism in a ripe sample of its own alogical style:

> Instead of a simple "either/or" structure, deconstruction attempts to elaborate a discourse that says neither "either/or," nor "both/and" nor even "neither/nor," while at the same time not totally abandoning these logics either. The very word *deconstruction* is meant to undermine the either/or logic of the opposition "construction/destruction." Deconstruction is both, it is neither, and it reveals the way in which both construction and destruction are themselves not what they appear to be.

However, de Man was a fake. Soon after his death in 1983, a young Belgian devotee of deconstruction discovered that de Man, rather than working with the Resistance and coming "from the left," had been a Nazi collaborationist who wrote anti-Semitic articles praising "the Hitlerian soul" for the pro-Nazi journal *Le Soir.* Moreover he was a swindler, a bigamist, and a liar, who had ruined his father financially then fled to America, promising to send for his wife and three sons when he found work but choosing instead to marry one of his Bard College students.

When these discomfiting facts emerged, Derrida came to his friend de Man's defense but in doing so inadvertently called attention to deconstructionism's moral abstruseness. "The concept of making a charge itself belongs to the structure of phallogocentrism," Derrida wrote dismissively. Nor was he alone in flying to the defense of de Man—who had presciently exonerated himself in advance, writing, in a 1979 study of Rousseau, "It is always possible to excuse any guilt, because the experience always exists simultaneously as fictional discourse and as empirical event and it is never possible to decide which one of the two possibilities is the right one." Years later, radical academics were still teaching de Man as if he were something other than a liar, a cheat, and a Nazi propagandist. As Alan Sokal and Jean Bricmont remind us, in their book *Fashionable Nonsense,* "The story of the emperor's new clothes ends as follows: 'And the chamberlains went on carrying the train that wasn't there.' "

The foundations of radical academic thought were further undermined when disquieting evidence came to light concerning Heidegger, the philosopher whose work stood at the headwaters of Derrida's deconstructionism, Jean-Paul Sartre's existentialism, and the structuralism and poststructuralism of Claude Levi-Strauss and Michel Foucault. Regarded in such circles as the greatest philosopher since Hegel, Heidegger celebrated irrationalism, claiming that "reason, glorified for centuries, is *the most stiff-necked adversary of thought.*" He flourished in wartime Germany, being named rector of Freiburg University just three months after Hitler came to power. The postmodernist party line was that Heidegger might have flirted with Nazism but had been a staunch defender of academic freedom. In 1987, a former student of Heidegger's found while searching the war records that Heidegger had joined the Nazi party in 1933 and remained a dues-paying member until 1945—signing his letters with a "Heil Hitler!" salute, declaring that *der Führer* "is the German reality of today, and of the future, and of its law," and celebrating what he called the "inner truth and greatness of National Socialism."

As might be expected of a dedicated Nazi holding a prominent university position, Heidegger worked enthusiastically to exclude Jews from academic life. He saw to it that the man who had brought him to Freiburg University—his former teacher Edmund Husserl, the founder of phenomenology—was barred from using the university library because he was Jewish. He broke off his friendship with the philosopher Karl Jaspers, whose wife was Jewish; secretly denounced his Freiburg colleague Hermann Staudinger, a founder of polymer chemistry who would win the Nobel Prize in 1953, after learning that Staudinger was helping Jewish academics hang on to their jobs; and scotched the career of his former student Max Müller by informing the authorities that Müller was "unfavorably disposed" to Nazism. Heidegger lied about all this after the war, testifying to a de-Nazification committee in 1945 that he had "demonstrated publicly my attitude toward the Party by not participating in its gatherings, by not wearing its regalia, and, as of 1934, by refusing to begin my courses and lectures with the so-called German greeting [Heil Hitler!]"—when in fact that is exactly how he started his lectures, for as long as Hitler remained alive. In a similarly revisionist vein, Heidegger rewrote his 1935 paean to the "inner truth and greatness of National Socialism," making it appear to have been an objection to scientific technology—an alteration that he repeatedly thereafter denied having made.

Particularly distressing, considering the subsequent spread of Heidegger's fame through the American academic community, was his attitude toward the relationship of universities to the political authorities. Upon becoming rector of Freiburg, Heidegger declared that universities "must be . . . joined together with the state," assuring his colleagues that "danger comes not from work for the State"—an astounding thing to say about the most dangerous state ever to have arisen in Europe—and urging students to steel themselves for a "battle" which "will be fought out of the strengths of the new Reich that Chancellor Hitler will bring to reality." Suiting his actions to his words, Heidegger changed university rules so that its rector was no longer elected by the faculty but rather appointed by the Nazi Minister of Education—the same Dr. Rust who in 1933 decreed that all students and teachers must greet one another with the Nazi salute. Heidegger was rewarded by being appointed *Führer* of Freiburg University, a step toward what seems to have been his goal of becoming an Aristotle to Hitler's Alexander the Great.

These and other revelations about the fascist roots of the academic left, although puzzling if analyzed in terms of a traditional left-right political spectrum, make better sense if the triangular diagram of socialism, conservatism, and liberalism is expanded to form a diamond:

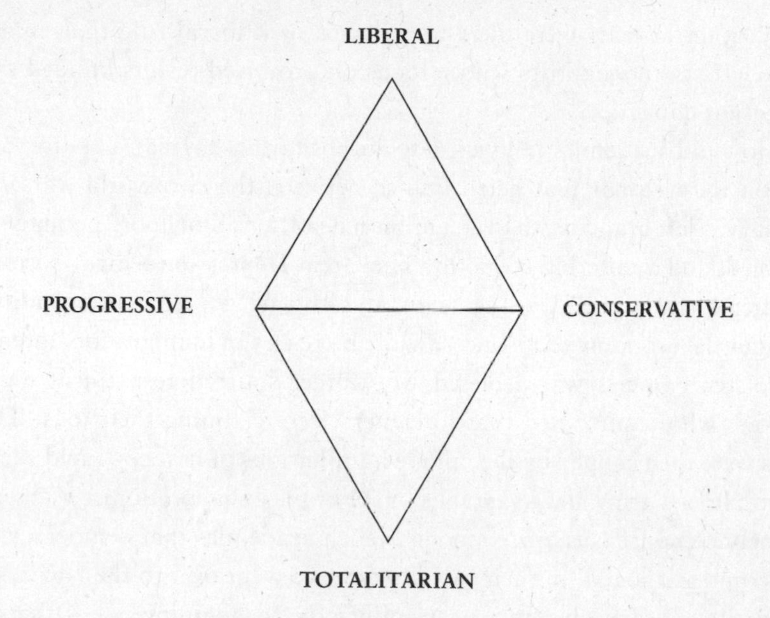

Such a perspective reflects the fact that liberalism and totalitarianism are opposites, and have an approximately equal potential to attract progressives and conservatives alike. (Try guessing who, for instance, said the following: "Science is a social phenomenon . . . limited by the usefulness or harm it causes. With the slogan of objective science the professoriate only wanted to free itself from the necessary supervision of the state." Lenin? Stalin? It was Hitler, in 1933.) The diagram also suggests why American liberals did at least as much as conservatives to expose the liabilities of communism, even when the USSR and the United States were wartime allies. In 1943, two liberal educators, John Childs and George Counts, cautioned their colleagues that communism "adds not one ounce of strength to any liberal, democratic or humane cause; on the contrary, it weakens, degrades or destroys every cause that it touches." The following year, the liberal political scientists Evron Kirkpatrick and Herbert McClosky sought to correct procommunist sympathies among leftist academics by pointing out similarities between the Nazi and communist regimes—e.g., that both prohibited free elections, freedom of speech, and freedom of the press; were dominated by a single political party whose views were broadcast by an official propaganda network and enforced by state police; and were dedicated to expansion of their power by force. Because liberalism is diametrically opposed to totalitarian rule, these and many other liberals were alert to fallacies of illiberal rule that escaped those leftists and rightists whose focus on promised results blinded them to present dangers.

How did academics fall victim to authoritarian dogma? The story centers on the traumas that befell France between the two world wars. The Great War left France with over a million dead, a million more permanently disabled, and a faltering economy. The Great Depression—made worse by Premier Pierre Laval's raising taxes and cutting government spending—brought the economy to its knees, while increases in immigration, intended to bolster France's war-depleted workforce, spurred resentment among workers who complained that foreigners were claiming their jobs. These woes were then capped by the unique humiliations of France's rapid surrender to Hitler's army and its establishment of the collaborationist Vichy government. A consensus arose among French academics that democracy was bankrupt and socialism their salvation. Some went over to the Nazis, who had the temporary advantage of seeming to be the winning side. Others adhered to communism, which became the winning side on the Eastern Front and had gained respectability by virtue of the fact that many members of the French Resistance were communists. The resulting clash of right-wing and left-wing French intellectuals drove political dialogue toward illiberal intemperance. As the social scientist Charles A. Micaud observes:

> A counter-revolutionary extreme right and a revolutionary extreme left, pitted against each other, emphasized the respective values of authority and equality at the expense of liberty. . . . Each coalition was kept together, not by positive agreement on a program, but by the fear inspired by the other extreme.

The result was what Micaud calls "a strong disloyal opposition" characterized by a contempt for liberalism and a romantic attachment to revolution. "Left and right alike felt a distaste for the lukewarm and were fascinated by the idea of a violent relief from mediocrity," writes the historian Tony Judt. The faith of French intellectuals in the power of revolution took on the trappings of a religion. Micaud:

> This psychological interpretation of the need for revolution explains reactions that would be incomprehensible at a more rational

> level—the impossibility of expecting reasonable and calm answers to certain pertinent questions concerning capitalism or socialism, reformism or communism; the impossibility of penetrating the charmed circle of a priori assumptions, based on faith, with arguments of fact and logic, the "moral and intellectual absolutism" characteristic of many left-wing intellectuals; a curious absence of any realistic analysis of economic and social problems; an extreme symbolism that abstracts reality to the point of meaninglessness or to what would seem, at times, intellectual dishonesty.

In terms of logic, postmodernism maintained that inasmuch as science involves induction, and induction is imperfect, science is imperfect. But since everybody already knew *that*—scientists claimed perfection for neither their results nor their methods—it was necessary to portray scientific research as not only imperfect but somehow subjective. This involved making much of there being no single, monolithic scientific method. Theoretical physicists work rather differently from molecular biologists, and both work differently from the anthropologists and sociologists upon whose examples the postmodernists tended to rely. (Many radical critiques of science acquire a certain coherence if read as describing *social* scientists, dispatched from the first world to study the third.) Hence the postmodernist argument took on the following odd, and oddly popular, form: There is no single scientific method, replicated in every detail in every science, and *therefore*—here comes quite a leap—scientific results are no more trustworthy than those obtained through any other procedure. Such arguments often began by trying to celebrate the prescientific beliefs of indigenous peoples, but soon ripened into attacks on science and logic so sweeping as to destroy the grounds on which the postmodernists themselves could critique any system of thought whatsoever.

The foundations of this endeavor were laid, to some extent unwittingly, by the Viennese philosopher Karl Popper and the American historian of science Thomas Kuhn.

Popper embraced Marxism as a youth but later evolved to a more liberal stance, eventually expressing dismay that liberal democracy was being "so often betrayed by so many of the intellectual leaders of mankind." His studies of science wrestled with the "problem of induction." Briefly put, this is the problem that if you observe a thousand white swans and then publish a paper declaring that all swans are white, you will be proven wrong should a black swan turn up. David Hume identified the problem of induction in 1739. His point was that if all knowledge is based on experience—as the empiricists maintain—then all conclusions, from the colors of swans to the laws of thermodynamics, are vulnerable to disproof.

It cannot be said that physicists lose much sleep over the problem of induction. They neither require, nor have they often resorted to, the claim that their findings are *absolutely* determined. Even the laws of science—the highest category of scientific near-certitude—are to some degree provisional. If you open a bottle of perfume and leave it in a sealed room, the perfume will eventually evaporate. Once that has happened and the perfume molecules are adrift throughout the room, they will never get back into the bottle unless a lot of work is exerted to bring about that result. This situation is expressed quantitatively in the laws of thermodynamics, which are rooted in calculations of probability. What those calculations say is not that it is *impossible* for the molecules to return to the bottle of their own accord, but that it is highly unlikely—so unlikely that if the entire observable universe consisted of nothing but open perfume bottles sitting in sealed rooms, not one bottle anywhere would yet have refilled itself with perfume. That's certain enough to satisfy most physicists.

In the social sciences, however, the limitations of induction come up more often, and have to do less with black and white swans than with shades of gray. Political opinion polling, for instance, is an empirical, highly competitive business—if your company fails to make reliable predictions of election results, it will lose clients to competing firms that do a better job—so pollsters pay a lot of attention to improving their sampling techniques. They don't need to be told that induction is less than perfect; they swim against the currents of inductive uncertainty all the time. But to claim that these limitations *invalidate* science is like saying that the sun is not spherical because it is not *perfectly* spherical. Science studies nature, and nothing in nature is ever a perfect anything.

Philosophers, however, tend to see things differently; in a sense, that's their job. Popper came up with the salient point that scientific hypotheses may be distinguished from other sorts of pronouncements by *virtue* of their being vulnerable to disproof. The statement that "man is born unto trouble, as the sparks fly upward" (Job 5:7) is not scientific, since it is too vague to be disproved, whereas the statement that "the charge of the proton is $1.602176487 \times 10^{-19}$ Coulombs to within an error bar *e*" is scientific, since it can be checked experimentally and might turn out to be wrong (although that would be a very black swan indeed). Popper's intention was mainly to debunk "pseudoscientific theories like Marx's [and] Freud's," but the postmodernists made a bonfire of its little spark, arguing that if scientific findings are falsifiable then they are no better than the findings of any other endeavor—about which more in a moment. But first, Thomas Kuhn.

Kuhn studied physics at Harvard before becoming a historian of science. He was fascinated by revolutions. His first book, *The Copernican Revolution*, examined how Copernicus had shown that a sun-centered model of the solar system could fit the observational data as well as did the earth-centered cosmology of Claudius Ptolemy that had preceded it. His second, *The Structure of Scientific Revolutions*, extended the concept to science more generally. In it, Kuhn stressed the provisional nature of scientific knowledge—as scientists themselves often do, referring for instance to Darwin's *theory* of evolution. Kuhn depicted the history of science as consisting of long, relatively docile periods of normalcy punctuated by spasms of revolutionary change. Kuhn called such changes "paradigm shifts." He defined paradigms as developments "sufficiently unprecedented to attract an enduring group of adherents away from competing modes of scientific activity [and] sufficiently open-ended to leave all sorts of problems for the redefined group of practitioners to resolve." As an example, he cited the discovery of the planet Uranus by the British astronomer William Herschel in 1781. On more than a dozen prior occasions, Kuhn pointed out, other astronomers had observed Uranus without recognizing that it was a planet. When Herschel spotted Uranus through his telescope, he noted that it looked like a disk rather than a starlike point. He then tracked it for several nights and found that it was slowly moving against the background stars. He thought that the object might be a comet, but further studies of its orbit revealed that rather than crossing the orbits of the planets, as most visible comets do, it remained outside the orbit of Saturn. Only at that point did Herschel conclude that it was indeed a planet. Writes Kuhn, "A celestial body that had been observed off and on for almost a century was seen differently after 1781 because . . . it could no longer be fitted to the perceptual categories (star or comet) provided by the paradigm that had previously prevailed." This interpretation is very forced. There was little or no such thing as a "star or comet" paradigm against which Herschel had to battle. What happened was that Herschel, an exceptionally sharp-eyed observer using a finely honed telescope that he had constructed with his own hands, was able to discern the disk of Uranus, as other astronomers who lacked his skill and his optics had not. (Viewed through inferior optics, the stars *all* look like blobs rather than points.) Having made that observation, Herschel checked to see if the object was moving against the stars, as a planet or comet would do. When he found that it was moving, he hypothesized that it was a comet. He had at least two good reasons for doing so: Comets were being discovered all the time, whereas nobody in all recorded history had yet discovered a planet; and to have claimed that it *was* a planet would have also been to insist that he, Herschel, should be accorded worldwide fame—not the sort of thing that a modest, responsible scientist would want to assert on his own behalf without stronger evidence. When, in the normal course of scientific investigation, the object's orbit revealed it to be a planet, it was so designated. Herschel's discovery was important, but few informed historians aside from Kuhn have found it useful to claim that it involved some sort of shift in the scientific gestalt or that it opened scientists' eyes to phenomena about which they had previously been blind.

Consider the subsequent discovery of Neptune, the next planet out from Uranus, in 1846. Neptune's existence was inferred from anomalies in the orbit of Uranus. Consulting these calculations, astronomers in Germany and England looked for a new planet in the particular part of the sky to which the mathematicians had pointed them. One such observer, James Challis in Cambridge, actually spotted Neptune through his telescope and noted that it seemed to have a disk—but he failed to examine it at a higher magnification or to follow up on his observation, leaving the discovery solely to the Germans. Evidently Challis hadn't received the Kuhnian memo: A half century after a paradigm shift was supposed to have changed the way astronomers looked at the sky, they were still apt to fall into the same lapses that, for Kuhn, ought by then to have become obsolete.

Like so many of the postmodernist "science studies" that followed in its wake, Kuhn's thesis was accurate only insofar as it was trivial. It may be useful to speak in terms of scientific revolutions, but they are rare and getting rarer (there has not been one that satisfies Kuhn's criteria in over a century), they do *not* result in the discrediting of all prior knowledge in the field (the Apollo astronauts navigated to the moon using Newtonian dynamics, having minimal recourse to Einstein's relativity), and it is usually extraneous to invoke new paradigms to explain why scientists thereafter are able to discern phenomena they had previously ignored. The physicist Paul Dirac in 1928 derived a theoretical equation for the electron which accurately described how electrons couple to electromagnetic fields. The Dirac equation also implied the existence of an unknown, oppositely charged particle, the positron, that was soon thereafter discovered. Dirac's name would be more widely known today had he predicted that positrons existed, but he downplayed that aspect of his own equation because he was unsure whether it was real or just a mathematical artifact. (Asked years later about his reticence, Dirac attributed it to "pure cowardice.") It is unclear how describing this as a paradigm shift adds to Dirac's own characterization of it, and anyway the facts differ substantially from Kuhn's prescription. Carl Anderson in 1932 discovered positrons *accidentally*, not because Dirac had predicted their existence, so once again the absence of a new "paradigm" failed to retard scientific progress.

Notwithstanding its severe limitations, Kuhn's thesis—which he immodestly described as "a historiographic revolution in the study of science"—was popular enough to make *The Structure of Scientific Revolutions* the most cited book of the twentieth century; even *Science* magazine called it "a landmark in intellectual history." Students intimidated by science were relieved to learn, as the historian Gerald Holton summarized their impression of Kuhn's argument, that "science provides truths, but now and then everything previously known turns out to have been entirely wrong, and a revolution is needed to establish the real truth." (Why bother to learn science, if it's all soon to change anyway?) Cultural critics were delighted to be reassured that sociological interactions among scientists had so great an influence; if scientific findings are but social constructions, then perhaps they are no better than any other. Through that imaginary breach flooded such oddities as "feminist algebra," "nonwestern science," and the assertion that it was parochial to prefer particle physics to voodoo. Kuhn's *Structure* became the indispensable starting point for academics out to diminish science without learning any.

The radicals wove Kuhn and Popper into an academic version of bipolar disorder: Everything real was depressing, everything desirable imaginary. Real science needed to be replaced by a "particularist, self-aware" science that would yield different answers depending on who—a woman, say, or a Sufi, or an Aborigine—was asking the questions. As the science critics could formulate no such particularized sciences, they fell back on celebrating their *efforts* to do so, filling their papers with self-congratulatory descriptions of how hard they were working to shuffle ideas around and create new projects. As the historian Robert Conquest observes, postmodernist theories "are complex, and deploying them is often a complicated process, thus giving the illusion that it is a useful one."

The Austrian philosopher Paul Feyerabend, inspired by Popper, crafted a critique of science that was for decades regarded as significant. A colorful figure—he was a German army veteran who walked with a limp, having been shot in the spine during a quixotic attempt to direct traffic in a combat zone on the Russian front, the wound rendering him permanently impotent although he went on to be married four times—Feyerabend was studying in postwar London when he happened to attend a lecture by Popper. "I am a professor of scientific method," Popper began, "but I have a problem: There is no scientific method." Popper was being hyperbolic, but Feyerabend inferred that having no one method meant having no method whatever. "Successful research does not obey general standards; it relies now on one trick, now on another," Feyerabend wrote, adding, in breathless italics, "*The events, procedures and results that constitute the sciences have no common structure. . . . Scientific successes cannot be explained in a simple way.*" But this was just an elaborate way of saying that scientific creativity, like all creativity, is somewhat mysterious.

Like many philosophers lacking firsthand experience in scientific research, Feyerabend thought of science as a species of philosophy, centered on logic and only incidentally entangled in the actual workings of the material world. (He boasts in his autobiography of having set a roomful of astronomers squirming by lecturing them on cosmology "without mentioning a single fact.") Starting from this flawed premise, he imagined that science needed to be freed "from the fetters of a dogmatic logic," and that this meant deconstructing scientific language. "Without a constant misuse of language there cannot be any discovery, any progress," he claimed, sounding rather like Lenin. "We can of course imagine a world where . . . scientific terms have finally been nailed down [but] in such a world only miracles or revelation could reform our cosmology." That is demonstrably false. Scientific terms get nailed down all the time, yet this has not inhibited the progress of cosmology. For example, astronomers studying the expansion of the universe discovered in 1997 that, to their astonishment, the cosmic expansion rate is accelerating. The enormous force responsible for the acceleration was subsequently, and rather unhelpfully, named "dark energy," but that was after the fact. The discovery that cosmic expansion is speeding up was *not* made by changing prior scientific terminology, but by employing existing terminology, tools, and concepts, a feat that Feyerabend claimed was impossible.

What Feyerabend was really attempting was to discredit Enlightenment values, which the radicals regarded as a mask for power politics. "I recommend to put science in its place as an interesting but by no means exclusive form of knowledge that has many advantages but also many drawbacks," he wrote.

> There is not one common sense, there are many. . . . Nor is there one way of knowing, science; there are many such ways, and before they were ruined by Western civilization they were effective in the sense that they kept people alive and made their existence comprehensible.

This charge should be examined closely, given that it has been widely entertained among Westerners lamenting the sins of colonialism. Feyerabend states that the ways of knowing of indigenous peoples "kept people alive." But people are kept alive by traditional ways of knowing only to the degree that such ways are themselves scientifically valid. If a medicine man gives you an herb that actually cures your disease, the cure is scientifically defensible—and scientists have learned a lot about herbs from just such individuals—but if the herb is of no empirical value, the only thing that may keep you alive is the placebo effect of your faith in the treatment, which improves the health of something like a fifth to a third of patients in Western doctors' offices and thatched huts alike. The 2004 Indian Ocean tsunami killed over 200,000 persons, but spared those whose traditions had alerted them to head for higher ground when an earthquake struck. The tradition worked because it was empirically valid, having been inducted by ancestors who inferred a causal relationship between earthquakes and tsunamis. Those whose traditions asserted something empirically invalid—say, that an earthquake means we've offended the sea gods, so we should all wade into the ocean and pray—would have been following bad advice, even though it too was a "way of knowing." Feyerabend's other claim for indigenous knowledge, that it "made their existence comprehensible," is flatly patronizing. Should doctors deny malaria vaccines to indigenous peoples on grounds that the local mythology claims that malaria is contracted by working too hard in the fields, or is a punishment wrought by the gods? Granted, some field researchers and physicians may have given short shrift to the knowledge and wisdom of the peoples they studied and treated. But it is a wild exaggeration to write, as Feyerabend did, that "the 'progress of knowledge and civilization'—as the process of pushing Western ways and values into all corners of the globe is being called—destroyed these wonderful products of human ingenuity and compassion without a single glance in their direction."

Although he represented himself as a friend to freedom and creativity, Feyerabend supported repressive governmental controls on scientific research. He even imagined that such interventions would open up science to new modes of thinking—a process that he called "proliferation":

> It often happens that parts of science become hardened and intolerant so that proliferation must be enforced from the outside, and by political means. Of course, success cannot be guaranteed—see the Lysenko affair. But this does not remove the need for non-scientific controls on science.

His sole example of such a success was a 1954 campaign by the British Ministry of Health to stress "traditional medicine," such as acupuncture, over Western "bourgeois science." Feyerabend credited this Marxist-inspired project with "making a plurality [actually a duality] of views possible." By these lights, Hitler could have increased scientific open-mindedness and creativity by ordering German astronomers to validate his pet theory that the earth was hollow. Feyerabend's offhand dismissal of the Lysenko affair, which involved the persecution of talented scientists whose views deviated from the party line, was consistent with his generally opaque ethical views. These were his remarks on the Chinese Cultural Revolution, in which scientists and intellectuals were murdered and cultural artifacts wantonly destroyed:

> Scientists are not content with running their own playpens in accordance with what they regard as the rules of scientific method, they want to universalize these rules, they want them to become part of society at large and they use every means at their disposal—argument, propaganda, pressure tactics, intimidation, lobbying—to

> achieve their aims. The Chinese communists recognized the dangers inherent in this chauvinism and they proceeded to remove it. In the process they restored important parts of the intellectual and emotional heritage of the Chinese people and they also improved the practice of medicine. It would be of advantage if other governments followed suit.

When Feyerabend's book *Against Method* was first published, in 1975, such demonstrations of ignorance and inadvertent cruelty—a philosopher *recommending* that governments follow the communist example of dispatching gangs of thugs to torment scientists—occasioned what might politely be called negative reviews. Feyerabend protested that the critics weren't getting his jokes, and that this showed them to be close-minded: " 'New' points of view," he sniffed, ". . . are criticized because they lead to drastic structural changes of our knowledge and are therefore inaccessible to those whose understanding is tied to certain principles." In a similar vein he complained that his ideas were being embraced by New Age mystics who failed to consider that science does, for some reason, work.

But by then, academics on hundreds of campuses were routinely disparaging science. "Science is not a process of discovering the ultimate truths of nature, but a social construction that changes over time," wrote the "ecofeminist" Carolyn Merchant of the University of Wisconsin, while the sociologist Stanley Aronowitz, of the City University of New York, declared that "neither logic nor mathematics escapes the 'contamination' of the social." The more hyperbolic the claim, the more celebrated its creator. A particularly astounding hypothesis was forwarded by the sociologist Bruno Latour, who imagined that "since the settlement of a controversy is *the Cause* of Nature's representation not the consequence, we *can never use the outcome—Nature—to explain how and why a controversy has been settled.*" If so, science would amount to no more than a solipsistic game. But it is not so. Prior to the soft landing of space probes on the moon, geologists feared that lunar dust, accumulated over 4.5 billion years of asteroid impacts, might be so deep that the Apollo astronauts would simply sink into it and disappear. The question, debated at length, was settled once robotic probes landed on the moon and did *not* sink into deep dust; further confirmation came when Neil Armstrong stepped onto the lunar surface and found it firm. But according to Latour this outcome cannot possibly explain "how and why" the controversy came to an end, since the discovery—in this case, that lunar dust is typically only a few inches deep—was somehow a "consequence" of the controversy itself. If science ran by postmodernist rules it could scarcely function at all—which, essentially, was the radicals' intent. Science, like liberalism, is anathema to ideologues.

The culture wars came to a head, though not to an end, with the Sokal Hoax of 1996. Alan Sokal, a physicist at New York University and one of the few scientists who paid close attention to the postmodernists, wrote a deliberately meaningless paper—rather like the ones created automatically by the online postmodernism generators—and submitted it to the journal *Social Text*, which published it. The article began with an apt summary of the radical consensus regarding science:

> There are many natural scientists, and especially physicists, who continue to reject the notion that the disciplines concerned with social and cultural criticism can have anything to contribute, except perhaps peripherally, to their research. Still less are they receptive to the idea that the very foundations of their worldview must be revised or rebuilt in the light of such criticism. Rather, they cling to the dogma imposed by the long post-Enlightenment hegemony over the Western intellectual outlook, which can be summarized briefly as follows: that there exists an external world, whose properties are independent of any individual human being and indeed of humanity as a whole; that these properties are encoded in "eternal" physical laws; and that human beings can obtain reliable, albeit imperfect and tentative, knowledge of these laws by hewing to the "objective" procedures and epistemological strictures prescribed by the [so-called] scientific method.

What followed was five thousand words of utter gibberish, duly adorned with citations from Kuhn, Feyerabend—and Derrida, rattling on about a (nonexistent) "Einsteinian constant." Sokal studded the paper with clues that he was joking, writing, among other deliberate howlers, that "the π of Euclid and the *G* of Newton, formerly thought to be constant and universal, are now perceived in their ineluctable historicity"—which if you stop to think about it, as the editors of *Social Text* evidently did not, means that even such universal constants as the ratio between the diameter and circumference of a circle, and the rate at which planets orbit stars, depend upon the historical circumstances of their discovery.

"The conclusion is inescapable," wrote Paul A. Boghossian, a professor of philosophy at New York University, "that the editors of *Social Text* didn't know what many of the sentences in Sokal's essay actually meant; and that they just didn't care." What they cared about was ideology. In the end, the postmodernist brew of socialism and cynicism was literally perverse—in the etymological sense of the word, as meaning to look in the wrong direction. Rather than pursuing knowledge about science, radical academics assumed a posture that the mathematician Norman Levitt calls "*knowingness*, an attitude that gives itself permission to avoid the pain and difficulty of actually understanding science simply by declaring in advance that knowledge is futile or illusory."

Postmodernism's lasting legacy—aside from having misled millions of students and made a laughingstock of the academic left—was its call to "democratize" science. This sounds nice, but in practice usually means wielding the power of the state to restrict scientific research. Hence it was not surprising that the "democratization" of science gained favor among authoritarian political thinkers of many stripes, from British socialists to right-wing Indian theorists to the Bush White House. When the mayor of Cambridge, Massachusetts, sought in 1976 to ban recombinant DNA research at Harvard, he exclaimed, "God knows what's going to crawl out of the laboratory!" The university prevailed, and as the biologist Carl Feldbaum reported, a generation later:

> What eventually "crawled out of the laboratory" was a series of life-saving and life-enhancing medications and vaccines, beginning in 1982 with recombinant insulin and soon followed by human growth hormone, clotting factors for hemophiliacs, fertility drugs, erythropoietin, and dozens of other additions to the pharmacopeia.

When confronted with fear tactics based on Faustian science-fiction fantasies, a useful exercise is to ask yourself which facts, adduced by science in the past, you would prefer had never been learned. The list is unlikely to be long.

How did it come to pass that so many teachers and students, in some of the freest and most scientifically accomplished nations in the world, entertained such an illiberal, illogical, and politically repressive account of the relationship between science and society? Part of the answer may be that universities generally, and humanities departments in particular, are more backward than is universally recognized. For most of their history, universities functioned primarily as repositories of tradition. It was professors, not priests, who refused to look through Galileo's telescope, and who drove genuinely progressive students like Francis Bacon and John Locke to distraction with their endless logic-chopping and parsing of ancient texts. Similarly it was twentieth-century humanities professors who, confronted with the glories of modern science and the triumph of the liberal democracies over totalitarianism, responded by denigrating virtually every political philosophy *except* totalitarianism. Campus enthusiasm for authoritarian rule today is predominantly leftist (one gets an impression that professors and students who live in university housing and eat at the commons think that everybody else should also enjoy life in a socialist paradise), but historically it has often come from the right as well. In 1927, years before Hitler came to power, 77 percent of the German student organizations were sufficiently in tune with the Nazis to ban non-Aryans from joining their clubs; by 1931, university support for Hitler was twice that of the German population at large. The Nazis book-burnings of 1933—where the works of Einstein, Thomas Mann, and H. G. Wells went up in flames—were staged by students and professors out to shape up Germany's youth for the trials ahead. The villains of the piece for right-wing socialists were Jews and bankers; for today's leftists they are big corporations and globalization. "Globalization has produced a world economic system and trade laws that protect transnational corporations at the expense of human life, biodiversity, and the environment [and by] greater levels of unemployment, inequality, and insecurity," reads a paper selected at random from the postmodernist literature. No facts are presented to support this claim, nor need they be. The statement is an assertion of shared belief, equivalent to the ritual incantations uttered to reassure the faithful at the outset of a religious service.

To say all this is not to condemn academics generally. The proponents of antiscientific, illiberal academic fads were mostly lackluster scholars, many of whom (e.g., Derrida, de Man, and Feyerabend) traded on their outsider status, while their defeat was wrought principally by liberal academics—debunkers like Paul R. Gross and Norman Levitt, whose book

Higher Superstition critiqued the postmodernists; Frederick Crews of the University of California, Berkeley, English department, who lampooned them in his bestselling *Postmodern Pooh*; and Allan Bloom of the University of Chicago, whose *The Closing of the American Mind* sold a million copies. (Bloom was pilloried by leftist critics as politically conservative, although he was a liberal Jewish homosexual who wrote his book at the behest of his friend the novelist Saul Bellow, and who protested, about as plainly as possible, "I am not a conservative.")

It may be useful to summarize the campaigns of the radical academics in terms of their principal errors. First, they ignored science as a source of knowledge and instead depicted it as a source of power, a distortion which ultimately led them to discount the validity of all objectively verifiable knowledge. Second, they thought that since scientific research is a social activity, the knowledge it produces must be nothing more than a social construct; this was like claiming that if various teams of mountaineers climb the north face of Eiger via different routes, there must be no *real* Eiger at all. Finally, they turned their back on learning. This may seem an odd thing to say of college professors—but academics, like most people, tend to form their political beliefs early on and then cease examining them, settling instead into blinkered judgments about how to realize their presumptively superior goals. The result is an academic shell game that was already well under way when Galileo, centuries ago, noted that it was the habit of many professors "to make themselves slaves willingly; to accept decrees as inviolable; to place themselves under obligation and to call themselves persuaded and convinced by arguments that are so 'powerful' and 'clearly conclusive' that they themselves cannot tell the purpose for which they were written, or what conclusion they serve to prove!"

Perhaps the most sympathetic thing to be said of the postmodernists and their sympathizers is that in witnessing the environmental degradation and decline of indigenous cultures that attended the postwar rise in global technology and trade, they feared that the wonderful diversity of our world was shrinking into something akin to the "single vision" of William Blake's Newton. Such concerns are not baseless, but the *worst* way to address them is by championing subjective ways of knowing over objective scientific knowledge. "Epochs which are regressive, and in the process of dissolution, are always subjective," noted Goethe, "whereas the trend in all progressive epochs is objective. . . . Every truly excellent endeavor turns from within toward the world [and is] objective in nature." A recent image transmitted from Mars showing a tiny blue crescent Earth adrift in the blackness of space reinforces the sentiment declared a half century ago by the eminent bluesman Huddie Ledbetter, known as Leadbelly:

> We're in the same boat, brother
> We're in the same boat, brother
> And if you shake one end
> You're going to rock the other
> It's the same boat, brother.

CHAPTER TWELVE

ONE WORLD

The world exists, as I understand it, to teach the science of liberty.

—RALPH WALDO EMERSON

Science and democracy are based on the rejection of dogmatism.

—DICK TAVERNE

One can live by dogma or by discovery. Dogmas (from the Greek for received opinions that "seem good") may seek to unify people (as is the implied intent of religious dogma, *religio* being Latin for "binding together") but insofar as a dogma must be taken on faith it winds up bifurcating humanity into a faithful *us* and a suspect *other*. Scientific discovery *might* have divided the world, but instead has found that all human beings are kin—to one another and to all other living things—in a universe where stars and starfish alike obey the same physical laws. So as humans move from dogma toward discovery, we increasingly find ourselves inhabiting one world.

This development raises the prospect that as the influence of science grows, people may overcome old prejudices and parochialisms and treat one another more liberally. To an extent this is already happening—the world today is more scientific and more liberal, better informed and less violent, than it was three centuries ago—but with such prospects have also come problems. Religious and political dogmatists react against science and liberalism with everything from denial and attempted suppression (of, for instance, the teaching of biological evolution) to terrorism. The liberal democracies have too often responded to such threats with insecurity rather than strength, reverting in times of trouble to illiberal practices little better than those of their adversaries. Meanwhile scientific findings challenge everybody's received opinions, while the growth of technology creates conundrums—with global warming currently at the top of the heap—that unless competently addressed threaten to reverse much of the progress our species has so recently made.

Dogmatists like to portray science as just another dogma—to the brazen all is brass—but science is a method, not a faith, and the unity of the universe was discovered by scientists who set out to demonstrate no such thing. When Newton identified the laws of gravitation he did not assert that they held sway everywhere, but wondered whether "God is able . . . to vary the laws of nature . . . in several parts of the universe." The physicist Ernest Rutherford, whose experiments exposed the structure of the atom, was so skeptical about drawing grand implications that he threatened to bar from his laboratory any scientist who so much as uttered the word "universe." When the astronomer Edwin Hubble established that the Milky Way was one among many galaxies, he called them "island universes" and questioned whether "the principle of the uniformity of nature" pertained across such enormous distances. This is the opposite of starting with a deeply held faith and accumulating evidence to support it. Scientists have a story of discovery to tell, dogmatists a story of obedience to authority.

The scientific discovery that everything—and everybody—is interwoven with everything else was a boon for liberalism, which took a unified view of humanity before such a stance could be justified empirically. When John Locke argued for human equality under the law, women were still considered unfit to vote and Europeans thought of their black and brown brothers as the benighted descendants of the biblically accursed Ham. The liberal claim that people ought to have equal rights was a theory, vulnerable to test by experiment and properly to be judged by the results. The experiment having since succeeded, while science determined that all human beings belong to the same species, we can now understand that we're all *us*; there is no *other*.

Darwin's discovery that biological evolution functions through random mutation and natural selection revealed the common ancestry of all human beings, but it did so at the cost of exposing the unsettling fact that we are here by virtue of chance. Genes mutate randomly, DNA/RNA copying errors altering the genetic inheritance of every organism. Changes in the environment—which are themselves random, to a first approximation—can alter circumstances in such a way that previously marginal mutants are better able to survive and reproduce than are those superlatively adapted to the prior order. The environmental changes involved may be as slow as the parting of continents or as sudden as an asteroid impact, but they never cease: Stasis is an illusion. Homo sapiens did not emerge because they were superior to other animals, but because their ancestors happened to be in the right place at the right time. This rather stark finding is difficult for humans to absorb; hence we are apt to regard ourselves as distinctly different from the other animals, and to imagine that we are here for a special purpose. To entertain this illusion is to approach biology the wrong way round.

The cognitive scientist Daniel Dennett, in his book *Darwin's Dangerous Idea*, offers a thought experiment to illuminate just how unaccustomed human beings are to thinking in Darwinian terms. In his scenario, you meet a gambler who claims that he can produce a man who, in your presence, will win ten consecutive coin tosses. You take the bet, knowing that the odds against anyone's winning ten straight coin tosses are 1,024 to 1. The gambler shows up in the morning accompanied by 1,024 men, who proceed to toss coins. At the end of the first round, half the men have lost. The same happens on the ensuing rounds, until only two men remain for the tenth and final round, the winner of which has indeed won ten straight coin tosses. He wasn't personally destined to win, any more than any given tennis player is destined to win next year's Wimbledon singles title, but somebody had to win, and it just happened to be him. "If the winner of the [coin-tossing] tournament thinks there has to be an explanation of why *he* won, he is mistaken," Dennett notes. "There is no reason at all why *he* won, there is only a very good reason why *somebody* won."

So to wonder, "Why am I here?" is to ask the wrong question. Nothing requires that you or I exist, or that the human species exists; it's just that so long as there is life on Earth *some* creatures will exist, and you and I happen, at present, to be among them. I may imagine that my existence is magically full of hidden meanings—just as amateur gamblers think they perceive patterns in the wholly random behavior of roulette wheels—but the silent majority of species that once thrived and are now gone would take a decidedly different view of the matter, were they around to be interviewed about it. Evolution reveals that humans got here the way everything else got here, through a long historical process of accident and selection.

This discovery offends secular as well as religious dogmatists.

Leftists have opposed evolution out of fear that genetics might reveal human beings to be other than John Locke's blank slates, whose faults were entirely attributable to their social milieu and who could, therefore, be redeemed entirely through political reforms. Many believed, as had Rousseau, that "Man is naturally good, and only by institutions is he made bad." Progressive constructs like behavioral psychology and the Standard Social Science Model ("Instincts do not create customs; customs create instincts. . . . The putative instincts of human beings are always learned and never native") were based on a genetics-free environmental determinism—a belief that, as the anthropologist Margaret Mead put it, "Human nature is almost unbelievably malleable." The young Charles Darwin, while sailing round the world aboard the scientific research vessel *Beagle*, wrote in his notebook, "He who understands baboon would do more towards metaphysics than Locke." Mead and many other progressives feared that Darwin was right, and made extravagant assertions to the contrary. "Give me a dozen healthy infants," claimed the psychologist Jon Watson, "and my own specified world to bring them up in and I'll guarantee to take any one at random and train him to become any kind of specialist I might select—doctor, lawyer, merchant chief . . . regardless of his talents." In studying the black and brown peoples of America and the South Pacific, Darwin, a steadfast abolitionist, was struck by "how similar their minds were to ours" and by "the close similarity between the men of all races in taste, dispositions and habits." Yet many progressives rejected whole swathes of his work rather than question their dogmatic belief that human behavior must be determined entirely by societal circumstances.

Rightists have depicted Darwinism as the engine driving socialistic efforts to engineer the human gene pool. In this narrative, the publication of Darwin's books in the latter half of the nineteenth century was soon followed by an upwelling of public enthusiasm for eugenics—the political commandeering of biological selection to attain political ends—that reached its apotheosis in the Nazi death camps and sterilization campaigns. This view of history is, while odd, far from fanciful; many early Darwinists did favor eugenics. In England, the science-fiction writer H. G. Wells, a prominent socialist, called for "the sterilization of failures" to bring about "an improvement of the human stock," while cheerfully proposing that state breeding methods be used reduce the world's "swarms of black and brown, and dirty-white and yellow people." His compatriot George Bernard Shaw argued that "the only fundamental and possible socialism is the socialization of the selective breeding of Man." John Maynard Keynes, who served on the board of the British Eugenics Society, called eugenics "the most important, significant and, I would add, *genuine* branch of sociology." In America, Justice Oliver Wendell Holmes maintained that by ruling in favor of "sterilizing imbeciles . . . I was getting near the *first principle of real reform*" (his italics) while the New Jersey governor and future president Woodrow Wilson created a state board to determine whether "procreation is inadvisable" for the "feebleminded, epileptics, and other defectives," including criminals and those living in poorhouses. Conservatives sensitive to this sad chapter in history may perhaps be forgiven for confusing Darwinian-sounding leftist dogma with evolutionary science, although their stance is more difficult to justify now that eugenics has been driven from the political arena.

Religious rather than secular convictions, however, motivate most of those Americans who reject evolution. Among the liberal democracies such staunch religious dogmatism is almost exclusively an American phenomenon. (The reason, economists theorize, is that innovative new churches found fresh parishioners in free-market America while European churches stayed put and lost their market share. If so, Christian fundamentalism is a product of what might be called social Darwinism.) In the United States, however, this latest spiritual awakening has been sufficiently influential that nearly a third of American teachers say they feel pressured to omit evolution from their lessons, or to mix in nonscientific concoctions such as creationism or intelligent design theory, while even science museums shy away from presenting films and exhibits about evolution.

The religious dogmatists who campaign against Darwin's theory of evolution seem to think that it involves little beyond the relatively narrow question of human origins—but its implications are much broader than that. In addition to being the only way to make sense of biology, evolution sheds light on phenomena ranging from universal behavior patterns (altruism, selfishness, "dad or cad" mating strategies) to the arts, where greatness equals universality (the Tokyo String Quartet playing Beethoven, Shakespeare performed in French) and so may harbor clues to our common evolutionary heritage. "If one reads accounts of the systematic nonintrusive observations of troops of bonobo," writes the novelist Ian McEwan, "one sees rehearsed all the major themes of the English nineteenth-century novel: alliances made and broken, individuals rising while others fall, plots hatched, revenge, gratitude, injured pride, successful and unsuccessful courtship, bereavement and mourning." Thinking itself may be evolution in action. According to the theory of neural Darwinism, the brain works by generating myriads of neuron patterns that jockey for preeminence, with the survivors mapping themselves onto cortical tissue in a contentious dynamic that works much like competition within and among biological species. The American neurobiologist Gerald Edelman, who originated the theory, maintains that "the key principle governing brain organization is a populational one and that in its operation the brain is a selective system" involving "hundreds of thousands of strongly interconnected neurons." Neuron groups that prevail are able to maintain and enlarge their territory. As Edelman puts it, "There is a kind of neuroecology occurring at several complex levels of development and behavior. . . . In certain respects, groups within a region [of brain tissue] resemble a species subdivided into many races, each interacting mainly within itself, but occasionally crossbreeding." Neural Darwinism may explain, among other things, why the human brain is able to compose music and do physics even though these activities do not seem to have been historically essential to human survival. If the brain functions in an evolutionary way, it *has* to be big, because most of its activities—thoughts, prethoughts, and unconscious impulses—amount to experiments, most of which are selected against. It may seem strange to think of evolution as taking place in the brain at lightning speed, but electrochemistry moves at the speed of thought, as thinking itself constantly demonstrates.

Dogmatism has been losing ground in the liberal-scientific world—even in the United States, where the ranks of the religiously devout are in general decline despite a fundamentalist resurgence. Meanwhile, though, religious and political extremism is on the rise in parts of the Middle East and the wider Muslim world, reaching a reduction to absurdity in the campaigns of Islamist terrorists. (The word *Islamist* designates Muslim extremists inclined toward violence. It came into general use with the 1979 hostage takings at the U.S. embassy in Iran. It relates to Islam in somewhat the way that *communist* relates to *commune*.) Islamism remains a minority dogma, with surveys indicating that Muslims generally support democracy and human rights while rejecting terrorism and theocracy, but terrorism always draws vastly more attention than its support merits; that's the point of it.

Following the 9/11 attacks, Americans asked, "Why do they hate us?"

and wondered whether Islam itself was to any extent responsible for Islamist terrorism. Many articles and books explored this question but mostly followed predictable political precepts. The left proclaimed Islam and Islamism to be almost equally innocent, and blamed terrorism on the evils of an allegedly neocolonialist West. The right often lumped Islam and Islamism together on one side of a "clash of civilizations"—the title of a widely read 1993 *Foreign Affairs* essay by the historian Samuel Huntington that also contributed the only marginally more illuminating claim, "Islam has bloody borders." From a liberal perspective, however, it is clear that Islam has little to do with Islamism—which is better understood as just one more totalitarian political dogma. All such dogmas justify present sins as the price to be paid to attain a future perfection. For communists, the wished-for state was Rousseau's heaven on earth, where nobody could own anything or evince anything short of complete faith in the government. For the Nazis it was a Wagnerian opera, with blond Nordic heroes playing all the leading roles. For Islamists it is the "Caliphate," their term for the domain carved out circa 632–750, when Islamic rule expanded from Arabia to Spain in the west and Afghanistan in the east—an epoch to which the Islamists would have all Muslims repair.

That regime was less an empire than a collection of loosely allied caliphates (from *khalifa,* meaning "descendant of Muhammed"), but it displayed the attitudes characteristic of empires: It cherished the heroic narrative of its own establishment, took pride in its considerable artistic and intellectual attainments, and regarded the outside world with complacent indifference.

Its establishment narrative was one of extraordinarily rapid military victories achieved from the very dawn of Islam. The prophet Muhammad began dictating his revelations in the year 610, when he was about forty years of age, and continued until three years prior to his death in 632. His words were subsequently collected as the Quran ("recitation"), a standardized edition of which appeared in 653. The Quran contains advice on almost every aspect of life, bound together by a call to return to traditional Arab values and to spread them across the known world—and spread they did, with remarkable alacrity. By 661, just one generation after Muhammad's demise, four major caliphates had extended the reach of Islam ("submission") from Persia to Egypt. By 750, Islam ruled Spain and had driven deep into the Byzantine Empire, where the tide turned when the Muslim forces—beset by Greek Fire, an early form of napalm—failed to conquer

Constantinople (today's Istanbul, which in 1453 would be conquered by the Turks and become central to the Ottoman Empire). Artistically and intellectually, the Islamic world could boast of exquisite architecture, mosaics, poetry, and calligraphy; significant contributions to mathematics (*algebra* is an Arabic word for an Arabic invention), and astronomy (many bright stars, such as Algol, Altair, and Rigel, bear Arabic names); plus libraries housing the Greek and Latin classics that would incite the European Renaissance. Though imposed by force, early Muslim rule was fairly liberal by the standards of its day. The caliphs generally paid respect to the teachings of Moses and Jesus, tolerated the religious observances of non-Muslims living under their governance, and accepted a degree of emancipation of women (whose veiling and sequestering was neither practiced nor advocated by Muhammad), although conservative Muslims maintained that a woman should leave home on only three occasions—when she took up residence with her bridegroom, attended to the death of her parents, and when her body was carried to the cemetery.

The caliphs were intrigued by the arts and crafts of the Far East that came through on the Silk Road, but paid scant attention to the rise of science and liberalism in a relatively primitive Europe. When Napoleon invaded Egypt in 1798, he might almost have come from Mars. The historian Abd al-Rahman al-Jabarti, who was living in Cairo at the time, noted wonderingly that should a Muslim such as himself approach the scholars on Napoleon's staff and display a desire for knowledge, "They showed their friendship and love for him, and they would bring out all kinds of pictures and maps, and animals and birds and plants. . . . I went to them often, and they showed me that." The European scientific revolution, notes Bernard Lewis, "passed virtually unnoticed in the lands of Islam, where they were still inclined to dismiss the denizens of the lands beyond the Western frontier as benighted barbarians. . . . It was a judgment that had for long been reasonably accurate. It was becoming dangerously out of date." By the time Sadik Rifat Pasha, the Ottoman ambassador to Vienna in 1837, warned that the Europeans were flourishing thanks to a combination of science, technology, and "the necessary rights of freedom," it was already too late. Western ships ruled the seas; British forces were extending their power inland from the coasts of India; and the U.S. Navy, having been created expressly for this purpose, had curtailed the plundering of American merchant ships and the hostage taking of their passengers by North African pirates hailing from what the

Americans called the Barbary Coast. By 1920 Arabia was encircled by the British Empire—while the Ottoman Empire, having sided with Germany in World War I, had seen its lands divided between Britain and France.

The technological exploitation of Arabian oil made matters worse. When the American explorer Joel Roberts Poinsett (who discovered the flower that bears his name) spotted a pool of petroleum in Persia in 1806 and speculated that it might someday prove useful as a fuel, nobody yet knew what to do with it. By the middle of the twentieth century a thirst for oil had drawn the Western powers ever more deeply into the Middle East. The states possessing large oil reserves—Iraq, Saudi Arabia, and to a lesser extent Kuwait, Oman, Bahrain, Qatar, and the United Arab Emirates—were transformed from an assortment of small fishing, herding, and trading villages into economically more vertical societies where the few who controlled the oil became rich and the rest stayed poor. Such inequities were offensive to Islam—which, like Christianity, had originated as a religion centered on the poor and devoted to social justice—but the attempts of Muslim leaders to redress them by resorting to wealth redistribution through state socialism failed.

Most Muslim intellectuals accommodated the incursions of the West by adopting what they found to be useful in Western thought and otherwise sticking to their own traditions. This was the approach taken by moderate nineteenth-century thinkers like the journalist and educator Muhammad Abdu, who maintained that science and democracy could strengthen Islamic societies, and Sayyid Ahmad Khan, who, finding no contradiction between science and Islam, established a school where Muslims could study science. But a minority reacted by reverting to political extremism.

The more retrograde among these extremists drew inspiration from the eighteenth-century fundamentalist Abdul Wahhab, who shrugged off a thousand years of scholarship to claim that the literal words of the Quran should govern political procedures and judicial values. (He had a particular enthusiasm for punishing women convicted of adultery by stoning them to death.) Assailed by many of his fellow Muslims, Wahhab came under the protection of a local chieftain, Muhammad ibn Saud, who in 1744 founded Saudi Arabia—today a police state that has spent nearly one hundred billion dollars indoctrinating students around the world in Wahhab's intolerant doctrines.

The progressives among the modern Islamic radicals, eager to find the wellsprings of Western power, tapped into the Parisian intellectual currents of the 1920s and 1930s—and there, seeking dreams, reaped nightmares.

Their desire for social justice (Muslims fast during Ramadan to stay mindful of the poor) resonated with communism. Their hopes of regaining the lost glories of the Caliphate drew them to fascism. Their thirst for philosophical verities attracted them to the pretentions of Rousseau, Nietzsche, and Hegel. Thus inspired, they set up Islamist organizations like the Young Egypt Society, the Arab Socialist Baath Party, and the Muslim Brotherhood. The members of Young Egypt called themselves Greenshirts, in emulation of the Nazi Brownshirts. The Baath Party—"We were racist," recalled Sami al-Jundi, one of its early leaders, "admiring Nazism, reading its books and the source of its thought, particularly Nietzsche"—empowered Saddam Hussein, the Iraqi ruler who modeled his career on Stalin's and murdered a quarter-million Iraqis. The Muslim Brotherhood spawned Hamas and Al-Qaeda. Its resident ideologue was Sayyid Qutb.

Born in 1906 in the northern Egyptian village of Musha, Qutb attended a provincial school, memorized the Quran, then studied education at Cairo University. He emerged as a young intellectual of the middling sort—hanging around after graduation to teach a few classes, writing novels that nobody read, and spouting quotations from Hegel, Heidegger, and the French eugenicist and Nazi sympathizer Alexis Carrel. By the late forties Qutb's views had become sufficiently irritating that the government shipped him off to the United States, where he took graduate courses in education and found little that met with his approval. American men, he wrote in his journal, were "primitive" though "armed with science"; the women in their tight-fitting sweaters were "live, screaming temptations"; and everyone was enamored of jazz, the "music that the savage bushmen created to satisfy their primitive desires." A dogmatic dualist, Qutb saw the world as divided between a perfect but as yet unrealized Islamist law and a scientific, technological culture that was materially powerful but devoid of "human values." The Caliphate had long ago done wonderful things like clothing naked Africans and delivering them "out of the narrow circles of tribe and clan [and] the worship of pagan gods into the worship of the Creator of the worlds." ("If this is not civilization, than what is it?") The West, in contrast, was preoccupied with science, which can discover "only what is apparent," makes overweening claims ("Darwinist biology goes beyond the scope of its observations . . . only for the sake of expressing an opinion"), and anyway was already in decline: "The resurgence of science has . . . come to an end," Qutb asserted, because science "does not possess a reviving spirit."

To thinkers of this stripe, the world is a nightmare from which humankind shall awaken only once the wished-for dogmas gain control.

On returning home Qutb became an advisor to Gamal Abdel Nasser, a former member of the fascist Young Egypt Society who had switched to a Stalinist-style pan-Arabism, becoming a Hero of the Soviet Union and winning the Order of Lenin. When Nasser shrank from subjecting Egypt to outright *sharia*, or Islamic religious law, members of the Muslim Brotherhood tried to assassinate him. Qutb's younger brother Muhammad escaped the ensuing government crackdown by fleeing to Saudi Arabia, where he became a professor of Islamic Studies. (One of his students was the wealthy and indolent, later to become austere and fanatically dedicated, Osama bin Laden, who went on to underwrite the 9/11 attacks, denounce freedom and human rights as "a mockery," and declare it Muslims' religious duty to murder millions of Americans.) Sayyid Qutb remained in Egypt. He was jailed and tortured but nonetheless managed to write dozens of books before being hanged in prison in 1966, at age sixty-one, on charges of advocating the violent overthrow of the government. The Quran forbids Muslims from attempting to depose any Muslim ruler, but Qutb got around this prohibition by claiming that a "society whose legislation does not rest on divine law . . . is not Muslim, however ardently its individuals may proclaim themselves Muslim."

Islamism thus resuscitates the totalitarian enthusiasms that nearly wrecked Europe. As a recent study puts it, "The line from the guillotine and the Cheka to the suicide bomber is clear." Nor were the shocks of 9/11 required for Americans to see that bright line. As early as 1954, the historian Bernard Lewis noted "certain uncomfortable resemblances" between communism and Islamism:

> Both groups profess a totalitarian doctrine, with complete and final answers to all questions on heaven and earth. . . . Both groups offer to their members and followers the agreeable sensation of belonging to a community of believers, who are always right, as against an outer world of unbelievers, who are always wrong. . . . The humorist who summed up the communist creed as "There is no God and Karl Marx is his Prophet," was laying his finger on a real affinity. The call to a communist *Jihad*, a Holy War for the faith—a new faith, but against the self-same Western Christian enemy—might well strike a responsive note.

Selling state power in the Arab world was made easier by the fact that much of the Middle East was already statist: Even today most of Saudi Arabia's adult males, and 95 percent of those in Kuwait, work for the government, while 80 percent of Iran's wealth is controlled by the government.

As is customary in such campaigns, the Islamists portrayed the West as a decaying shell and Western man as, in Qutb's words, "suffering from affliction [and] killing his monotony and weariness by such means as . . . narcotics, alcohol and . . . perverted dark ideas." (Westerners *must* be deluded or debased; otherwise there would be no accounting for their support of liberal democracy.) The Islamist remedy was to enforce total obedience to an ideology capable of inspiring peoples' imaginations and dictating their every action—an ideology that combined the tactics of fascism and communism with faith in the Quran as a guide to good government. "Religion and politics are one and the same in Islam," stated Gulam Sarwar, whose books have influenced *madrasa* students in Britain and elsewhere. "Just as Islam teaches us how to pray, fast, pay charity and perform the *Haj*, it also teaches us how to run a state, form a government, elect councilors and members of parliament, make treaties and conduct business and commerce." Once Quranic law was imposed by all-powerful governments, it would be possible to resurrect the Caliphate.

To this end idealistic students were recruited and organized into cells. In his book *The Islamist,* Ed Husain recounts how he became beguiled by radical Islamist literature while studying medicine in London. When his father, a devout Muslim, saw in his room a quotation from Hasan al-Banna, the founder of the Muslim Brotherhood—"The Quran is Our Constitution; *Jihad* is Our Way; Martyrdom is Our Desire"—he reacted with apprehension:

> My son, the Prophet is not our leader, he is our master, the source of our spiritual nourishment. Leaders are for political movements, which Islam is not. The Quran is his articulation, as inspired by God, not a political document. It is not a constitution, but guidance and serenity for the believing heart. How can you believe in these new definitions of everything we hold so dear?

Undeterred, Husain and his fellow Islamist students circulated posters proclaiming, "Islam: The Final Solution." "We failed to comprehend the totalitarian nature of what we were promoting," he writes.

"Democracy is *haram*! Forbidden in Islam," a cell member scolded Husain. "Don't you know that? Democracy is a Greek concept, rooted in *demos* and *kratos*—people's rule. In Islam, we don't rule; Allah rules. . . . The world today suffers from the malignant cancers of freedom and democracy." Husain eventually fell in with members of Hizb ut-Tahrir, the political party founded by Taqiuddin al-Nabhani, whose works enthralled Husain until he discovered that they were "based on the writings of Rousseau [who] called for God to legislate, because man is incapable of legislation," and Hegel, the author of stirring totalitarian edicts such as, "The state is the march of God through the world."

Like the European totalitarians who preceded them, the Islamists preach an ideology of purification through total mobilization, violent struggle, and death. "Combat is today the individual duty of every Muslim man and woman," asserts the Abu-Hafs al-Masri Brigades, an affiliate of Al-Qaeda. "History does not write its lines except with blood," declares Abdullah Azzam, a disciple of Qutb who fought with bin Laden in Afghanistan. "Glory does not build its lofty edifice except with skulls. Honor and respect cannot be established except on a foundation of cripples and corpses." Posters decorating the walls of Palestinian kindergartens proclaim, "The Children are the Holy Martyrs of Tomorrow." Nor have such injunctions remained on the level of street marches, or of teenagers blowing themselves up in marketplaces on the promise of eternal salvation. When the Islamist scholar Hassan al-Turabi (B.A., Khartoum University; M.A., London School of Economics; Ph.D., the Sorbonne) seized power in Sudan and instituted *sharia* and *jihad*, his reign resulted in the deaths of more than a million Sudanese. The Islamists would bring such campaigns to the world at large: "Islam is a revolutionary doctrine and system that overthrows governments [and] seeks to overturn the whole universal social order," writes Abul Ala Mawdudi, a Pakistani journalist who promotes Islam as a political ideology. "Islam wants the whole earth and does not content itself with only a part thereof. It wants and requires the entire inhabited world. . . . It is not satisfied by a piece of land but demands the whole universe [and] does not hesitate to utilize the means of war to implement its goal."

To justify such ambitions, Islamists fan the flames of resentment over colonialism and the intrusion of Western wealth, popular culture, and technological power into the Middle East. "The problem of modern Islam in a nutshell," writes Omar Nasiri, a Moroccan who once advocated global *jihad*, is that "we are totally dependent on the West—for our dishwashers, our clothes, our cars, our education, everything. It is humiliating and every Muslim feels it. . . . We were the most sophisticated civilization in the world. Now we are backward. We can't even fight our wars without our enemies' weapons." In a widely quoted passage, David Frum and Richard Perle describe the situation this way:

> Take a vast area of the earth's surface, inhabited by people who remember a great history. Enrich them enough that they can afford satellite television and Internet connections, so that they can see what life is like across the Mediterranean or across the Atlantic. Then sentence them to live in choking, miserable, polluted cities ruled by corrupt, incompetent officials. Entangle them in regulations and controls so that nobody can ever make much of a living except by paying off some crooked official. Subordinate them to elites who have suddenly become incalculably wealthy from shady dealings involving petroleum resources that supposedly belong to all. Tax them for the benefit of governments that provide nothing in return except military establishments that lose every war they fight: not roads, not clinics, not clean water, not street lighting. Reduce their living standards year after year for two decades. Deny them any forum or institution—not a parliament, not even a city council—where they may freely discuss their grievances. Kill, jail, corrupt, or drive into exile every political figure, artist, or intellectual who could articulate a modern alternative to bureaucratic tyranny. Neglect, close, or simply fail to create an effective school system—so that the minds of the next generation are formed entirely by clerics whose own minds contain nothing but medieval theology and a smattering of third world nationalist self-pity. Combine all this, and what else would one expect to create but an enraged populace ready to transmute all the frustrations in its frustrating daily life into a fanatical hatred of everything "un-Islamic."

The Islamist diagnosis is that Muslims, though they rank among the world's most religiously devout peoples, just aren't devout enough—that only by bringing back old-time religion plus totalitarian governance will Muslims surpass the West. Somehow they will do this without resorting to science, since, as the London-based Islamist writer Ziauddin Sardar asserts, "Western science is inherently destructive and does not, cannot, fulfill the needs of Muslim societies." Impatient with compatriots who blanch at their extremism, Islamist terrorists are as content to murder Muslims (the "near enemy") as they are Christians (the "distant enemy"). "Fighting the near enemy is more important than fighting the distant enemy," declared the Egyptian Islamist Abd al-Salam Faraj. "In *jihad* the blood of the Muslims must flow until victory is achieved." Their actions have matched their words: To date, the majority of their victims have been Muslims. Islamists have murdered Muslim real estate agents, barbers, and ice vendors for selling services that were not available when the Prophet was alive. To discourage the education of women, the Taliban throws acid in the faces of Muslim schoolgirls and blows up their schools. Such conduct is so far removed from Muslim morality that the terrorists, when called upon to explain themselves, soon abandon their religious rhetoric and revert to the language of totalitarian *realpolitik*. Fouad Ali Saleh, the leader of an Islamist student network that detonated bombs in Paris in the eighties, was confronted at his trial by a man who had been badly burned in one of the attacks. "I am a practicing Muslim," said the victim. "Did God tell you to bomb babies and pregnant women?" Saleh replied not with a quotation from the Quran, but with an appeal to ethnic grievances worthy of Lenin: "You are an Algerian. Remember what [the French] did to your fathers."

A liberal-scientific diagnosis of what has gone wrong in the Middle East is that the problem arises from a paucity of science and liberalism.

Regarding science: Arab investment in R&D is only a tenth of the world average and a third of what other developing countries spend. Muslim nations produce 8.5 scientists, engineers and technicians per 1,000 persons—the world average is 40—and contribute less than 2 percent of the world's scientific literature. When the Turkish-American physicist Taner Edis was asked, "How would you assess the state of scientific knowledge in the Islamic world?" he replied, "Dismal. Right now, if all Muslim scientists working in basic science vanished from the face of the earth, the rest of the scientific community would barely notice."

Regarding liberalism: Only a quarter of the world's Muslim-majority nations are electoral democracies, compared to almost three quarters of the non-Muslim nations. No Arab leader has yet lost power in a general election, nor are many likely to expose themselves to any such peril, considering the wealth that customarily comes with the position: Seven of the world's ten richest heads of state are Arabs. Saudi Arabia has more military fighter jets than trained pilots, presumably because at least a third of the jets' purchase price is kicked back to the royal family. A public appearance by a Middle Eastern ruler, writes the Tunisian historian Mohammed Talbi, "always triggers thunderous applause. The most zealous vow to sacrifice their blood for him and shriek their undying loyalty until their voices are hoarse and their bodies exhausted. . . . It is a gripping spectacle, greatly enjoyed by the leader, whose passage through the crowds has the effect of a huge collective brainwashing." That so many such dictators have enjoyed American support, and that so much of their wealth has come from oil, fosters Islamist condemnations of the West as brutal and untrustworthy.

"The Middle East has become dominated by a totalitarian model that destroyed traditional freedoms and stifled economic growth, while educating generations of Arabs to oppose commerce, pragmatic compromise, and western science," notes the political scientist Colin Rubenstein. Sexism pervades society, from education (four-fifths of adult women in Bahrain, Jordan, Kuwait, Qatar, and the United Arab Emirates are illiterate) to the law (Iranian law weighs the testimony of female witnesses as half that of males), and even in the terrorist cells, which pay women half what men are paid to strap on a bomb and blow themselves up on a bus or in a marketplace. Young Arabs find their prospects crimped by mediocre schools, high unemployment rates, and repressive social mores that make it difficult to meet, much less romance, members of the opposite sex. Iranians and Saudis who do manage to get together are discouraged by the morals police from holding hands, dancing, or attending mixed-gender theatrical events. The three-thousand-seat King Fahd Cultural Center, built in Riyadh, Saudi Arabia, in 1989 at a cost of $140 million, sat empty for a decade owing to the risks involved in actually staging a play. Small wonder that a majority of Arab youths say they want to emigrate.

Islamism is opposed by most Muslims—in Pakistan, Egypt, Morocco, Indonesia, and many other Islamic nations. Thousands of young Iranians defied their government and took to the streets on the night of 9/11 to hold candlelight vigils for the American victims. Reformers like the Iranian chemist and philosopher of science Abdul Karim Soroush ("science operates in a matrix of freedom of research and adversarial dialogue of ideas, a practice incompatible with political repression") and the Tunisian philosopher Rachid al-Ghannouchi ("the only legitimacy is the legitimacy of elections") have continued to speak out even when jailed by the authorities or beaten up by Islamist goons. A majority of Arabs polled say they approve of "American freedom and democracy."

Can such sentiments be translated into genuinely democratic governments? Doubters argue that Muslims are not ready for democracy or that liberalism is not part of their culture, but such claims are too vague to be disproved and there are factual grounds for hope. Millions of Muslims already live in democratic nations, 200 million of them in Indonesia alone. Economically, a number of Middle Eastern nations are wealthy enough to be promising candidates for democracy, were the wealth better distributed. They include the United Arab Emirates, Qatar, Bahrain, and Iran (which has a pseudodemocratic government kept firmly under Islamist rule). Extremism may be brazen in the Middle East and the voices of moderation muted, dictatorship the rule and liberal governance the exception, the Islamists full of passionate intensity and moderates slow to respond, but those who blame the situation on Islam itself are invited to compare the torpid, muddled reactions of many European and American politicians, preachers, and scholars to the threats posed by communism and fascism just decades ago—or, for that matter, the cynicism with which science, liberalism, and democracy continue to be viewed by many in the West today. If rising income levels tend to foster liberalism, science, and democracy, then to claim that the troubles afflicting the Islamic world are cultural rather than material is to make the rather tortured case that Muslims are somehow unique.

As anguished and confused as Americans may have been over the 9/11 attacks, their responses were on the whole admirably liberal, and reported instances of public hostility toward Muslims were rare. But while the George W. Bush administration rightly drew a clear distinction between Islam and Islamism, it also demonstrated a lamentable insecurity regarding the strength of America's liberal-democratic institutions. Foreign nationals were kidnapped and tortured while others were held without bail or legal representation in a makeshift prison at the Guantanamo naval base, President Bush going so far as to assert that he had the power to imprison American citizens, without legal counsel, if he deemed them to be a security threat. Such measures were imposed dogmatically, as if it were self-evidently true that liberal democracies are incapable of dealing with terrorism through legal due process. The administration ignored the many foreign-policy and intelligence professionals who warned that, as one put it, they had made the United States look like "a fearful superpower that has relaxed its own standards of openness and the rule of law at home."

While these steps were being taken by a conservative American administration, progressives in England promoted illiberal campaigns aimed at scotching anti-Islamic "hate speech." As so often happens when civil liberties are suppressed, the intentions were good but the consequences lamentable. A woman picnicking in London's Parliament Square was arrested for having the word "Peace" inscribed in icing on a cake, another for reading aloud the names of British soldiers killed in Iraq. A conservative member of the Dutch parliament, Geert Wilders, was turned back at Heathrow Airport to prevent his accepting an invitation to show the House of Lords a film he had made depicting the Quran as a fascist book—the Foreign Office advising that his presence in England might "threaten community harmony and therefore public security." For one liberal democracy to bar another's elected representative from showing its own legislature a film that could be viewed by anyone with Internet access was so odd that even many British Muslims were moved to remind their compatriots of the virtues of liberalism. "Freedom of speech should be protected—so long as people do not use this freedom to call for violence against others," said the Quilliam Foundation, an antiterrorism Muslim organization funded in part by the government, adding that it would have been better to admit Wilders and evaluate his views "through debate and argument."

Across a wide political spectrum—from conservatives who would abridge the legal rights of suspected terrorists to progressives who would silence politically incorrect opinions about them—liberalism was mistaken for weakness. "It seems we need to fight the battle for the Enlightenment all over again," lamented the author Salman Rushdie, who opposed hate speech legislation even though he had spent years in hiding after the Ayatollah Khomeini of Iran issued a *fatwa* in 1989 urging his murder. When Theo van Gogh, who had also made a film critical of Islam, was shot while riding his bicycle and lay dying on an Amsterdam sidewalk, he pleaded with his assassin, "Surely we can talk about this." The young thug instead slit van Gogh's throat, kicked him, and walked away. (Later arrested, he was sentenced to life in prison without parole.) What did van Gogh die for, if not free speech? "If liberty means anything at all," observed George Orwell, "it means the right to tell people what they do not want to hear."

Many Muslims share with other religious believers a conviction that religion is the sole or at least the most effective defender of morality. It is not. If it were, religious believers ought at the very least to commit fewer serious crimes than do atheists and agnostics, but such is not the case. As many surveys have shown, atheists and agnostics are, if anything, *less* apt to commit serious crimes—and they persist in their erstwhile ethicality even though they belong to the most distrusted minority in the modern world. What is called secularism—meaning atheism, agnosticism, or simply having no interest in religious faith—is on the rise in the United States, having jumped from 8 percent of the population in 1990 to 15 percent in 2008. The trend is geographically widespread—secularism is growing in all fifty states—and likely to accelerate. While only 5 percent of Americans born before 1946 describe themselves as nonbelievers, that number more than doubles for those born in the years 1946–1964 and reaches nearly 20 percent for Americans born since 1977. Yet the American violent crime rate remained flat from 1990 to 1993, and has since been declining. Indeed, crime correlates *inversely* with levels of religious conviction, if it correlates at all. While 15 percent of all Americans identify themselves as having no religious beliefs, the Federal Bureau of Prisons reports that nonbelievers make up only two-tenths of 1 percent of its inmates. (Christians constitute about 80 percent of the American population and 75 percent of its prisoners.) A ten-year study of death-row inmates at Sing-Sing found that 91 percent of those executed for murder were Christians, less than a third of 1 percent atheists. Similar anticorrelations between religion and crime are found internationally. Only 20 percent of Europeans say God plays an important part in their lives, as opposed to 60 percent of Americans, but Europe's crime rates are lower than America's. Denmark and Sweden rank among the most atheistic nations in the world—up to three quarters of their citizens identify themselves as nonbelievers—yet these Godless souls somehow enjoy admirably low levels of corruption and violent crime while scoring near the top of the international happiness indices.

Religious fundamentalists are often surprised to hear this, just as their forebears were surprised to learn from explorers' reports that upright Hindus and Buddhists living in faraway lands comported themselves as ethically as did Anglican bishops. But the basis of such confusion disappears when the genesis of morals is examined empirically. The basic tenets of morality, such as prohibitions against murder and incest, are common to most peoples and most religions. This makes sense if the moral precepts evolved over time, socially and perhaps biologically, because they promoted human survival—as they obviously do—and are reflected in religious texts rather than having been handed down from heaven. If morality evolved, rather than having been independently invented by thousands of gods, people should behave at least as ethically without religion as with it—as, evidently, they do.

Neither liberalism nor science need quarrel with religion. Liberals defend religious rights and support the separation of church and state—a stance that benefits churches as well as states, as many wise religious leaders have pointed out. President Jefferson's 1802 letter advocating "a wall of separation between church and state" was written at the instigation of the famous Baptist minister John Leland, who presented Jefferson with a 1,235-pound wheel of cheese to express his gratitude for Jefferson's keeping religion safe from politics. American Baptists heeded Leland's doctrine for almost two centuries thereafter, only to repudiate it in 1998 in hopes of gaining political leverage—a step that awakened widespread opposition to the Christian right and sparked a spate of widely read books assailing religion. Liberals caution the faithful that seeking political privileges for their particular church is unlikely to benefit it or any other religious group, in the United States or any other liberal democracy.

Scientists originally were as religious as the rest of the population, but the scientific process and the knowledge it obtains are so different from religious practices and doctrines that it is becoming increasingly difficult, as science progresses, to accommodate both within a single worldview. Religions value faith but scientists have found, often to their own embarrassment, that having faith in an idea has no bearing on whether the experimental evidence will verify it. (Nobody asked for, much less prayed for, irrational numbers or quantum nonlocality, but they became part of science anyway; nature is as it is, regardless of what we wish for.) Scientific theories stand or fall on their ability to make accurate predictions; religions have such a poor record in this regard that to champion divine prophecy is to risk being thought supercilious or deranged. Religions account for natural phenomena by positing the existence of an invisible and miraculously complex agency, science by sticking to discernable phenomena that are simpler than what they seek to explain. In that sense, God is literally the last thing a scientist should look for when studying nature.

Global warming, discovered by various scientists over more than a century of research, has opened an arena of contention among dogmatists, discoverers, and everyone in between that is likely to be with us for many years to come. The phenomenon may be threatening but the ongoing matter of dealing with it, involving as it must all the peoples of the world and raising scientific and ethical questions that resound down through future generations, could have a healthy effect on global discourse. Extremists on the left see global warming as a condemnation of capitalism and globalization and an invitation to impose stricter state controls. Their counterparts on the right dismiss it as a hoax or an exaggeration and express a touching faith that God or nature will see us through—the same God or nature that, in spasms like the Permian mass extinction of 248 million years ago, has repeatedly and unapologetically killed off most living species. Clearly, dogma won't help. Global warming will require ongoing scientific investigation, quantitative analysis, and open, liberal discussion and debate on a worldwide scale.

Its essentials are not terribly complicated. The earth's atmosphere is thin—thinner, relative to the planet's diameter, than the layer of moisture covering one's eyes. Through this membrane passes sunlight, which warms the earth's surface. Much of the resulting heat is radiated back into the atmosphere as invisible, long-wavelength ("infrared") light. Nitrogen and oxygen comprise 99 percent of the atmosphere. These gases are transparent to infrared light, so if they were the whole story the heat would simply be radiated back into space, in which case the earth would be considerably colder than it is. But the remaining 1 percent of the atmosphere includes several "greenhouse" gases—notably water vapor, carbon dioxide (CO_2), and methane—that absorb the infrared light rather than letting it escape. They act like the closed windows of a parked car, trapping heat and warming the planet. Since the dawn of the Industrial Revolution, humans have been pumping carbon dioxide and other greenhouse gases into the atmosphere at an accelerating rate, through such activities as burning coal in power plants and gasoline in automobiles. It is estimated that nearly a third of the CO_2 in the atmosphere today is anthropogenic.

Meanwhile the earth's average temperature has been slowly rising, up about 0.74 degrees Celsius from 1905 to 2005. Some of the warming may be due to other causes, such as fluctuations in the amount of energy released by the sun or by the climatic cycles that, over the eons, have plunged the earth into a series of ice ages. Conceivably, *all* the warming might be of nonhuman origin, but the present trend is historically unusual—the world today evidently is hotter than it has been for many centuries—and anthropogenic greenhouse gases constitute the most plausible explanation for it.

If temperatures continue to rise, the results could be extremely unpleasant and dauntingly expensive to reverse. An increase of only one to two degrees Celsius over preindustrial levels could drive a third of the world's living species to extinction. An increase of two to three degrees could sharply reduce the world's supplies of drinking water and acidify the oceans, with unknown consequences for fish stocks and other marine ecosystems. A three- to four-degree increase could melt the Antarctic and Greenland ice caps, raising sea levels and driving hundreds of millions of people from their homes. (Sea levels have risen nearly a foot already, impacting coastal real estate markets and obliging the 11,000 citizens of low-lying Tuvalu, a tiny Pacific nation where the intruding ocean is already contaminating drinking water and destroying farmland, to apply to New Zealand for permission to emigrate there *en masse.*) At higher levels the threat arises of runaway warming, in which greenhouse gases currently locked into permafrost are released in such quantities that global temperatures spiral upward, utterly out of control.

On which point it may be useful to contemplate Venus, the brightest planet in the sky. Venus is virtually Earth's twin—the two planets have the same diameter and the same mass—but while much of the earth's carbon is bound up in its oceans and plants and in fossil fuels like coal, oil, and natural gas, the carbon on Venus resides in its atmosphere. The surface temperature of Venus is 457 degrees Celsius, hot enough to melt lead. Should the earth be pushed into runaway greenhouse warming, it might wind up resembling the Venus of today.

That, in essence, is the problem. The current rise in both anthropogenic greenhouse gases and global temperatures *might* be a coincidence, but how much do you want to bet on it? The cost of bringing global warming under control is about 1 percent of global GDP, estimates the economist Nicholas Stern, whereas the cost of inaction, should global warming continue, "will be equivalent to losing at least 5 percent of global GDP each year, now and forever."

You can dispute Stern's figures, but unless you can demonstrate that a significantly hotter world would be a boon for humanity and that there is no reason to worry about a runaway greenhouse effect—a very long row to hoe—you come up against questions of what to do about it, now or later. Those who doubt that human activity has anything to do with global warming often point out that scientists have made mistakes in the past (which they certainly have) and that their models of global climate are inadequate (which they may be). But nothing can be accomplished if those involved are required to know everything about the relevant systems and to predict their every possible future. If you examine a pond where fish are dying and discover a pipe through which raw carbolic acid is gushing into the water, you are not obliged to ascertain every specific of the pond's ecosystem before plugging the pipe to see whether that allows the fish population to recover. Just such empirical approaches have succeeded in mitigating previous air-pollution problems: Urban smog levels were reduced by fitting automobiles with catalytic converters, acid rain diminished by fitting sulfur-reducing scrubbers on the smokestacks of coal-burning power plants, and ozone depletion addressed by banning chlorofluorocarbons, which had been employed in refrigeration systems. Had the industrialized nations instead waited until their scientists could accurately model the atmosphere in every detail, people today would be choking on auto emissions while forests and lakes were dying from acid rain and sterilizing ultraviolet sunlight poured through an ever-widening ozone hole.

Global warming was discovered in a haphazard manner that illustrates the benefits of having an open scientific community where all sorts of individuals are free to pursue their interests, report their findings, and make themselves heard. In the eighteenth century a wealthy Swiss amateur physicist named Horace Saussure—a pioneering alpinist who climbed with barometers, thermometers, and hygrometers in his kit, paying close attention to the play of sunlight at altitude and the way different materials reflected or absorbed sunlight—noted that a black box in direct sunlight got warmer when a thin pane of glass was placed on top of it. The reason, a scientist would say today, was that the glass, like the atmosphere, was opaque to infrared light—but that much would not be understood until William Herschel discovered infrared light decades later.

Meanwhile the French scientist Joseph Fourier realized that Saussure's result might explain why the earth is so warm. Fourier had already calculated that the earth would be cold enough to freeze the oceans unless some unknown mechanism was retaining part of the sun's heat. Now he theorized that the atmosphere was responsible—that it acted like the sheet of glass atop Saussure's black box. The question of just how this worked remained for John Tyndall, in England, to investigate in 1859. Measuring the temperature of infrared light after it passed through various atmospheric gases, Tyndall established that methane and carbon dioxide are opaque in the infrared and so retain solar heat. With that it became apparent that the greenhouse effect keeps the earth warm.

Global warming seemed to be strictly of scientific interest, although two worrisome possibilities had appeared in the scientific literature by the end of the nineteenth century. The first was the prospect of climatic feedback loops, identified by the Swedish physicist Svante Arrhenius. People had been thinking of the climate as a linear system—if you pumped in some additional greenhouse gas you'd get back a directly related amount of global warming—but Arrhenius realized that one such change might amplify others. Suppose the amount of carbon dioxide in the atmosphere increases, warming the air. Warm air holds more water vapor than does cold air, and water vapor is a major greenhouse gas, so the slight initial increase in one greenhouse gas can spur the increase of another. Intrigued, Arrhenius consulted a colleague, Arvid Högbom, who had been studying how carbon dioxide erupts from volcanoes, circulates in the air, and is absorbed by trees and seas. Högbom had considered that coal-burning in factories might contribute to the greenhouse effect, but anthropogenic emissions were still negligible in the 1890s, so few paid much attention. By 1908, though, the rate of coal burning in the industrialized world was rising rapidly, and Arrhenius speculated in print that industrialization might eventually contribute enough CO_2 to cause significant global warming. The problem was thought to be remote—nobody anticipated that the human population would quadruple by the end of the twentieth century or that its per capita energy use would also quadruple, resulting in a sixteenfold increase in CO_2 emissions—and anyway, as the historian of science Spencer Weart remarks, "Warming seemed like a good thing in chilly Sweden."

The first study connecting the rise in atmospheric greenhouse gases with rising global temperature was presented in the 1930s by an English steam-power engineer and amateur meteorologist named Guy Stewart Callendar, who by checking old weather reports confirmed rumors that average temperatures were increasing. He then compiled measurements of atmospheric

CO_2 and found that it, too, was increasing. Addressing the Royal Meteorological Society in London in 1938, Callendar suggested that the two trends were related—that industrial dumping of carbon dioxide into the air contributed to the warming trend. "We owe much to Callendar's courage," writes Weart. "His claims rescued the idea of global warming from obscurity and thrust it into the marketplace of scientific ideas." It helped that more than a few scientists were willing to consider evidence presented by a rawboned engineer for whom scientific research was only a hobby.

In 1956, Gilbert Plass of the U.S. Office of Naval Research ran some calculations on an early digital computer and predicted that the ongoing release of greenhouse gases by human activity would raise earth's average temperature by 1.1 degrees Celsius per century. His work attracted the attention of the Caltech chemist Charles David Keeling, who established carbon dioxide monitoring stations in California, Antarctica, and atop Mauna Loa in Hawaii. Despite a break in data gathering imposed when his funding ran out, Keeling's results clearly showed a substantial rise in global CO_2 levels, from an average of around 315 parts per million (ppm) in the late fifties to over 370 ppm by 2002. (Updated estimates find levels of 280 ppm in 1850 and 380 ppm in 2008.) Moreover, the rate was accelerating as industrialization spread. The world in 2008 produced three times as much CO_2 as in 1957. Obviously this could not continue indefinitely—project the current rate of rise far enough into the future and you get Venus—but too little was understood about the atmosphere to establish how much CO_2 it could absorb safely. "The atmosphere is not a dump of unlimited capacity," warned the National Academy of Sciences in 1966, but, "We do not yet know what the atmosphere's capacity is."

As more scientists got into global-warming research, the geological history of global temperature change became better understood. Cesare Emiliani, a geologist who collected core samples from the ocean floor (creating a mathematical formula to determine exactly how many bottles of beer could be stored in the empty tubes, based on the crew's anticipated consumption rate and the scheduled dates when each tube would be crammed with ancient mud and sand) compiled a temperature record dating back three hundred thousand years. Emiliani found that there have been dozens of ice ages—not just four, as had been thought—and that their chronology fits the theory that ice ages result from the slow oscillation, or precession, of the earth's axis. By the nineteen eighties, sufficient core-sample data were on hand to show that global carbon dioxide levels historically have been low during the ice ages and high when the earth was warm. The other greenhouse gases appear to have behaved similarly. Methane, for instance, is mostly locked up in peat moss and permafrost, but could be released should the world warm sufficiently. Molecule for molecule, methane is twenty times as potent as carbon dioxide in fueling the greenhouse effect. The world's wetlands and permafrost are already melting, with studies indicating that methane levels are increasing by more than 10 percent per decade. They contain more greenhouse gases than have yet been released by all human activities combined.

Each ice age typically was preceded by a period of global warming. One way to visualize how this can happen is by considering the phenomenon known as glacial surge. Giant glaciers like those in Antarctica, Greenland, and Alaska date back to the last ice age, when New York and Chicago were buried under a mile of ice. Glaciers normally move with trademark slowness but occasionally slip, accelerating to velocities as high as three hundred meters per day. A feedback loop is thought to be responsible: Water under the glacier melts sufficiently to create a slippery surface, the glacier starts speeding up, and friction from its increased speed liquefies more of the ice, so the glacier accelerates. In recent decades, geologists have found that the mechanisms holding glaciers in place can be so delicate that a small rise in global temperatures suffices to release massive amounts of glacial ice: The glaciers could slide off the landmasses that supported them and splash into the sea. They contain enough water to raise global sea levels by several meters—enough to flood New York and many other cities, much of Florida and Bangladesh, and the entirety of inhabited atolls like Tuvalu. Geologists caution that such a catastrophe might be followed not by further warming but by the onset of a new ice age. The reason is that the glaciers create huge icebergs, which, combined with the remaining ice left behind in Iceland and Antarctica, makes the earth a much whiter planet, one that reflects much more sunlight than before and consequently cools down rather rapidly. That is why global warming is sometimes called global climate change, to emphasize that it can invoke cold as well as heat.

By 1981, enough scientists were convinced that increased levels of greenhouse gases could set off rapid climate change to put the story on the front page of the *New York Times*. Public response was torpid, until the summer of 1988 brought heat waves not experienced in the United States since the

Dust Bowl of the 1930s. Four in five Americans thereafter had at least heard of the greenhouse effect. A turning point was reached in 1990, when 170 scientists comprising the Intergovernmental Panel on Climate Change (IPCC) released the first in a series of reports confirming that the world was getting warmer and stating that human activities were at least partly responsible. "Warming of the climate system is unequivocal," warned the IPCC in 2007, adding that global sea levels were rising and northern-hemisphere snow covers shrinking, that anthropogenic greenhouse gas emissions had increased 70 percent between 1970 and 2004, and that "most of the global average warming over the past 50 years is *very likely* due to anthropogenic GHG [greenhouse gas] increases."

The IPCC uses the words *very likely* to mean "having a probability of over 90 percent." So when it reports that "most of the observed increase in global average temperatures since the mid-twentieth century is *very likely* due to the observed increase in anthropogenic GHG concentrations," it is assigning a 90 percent probability to the theory that anthropogenic greenhouse emissions are warming the planet. Other conclusions assigned that probability in the 2007 report are that "the observed widespread warming of the atmosphere and ocean, together with ice mass loss . . . is not due to known natural causes alone," that human activities "contributed to sea level rise during the latter half of the 20th century," and "that climate change can slow the pace of progress toward sustainable development," retarding or reversing the recent gains made in alleviating poverty and hunger.

Dogmatists accustomed to the rhetorics of religious or political certitude may regard such quantifications as shilly-shallying, but *all* predictions are statements of probabilities. The odds are good that the sun will rise tomorrow morning but they are less than 100 percent, as are the odds that the book you are reading will still be here ten seconds from now. (It is possible, though unlikely, that the sun will explode tonight, or that every atom in the book will simultaneously perform a quantum leap to the asteroid belt.) Lyman Bryson, writing a half century ago about the "moral atmosphere in which [scientists] live and for which its acolytes are trained," cited as an example of scientific ethics "that no judgments will be given as absolutes." To dismiss the IPCC conclusions as merely probabilistic is like playing Russian roulette with a ten-shot revolver containing nine live rounds.

Global warming poses ethical issues of long duration and universal human concern—issues which demand quantitative analysis, and cannot be addressed satisfactorily by invoking the on-off switches (Thou shalt do this, Thou shalt not do that) that characterized the moral precepts of old. With the rise of science and mathematics it became commonplace to bifurcate human thought into two realms—quantitative thinking (employed in science, finance, and the like) and something called qualitative thinking, alleged to be the realm of creative pursuits like painting and poetry. This distinction is probably false—most likely all thought is quantitative, whether we know it or not—but in any event the ethics of global warming demand quantitative analysis on virtually all levels. Rich nations produce far more greenhouse gases per capita than poor nations do—but *how much* more, weighed against the rich world's contribution to global productivity and growth? Greenhouse gases put into the atmosphere today will stay up there for decades: Is it ethical to bequeath them to our descendants, and if so *to what degree?*

Risks are calculated by multiplying the odds of something happening times the cost if it does. The potential costs of global warming are high, so the risk must be taken seriously even if the odds of its occurring are low. The question of *when* to address them is also important, but more subtle than is generally realized. All else being equal, it should be cheaper to head off catastrophic global warming now than later; only a slight course correction is required to avoid an iceberg miles ahead, whereas if you wait till the last moment, the energy required to evade the iceberg may exceed the capacity of your ship. But all else is not equal. Our descendants, decades from now, will probably still be dealing with global warming, but they will have the advantages of advanced technology and a stronger economy—or of neither, should climate change badly erode the economy. Any responsible policy has to take these prospects into account.

Suppose that within the next few decades the world has taken all the relatively easy steps indicated to abate global warming—improving energy efficiency, reducing carbon emissions through cap-and-trade or some such approach, and promoting the development of new technologies ranging from solar cells to genetically engineered, carbon-devouring trees—and has now come up against the harder, more costly measures required to bring the problem fully under control. Should more money be spent immediately, or later? At one extreme all the money in the world could immediately be allocated to address it, but that would wreck the economy, inflicting more human suffering immediately than global warming would have wrought in the long term. At the other extreme nothing more would be spent, but then the problem might just get worse. The point is that light-switch ethics—pounding the table and insisting, "We must save the lives of our grandchildren," or, alternately, "We must leave free enterprise alone"—are inadequate when dealing with predictions, since predictions are inherently probabilistic. You can ignore the odds, just as you can close your eyes while driving a speeding automobile, but to do the job responsibly means staring at a lot of curving lines on graphs and trying to find the sweet spot. One line predicts the growth of the world economy—which increases the potential ability of our descendants to deal with the problem. Another projects a long-term economic decline occasioned by global warming—which weakens our descendants' spending power, though it may also reduce pollution. A third line predicts the "discount rate," which takes into account the declining purchasing power of a dollar over time—and so forth. Responsible stewardship requires dealing in quantities and probabilities, with little recourse to comforting certitudes.

A quantitative ethics involving intersecting lines on graphs may be less than ideal, but traditional ethics are not ideal, either. One way to evaluate ethical systems is to ask which are invariant, meaning that they are accepted and practiced across virtually all cultures and nationalities.

Religious ethical systems tend to mix invariant prescriptions with others more peculiar to the cultures from which each religion emerged. Six of the Ten Commandments, for example, are invariant—honor your parents; do not commit murder, adultery, or theft; and do not bear false witness or covet your neighbor's property. The other four are parochial: To observe the Sabbath and to worship only Yahweh, making no graven images of him and never taking his name in vain, are precepts neither honored nor esteemed in all cultures. Nor is invariance characteristic of hundreds of other religious injunctions, such as the Bible's advice that if parents find their son to be stubborn and rebellious, "All the men of the city shall stone him to death," or the Quran's advising women to "stay in your houses and not display your finery." The Golden Rule—"Do unto others as you would have others do unto you," or, as Kant put it, "Act only on those maxims that you can at the same time will should become a universal law"—is invariant across the human species, and hence is widely regarded as an objective basis for ethics generally.

Among political philosophies, conservative values of patriotism and honoring military service are embraced by many cultures but can break down over national differences: American conservatives supported the mass demonstrations in which hundreds of thousands of Iranians protested the rigged elections of 2009, but Iranian conservatives did not; American conservatives are anticommunist, but many Chinese conservatives belong to the Communist Party. Progressivism lacks invariance to the extent that leftists in various societies differ in their goals and in their estimations of how best to achieve them: Progressives like to blame social problems on wealthy elitists, yet many wealthy elites *are* progressive. The one genuinely invariant political philosophy is liberalism. Its prescription of personal freedom and universal human rights works without amendment in all cultures. When people express admiration for American freedom and democratic values, they are admiring not the fact that Americans are patriotic, or that they have Medicare and Social Security, but that thanks to liberal democracy the American electorate has the power to go from a President George W. Bush to a President Barack Hussein Obama—or the other way around.

Science is likewise invariant: Scientific research is practiced in much the same ways everywhere, deriving results that apply throughout the known universe. Therefore it would be surprising if science did not eventually imply and evince invariant ethical standards. Some of these emerging values may strike us as odd—if, for instance, humans are animals, and differ from other animals only by degrees, then shouldn't the other animals have some rights?—but the teachings of Jesus seemed odd at first, too, and yet have proved lastingly popular. What might a scientific ethics look like? Two promisingly invariant precepts were suggested recently by the geneticist Sydney Brenner, who when asked by a student at the Salk Institute what commandments should govern the behavior of scientists, replied, "To tell the truth," and, "To stand up for all humanity."

People everywhere wish for a better world—a more peaceful and prosperous world, where their children can live healthy, happy lives—and they have long sought the right intellectual tools with which to pursue this goal. Religion works best when it emphasizes common decency, philosophy when stressing our ignorance, art when exposing us to visions larger than ourselves, history by drawing lessons from the past—but the most effective tools are liberalism and science. They may on occasion lead to harmful results, as may anything else: You can poison a prophet with a Girl Scout cookie. But science and liberalism have an unequaled capacity for doing good—for reducing cruel ignorance and villainous certitude, encouraging freedom and effective government, promoting human rights, putting food in the mouths of the hungry and attainable prospects in their future. If we keep our heads, use our heads, nourish learning, tend the fires of freedom, and treat one another with justice and compassion, our descendants may say of us that we had the vision to do science, and the courage to live by liberty.

ACKNOWLEDGMENTS

For my having been able to write this book I am grateful to more individuals than can be named here, going back to the public school teachers who showed me that one can be both educated and sensible. Special thanks to Tim Duggan, Patrick Ferris, Owen Laster, Tom Leonard, Jay Mandel, Alice Mayhew, Menno Meyjes—and to my wife, Carolyn, thanks to whose unwavering support I have been able to do my best, which really is all a writer can ask.

NOTES

CHAPTER ONE: SCIENCE & LIBERTY

3 "Everything of importance": Jefferson Hane Weaver, *The World of Physics* (New York: Simon and Schuster, 1987), 683. The same point was made by the Islamic Indian poet Humayun Kabir: "There have been brilliant scientists and speculators in ancient India, China, Egypt and Greece, but like solitary stars in the firmament they shone in splendid isolation." Humayun Kabir, *Science, Democracy and Islam* (London: George Allen, 1955), 10. As the twentieth-century physicist Hans Bethe noted, "Most philosophical questions were quite well answered by the old Greeks, and even better by people from 1500 to 1800 [but] science is always more unsolved questions, and its great advantage is you can prove something is true or something is false." (Associated Press obituary of Hans Bethe, March 7, 2005.)

3 eighty-nine democracies: Freedom House "Freedom in the World 2009 Survey," http://www.freedomhouse.org/template.cfm?page=445 (accessed March 31, 2009).

5 "There is but one state of learning": Francis Bacon, *The Great Instauration*, preface.

5 "Freedom of inquiry": *Encyclopaedia Britannica*, 15th ed., 6:986.

6 "the three greatest men that have ever lived": Thomas Jefferson, letter to John Trumbull, February 15, 1789.

9 "The society of scientists is simple": Jacob Bronowski, *Science and Human Values* (New York: Harper & Row, 1965), 68.

9 "Men have asked for freedom": Ibid., 70.

10 One factor is wealth: See Adam Przeworski and Fernando Limongi, "Modernization: Theories and Facts." *World Politics* 49.2 (1997) 155–83, and a discussion in Fareed Zakaria, *The Future of Freedom: Illiberal Democracy at Home and Abroad* (New York: Norton, 2003), 69ff.

11 "the role of thought is to explain and transmit": *The Economist*, July 6, 2002, 27.

11 "I approve of democratic ideals": World Values Survey, in Mark Leonard, Rouzbeh Pirouz, "Iraqis Don't Need More Propaganda," *International Herald Tribune*, February 6, 2004.

13 "a monarchy is a merchantman": Fisher Ames, address to the House of Representatives, 1795.

14 "Perfect democracy is an illusion": Stein Ringen, *Democracy, Science, and the Civic Spirit: An Inaugural Lecture* (Oxford: Clarendon Press, 1993), 14. The lecture was originally delivered before the University of Oxford on October 27, 1992.

15 "Nature is only subdued by submission": Bacon, *Novum Organum*, I:1 (Chicago: University of Chicago Press, 1952), 107.

CHAPTER TWO: SCIENCE & LIBERALISM

16 "one very simple principle": J. S. Mill, *On Liberty* (London: Penguin, 1985), 28. Two centuries earlier, in chapter 21 of his *Leviathan*, Thomas Hobbes defined liberty as "freedom . . . the absence of opposition (by opposition, I mean external impediments of motion)."

17 "There are no aims of education": Andrew Abbott, "Welcome to the University of Chicago," address to incoming freshmen, September 26, 2002. Italics added.

17 "Those are not at all to be tolerated": John Locke, *A Letter Concerning Toleration* (Stilwell, KS: Digireads, 2005), 172. Available online at http://odur.let.rug.nl/~usa/D/1651-1700/locke/ECT/toleraxx.htm (accessed November 16, 2004).

18 "I beseech ye": Learned Hand, address to 150,000 newly naturalized American citizens, New York City's Central Park, May 21, 1944; Oliver Cromwell, letter to the general assembly of the Church of Scotland, August 1650.

18 "The starting point of liberal thought": Ludwig von Mises, *Liberalism: The Classical Tradition* (Indianapolis: Liberty Fund, 2005), 76.

19 "Where there is no law there is no freedom": John Locke, *Second Treatise of Government*, ed. C. B. Macpherson (London: Hackett, 1980), 32.

19 "Opinion tends to encroach more on liberty": J. S. Mill to Harriet Taylor, January 15, 1855; in J. S. Mill, *On Liberty* (London: Penguin, 1985), 23.

20 "The application of the teachings of science": Ludwig von Mises, *Liberalism*, xix. Materialistic criteria can inform decisions often assumed to be primarily or exclusively matters of ethics and morality. It was possible, for instance, to oppose slavery simply by testing the postulate that slave economies perform poorly by comparison to free economies. The economic calculus goes like this: A slave will not work as hard to tend a master's field as he would his own, unless he is constantly harassed by an overseer—and that requires the presence of two men where one would do, were that one man free. Hence slavery is unproductive. Universal health care can similarly be defended simply on grounds that it improves both the health (and hence the productivity) of the workforce—and its mobility, the latter by freeing those who previously got their health insurance through their employers to seek work wherever they please. People may introduce moral or emotional arguments as they wish, but the "cold and heartless" quantitative approach is more revealing than it is given credit for.

20 "We are socialists": John Toland, *Adolph Hitler* (New York: Anchor, 1976), 224–25. Toland attributes this quotation to Hitler, others to Hitler's Nazi Party rival Gregor Strasser.

21 "Whether democracy should be defined": Reinhold Niebuhr, "Russia and the West," *The Nation*, January 16, 1943, 83.

21 Only 20 percent of Americans: National Election Study, Pew, Harris, and Democracy Corps polls; Patricia Cohen, "Proclaiming Liberalism, and What It Now Means," the *New York Times*, June 2, 2007. Cohen notes that the 20 percent figure has held steady "since at least 1992."

23 The divisions imposed on Korea and China: *CIA World Factbook*, 2008, expressed in purchasing-power parity (ppp). The contrast is even more imposing when it is taken into account that China and North Korea have considerably more natural resources than do Taiwan and South Korea.

25 "highest quality of life": *The Economist Pocket World in Figures* (London: Profile, 2007).

26 "All institutions of freedom": Friedrich A. Hayek, *The Constitution of Liberty* (Chicago: University of Chicago Press, 1960), 30.

26 "science unifies humanity": Robert Nozick, *Invariances: The Structure of the Objective World* (Cambridge, MA: Harvard University Press, 2001), 63.

26 "Science could contain values": Ibid., 95.

27 "In an advancing society": H. B. Phillips. "On the Nature of Progress." *American Scientist* 33 (1945), 255.

27 The value of individual creativity: It was once widely assumed that scientific and technological innovations would increase the power of big governments and big corporations at the expense of the average individual. Students in the 1960s, obliged to register for courses by filling out punch cards to be read by the big mainframe computers in the college administration building, demonstrated, carrying signs reading, "I am a human being. Please do not fold, mutilate, or staple me!" One visionary who glimpsed a different technological future was the science writer James Gerald Crowther, who wrote in the first half of the twentieth century, "At present science is developing in the direction of big instruments and organizations [but] it seems probable that it will evolve through this phase, and arrive at a new and higher one in which its instruments will again be small and compact. Science may show how a man can provide all his needs . . . from very small instruments and concentrated supplies carried in his pockets. . . . If science is developed to this stage, it would provide new concrete bases for freedom." J. G. Crowther, *The Social Relations of Science,* xxvii, in Joseph Needham, *Time: The Refreshing River* (London: George Allen, 1943), 165–66. Crowther had trouble imagining what such "instruments" would be like—human imagination seldom being capable of foreseeing such a thing as cell phones with more computing power than the Apollo lunar lander.

28 "not the work of philosophers": Mill, "Bentham," in his *Utilitarianism and Other Essays* (New York: Penguin, 1987), 134.

28 "all theory was against it and all experience for it": Ibid.

28 The people as a "beast," etc.: Ibid., 19; Ed Husain, *The Islamist: Why I Joined Radical Islam in Britain, What I Saw Inside and Why I Left* (London: Penguin, 2007), 162; Plato, *Republic*, VI, 493B; Sir Thomas Browne, *Religio Medici*, 1643, pt. II, sec. 1; Wentworth Dillon, *Essay on Translated Verse*, l. 96; Stendhal, *De l'Amour*, 1822, Fragments; Tennyson, *The Princess*, Conclusion, l. 54; Thoreau and Carlyle, in James Surowiecki, *The Wisdom of Crowds* (New York: Anchor, 2005), xv, xvi.

29 "You're nobody if you don't get booed": Robert Hilburn, "Some Q&A with Bob Dylan," *Los Angeles Times*, October 12, 2001, 20.

30 "I surveyed them in every way": Edwin Black, *War Against the Weak: Eugenics and America's Campaign to Create a Master Race* (New York: Four Walls Eight Windows, 2003), 14.

30 "peculiarities in the silkworm": Charles Darwin, *Origin of Species* [1859] (New York: Modern Library, 1998), 19, 32, 36.

30 "the stupidity and wrong-headedness of many": Francis Galton, *Memories of My Life* (London: Methuen, 1908), 246.

30 "stern compulsion ought to be exerted": Ibid., 311.

31 "the freedom to marry": *Loving v. Virginia*, 388 U.S. 1 (1967).

31 "The sixpenny fee deterred practical joking": Francis Galton, "Vox Populi," *Nature*, March 7, 1907, 450.

31 "more creditable to the trustworthiness of a democratic judgment": Ibid. The weight estimate Galton published was the statistical mean of all the legible tickets.

31 "random crowds of people with nothing better to do": Surowiecki, *The Wisdom*

of Crowds, 4. Contestants on this particular quiz show had the option of polling the audience, who at most shows leave the studio without having been asked for their opinions.

34 "Being an informed participant citizen": Gerald Holton, "How to Think about the 'Anti-Science' Phenomenon." *Public Understand. Sci.* 1:103–128 (1992).

CHAPTER THREE: THE RISE OF SCIENCE

35 The European region that best fit those criteria was Italy: This point is made, with some exaggeration, by Harry Lime, the villain in Orson Welles's film *The Third Man*: "In Italy for thirty years under the Borgias they had warfare, terror, murder, and bloodshed, but they produced Michelangelo, Leonardo da Vinci, and the Renaissance. In Switzerland they had brotherly love—they had five hundred years of democracy and peace—and what did that produce? The cuckoo clock!" (Script by Graham Greene, Alexander Korda, Carol Reed, and Orson Welles.)

36 "our liberty, which we have inherited from our forefathers": Quentin Skinner, *The Foundations of Modern Political Thought. Volume 1: The Renaissance* (New York: Cambridge University Press, 1978), 7.

36 "God's most precious gift to human nature": Dante, *Monarchy*, in Skinner, *Foundations*, 16.

36 "the eldest child of liberty": William Wordsworth, "On the Extinction of the Venetian Republic." In the fifteenth century the Venetians tried an elaborate system that combined ballot voting with chance. A thousand or more Venetian patricians lined up and each picked one marble, thirty of which were gilded. Each time a patrician happened to get a gilded marble, his relatives were all escorted from the hall and excluded from further selection, to prevent undue influence by any one family. Once the thirty had been chosen, twenty-one of them were eliminated by another random draw. The remaining nine then voted for a total of forty electors, whose numbers were reduced to twelve by another round of drawing by gilded marbles. The twelve then elected twenty-five more electors, of whom sixteen were randomly eliminated . . . and so on.

36 "There is no city in the world where merit is more recognized": Domenico Sella, *Italy in the Seventeenth Century* (London: Longman, 1997), 70.

36 "It is impossible to judge of equality": Casanova, *History of My Life*, William R. Trask, trans. (New York: Knopf, 2007), 233–34.

37 "being unarmed . . . causes you to be despised": Machiavelli, *The Prince*, chapters 8 and 14.

37 "the principal matter . . . is for a courtier": Baldassare Castiglione, *Il Cortegiano*, in Geoffrey R. Elton, ed., *Renaissance and Reformation 1300–1648*, Thomas Hoby, trans. (New York: Macmillan, 1968), 75–78.

37 Michelangelo's larger-than-life sculpture *David*: This is not to say that Michelangelo gave much weight to politics in accepting commissions. His *Pietà* was sculpted on commission by a cardinal in Rome, and he cast a gigantic bronze of Pope Julius II (it was later destroyed) in Bologna to commemorate and cement the pope's recent conquest of that city. The fresco to cover his remarkable cartoon the *Battle of Cascina* in Florence was never executed, Michelangelo having decamped for Rome in response to an offer of more money.

38 Leonardo's "thought seems always to be moving": John Herman Randall Jr. "The Place of Leonardo da Vinci in the Emergence of Modern Science." *Journal of the History of Ideas* vol. 14, no. 2 (April 1953): 195. Randall adds that Leonardo, given his fascination with hands-on experimentation and inventing, might "have found Edison more congenial than Einstein."

39 "I am fully aware": Irma A. Richter, ed., *The Notebooks of Leonardo da Vinci* (Oxford: Oxford University Press, 1952), 2. Trans. slightly altered by T.F. See also

Jacob Bronowski, "Leonardo da Vinci," in Richard M. Ketchum, ed., *The Horizon Book of the Renaissance* (New York: Doubleday, 1961), 187–88.

39 "You designed a horse to be cast in bronze": George Bull, *Michelangelo: A Biography* (London: Penguin, 1996), 47.

40 "a world on paper": Stillman Drake, *Galileo at Work: His Scientific Biography* (New York: Dover, 1987), 15.

40 invoking an irrational number: Vincenzio was ahead of his time on this issue. By the end of the seventeenth century, a new system of "tempered" tuning was developed—based, indeed, on an irrational number, the twelfth root of two. Johann Sebastian Bach wrote the *Well-Tempered Clavier* to demonstrate its advantages. Galileo's biographer Stillman Drake speculates that this work of Bach's "would never have been written had Zarlino won the argument against Galileo's father." Drake, *Galileo at Work*, 16.

41 "a hundred times heavier": Laura Fermi and Gilberto Bernardini, *Galileo and the Scientific Revolution* (New York: Basic Books, 1961), 15.

43 "It is impossible to obtain wages from a republic": Stillman Drake, *Discoveries and Opinions of Galileo* (New York: Anchor, 1957), 65.

43 "Here the freedom and the way of life of every class": Giovanni Sagredo to Galileo, 1611.

43 "I deem it my greatest glory": Drake, *Discoveries and Opinions of Galileo*, 62.

43 "to punish me for my contempt for authority": Banesh Hoffman, *Albert Einstein: Creator and Rebel* (New York: Viking, 1973), 24.

44 "the prohibition of the printing and sale of my *Dialogues*": Galileo to Cardinal Barberini, October 13, 1632.

44 "errors and heresies": Galileo's abjuration, in Maurice A. Finocchiaro, ed. and trans., *The Galileo Affair: A Documentary History* (Berkeley: University of California Press, 1989), 292.

44 "see that just as nature has given to them": Letter to Paolo Gualdo, in Drake, *Discoveries and Opinions of Galileo*, 84.

44 "since if they had seen what we see": Ibid., 135.

44 "In the discovery of secret things": William Gilbert, *De Magnete*, P. Pleury Mottelay, trans. (New York: Dover, 1958), 1, 14–15.

45 "dampened the glory of Italian wits": Domenico Sella, *Italy in the Seventeenth Century* (London: Longman, 1997), 233.

45 "that there had been men before Adam": Margaret Jacob, *Scientific Culture and the Making of the Industrial West* (New York: Oxford University Press, 1997), 28.

45 rioters who sacked his house: The Priestley Riots, as they are known to history, were provoked by Priestley's outspoken skepticism about certain articles of religious faith, among them the doctrines of the Trinity and of virgin birth. (He was a "dissenting" theologian.) The proximate cause was troublemaking by a clique seeking to curry favor with the Church of England. Having fomented the riots by circulating various lies and exaggerations about Priestley, they celebrated by drinking till dawn at a pub called the Swan. When the sun rose, one of the plotters—Benjamin Spencer, the Vicar of Aston—rode out to survey the smoking ruins of Priestley's home and looted them of Priestley's scientific papers. They were never recovered.

45 "All depends on keeping the eye steadily fixed": Francis Bacon, *The Great Instauration*, "The Plan," James Spedding, Robert Leslie Ellis, and Douglas Denon Heath, trans. In *The Works of Francis Bacon* (Boston: Taggard and Thompson, 1863), http://www.constitution.org/bacon/instauration.htm.

45 "I showed what the laws of nature were": Descartes, *Discourse on Method*, part V, Robert Stoothoof, trans. In John Cottingham et al., trans., *Descartes: Selected Philosophical Writings* (New York: Cambridge University Press, 1988), 41.

46 "becalmed ships": Basil Montagu, *The Life of Francis Bacon, Lord Chancellor of England* (London: Pickering, 1833), 10.

46 came as a guest and not as an enemy: Thomas Babington Macaulay, *The Life and Writings of Francis Bacon, Lord Chancellor of England* (Edinburgh: Edinburgh Review), 1837.

46 He dismissed Copernicus': Bacon's objection typically was that these researchers had induced too wide a range of concepts from their few experiments. "The race of chemists, again out of a few experiments of the furnace, have built up a fantastic philosophy, framed with reference to a few things; and Gilbert also, after he had employed himself most laboriously in the study and observation of the loadstone, proceeded at once to construct an entire system in accordance with his favorite subject." Bacon, *Novum Organum*, Book I, Aphorism LIV.

46 "Bacon . . . possessed in a most eminent degree the genius of philosophy": H. R. Fox Bourne, *The Life of John Locke* (New York: Harper, 1876), 69.

47 "It will be seen amidst the erection of temples": Bacon, *Novum Organum*, in Montagu, *Life of Francis Bacon,* 465.

47 "shut up in the cells of a few authors": Montagu, *Life of Francis Bacon,* 7; "For inquiries in new and unlabored parts of learning" Ibid., endnote K; "Spin cobwebs of learning" Ibid., 7.

48 "to escape upwards from this Satan-ridden earth": Willey, Basil, *The Eighteenth Century Background: Studies on the Idea of Nature in the Thought of the Period* (New York: Columbia University Press, 1961), 32.

48 "the ostentation of dispute": Bacon, *Novum Organum,* 1:20.

48 "They certainly have this in common with children": Ibid., 1:71.

48 "The reverence for antiquity": Ibid., 1:84.

48 "the discovery of things is to be taken from the light of nature": Ibid., 1:122.

48 you find them to be not empty notions but well defined: Ibid., 1:14; ibid., Preface; *The Great Instauration,* "The Arguments," in Sidney Warhaft, ed., *Francis Bacon: A Selection of His Works* (Indianapolis: Bobbs-Merrill, 1982), 314–15.

49 "familiarity between the mind and things": Bacon, *Magna Instauratio,* opening sentence.

49 "a sort of soft and stinking mud": Peter Ackroyd, *London: The Biography* (New York: Doubleday, 2000), 100.

49 "in every street, carts and coaches": Stephen Inwood, *A History of London* (New York: Carroll & Graf, 1998), 204.

50 "I don't think much of my way of life": Ackroyd, *London*, 133.

50 "None but the clergy": Joseph Priestley, "Essay on a Course of Liberal Education for Civil and Active Life," 1765.

50 "It is by instruments and helps that the work is done": Bacon, *Aphorisms Concerning the Interpretation of Nature and the Kingdom of Man,* II.

50 "Nature to be commanded must be obeyed": Ibid., III; "Knowledge itself is power" Bacon, *Meditationes Sacrae* 11, "de Haeresibus," 1597.

50 "New artificial metals": Bacon, *New Atlantis* (Philadelphia: Franklin Library, 1982), 382ff.

51 "He wrote like a philosopher and lived like a prince": Will and Ariel Durant, *The Age of Reason Begins* (New York: Simon and Schuster, 1961), 170.

51 "to provide instruction to sailors and merchants": Peter Dear, *Revolutionizing the Sciences: European Knowledge and Its Ambitions, 1500–1700* (Princeton, NJ: Princeton University Press, 2001), 52.

51 "all knowledge is to be limited by religion": Bacon, *Valerius Terminus,* ch. 1, in *Philosophical Works,* J. M. Robertson, ed. (London: Longmans, 1905), 186.

51 "the scaffold with which the new philosophy was raised": Voltaire, *Philosophical Letters,* letter 12, "On Chancellor Bacon."

51 "Bacon, like Moses, led us forth at last": Abraham Cowley, "To the Royal Society," lines 93–98.

51 "I have only taken upon me to ring a bell": Bacon, letter to Dr. Playfer, in I. Bernard Cohen, *Revolution in Science* (Cambridge, MA: Harvard University Press, 1985), 152.

52 "men's thoughts are frozen here": Descartes, letter of January 15, 1650, in *Oeuvres De Descartes*, Charles Adam and Paul Tannery, eds. (Paris: Librairie Philosophique J. Vrin, 1983), vol 5, 467; and *The Philosophical Writings of Descartes*, John Cottingham, Robert Stoothoff, and Dugald Murdoch, trans. (Cambridge: Cambridge University Press, 1988), vol. 3, 383.

52 "purely mathematical type of mind": D. E. Smith, *History of Mathematics* (New York: Dover, 1951), vol. 1, 376.

53 "The only principles which I accept, or require": Dear, *Revolutionizing the Sciences*, 88.

53 explain things but predict next to nothing: Peter Dear puts this aptly in *Revolutionizing the Sciences:* "An Aristotelian world was not one in which there were countless new things to be discovered; instead, it was one in which there were countless things, mostly already known, left to be *explained*. That Aristotle himself does not seem to have believed this is beside the point; it was nonetheless the lesson that his scholastic followers in medieval and early-modern Europe tended to draw from those of his writings that they found most interesting and most teachable." (132)

53 "good sense is of all things": Descartes, *Discourse on the Method of Rightly Conducting the Reason and Seeking for Truth in the Sciences*, Elizabeth Haldane, trans., Part I.

53 "the increasing discovery of my own ignorance": Ibid., 46.

53 "preoccupation with the indubitable": "René Descartes," in Paul Edwards, ed., *The Encyclopedia of Philosophy* (New York: Macmillan, 1967), vol. 2, 346.

55 "I have explained the phenomena of the heavens": Newton, *Principia*, I. Bernard Cohen and Anne Whitman, trans. (Berkeley: University of California Press, 1999), 943.

55 "although the whole of philosophy is not immediately evident": Ibid., 61.

56 *cogito ergo sum*: If Descartes' famous "I think, therefore I am" is reworded to repair the logical circularity invoked by assuming the existence of an "I" in order to establish its existence, it devolves into the rather inconsequential statement, "Thinking transpires, therefore somebody must be thinking!" For an able discussion concerning English translations of Newton's "fingo," see I. Bernard Cohen, "The First English Version of Newton's Hypotheses Non Fingo," *Isis*, vol. 53, no. 3 (September 1962): 379–88.

56 "a largely unintelligible book": I. Bernard Cohen, "The Newtonian Revolution in Science," in Paul Theerman and Adele Seeff, eds., *Action and Reaction: Proceedings of a Symposium to Commemorate the Tercentenary of Newton's* Principia (Newark: University of Delaware Press, 1993), 77.

56 "a conqueror rather than a rebel": Hunter Crowther-Heyck, review of A. Rupert Hall, *The Scientific Revolution, 1500–1800: The Formation of the Modern Scientific Attitude* (Boston: Beacon Press, 1966), *H-Ideas*, February 2001, 2. Crowther-Heyck is paraphrasing Hall.

CHAPTER FOUR: THE SCIENCE OF ENLIGHTENMENT

58 "Newton rose at once to the highest pinnacle of glory": Thomas Thomson, *History of the Royal Society, from Its Institution to the End of the Eighteenth Century* (London: Baldwin, 1812), 342.

58 "the greatest and rarest genius": Peter Gay, *The Enlightenment: The Science of Freedom* (New York: Norton, 1996), 130.

58 "God said 'Let Newton be' ": Alexander Pope, "Essay on Man" [1734].

58 "people were beginning to dispute": Gay, *Enlightenment*, 129.

58 "We are all his disciples now": Ibid.

58 "one of the architects of our civil liberties": Freeman Dyson, "A New Newton," *New York Review of Books*, Volume 50, Number 11 (July 3, 2003).

59 "he was perpetually unbuttoning the stores of his royal wisdom": G. M. Trevelyan, *A Shortened History of England* (London: Penguin, 1987), 279.

59 "kings exercise . . . divine power on earth": John Cannon and Ralph Griffiths, *The Oxford Illustrated History of the British Monarchy* (New York: Oxford University Press, 1997), 356. The divine-right doctrine was upheld by contemporary royalists like Sir Robert Filmer (1590–1653). Knighted by a like-minded Charles I, Filmer argued that nations are like families and monarchs their fathers and mothers, and that therefore (!) regents are above all earthly law, just as parents are not to be ordered around by their children. (They enjoy what Filmer called the "Fatherly Right of Sovereign Authority.") Locke's *Two Treatises of Government* was written to refute Filmer's *Patriarchia, or the Natural Power of Kings* (published posthumously in 1680). Otherwise Filmer is of interest today chiefly because his stance, if inverted, becomes perhaps the strongest argument for having a constitutional monarch with little real power rather than having no regent at all. This argument states that (1) Many people are naïve enough to regard their chief of state as a father figure; (2) A figurehead monarch serves as a kind of lightning rod for this misplaced affection, and so helps citizens evaluate their real leaders in a clearer light, free from a confusing nimbus of sentimentality; (3) The nation's actual leader—the prime minister, say—is spared presiding over ribbon cuttings and other ceremonial duties handled by the monarch, and so has more time to get work done—or go fishing, during which recreation he or she may conceivably come upon a useful idea.

59 "Literature and science were . . . surrounded with pomp": Lord Macaulay, *The History of England* (London: Penguin, 1986), 180–181.

60 "He was the steady friend of civil liberty": Ibid., 184.

60 "Go your way and sin no more": Ibid., 185.

60 "Go home": Ibid., 196.

61 Whigs . . . and Tories: The word *Whig* originated as a slur—as have many other terms denoting political philosophies, including *Tory*, for "outlaw." Originally reserved for country bumpkins, the term evidently came from a buttermilk-like drink favored by the bucolic poor. By the seventeenth century it was being applied to political rebels, particularly the Protestant insurgents who wrested control of western Scotland from the Royalists in 1648, and American revolutionaries who supported independence from the Crown. After 1689 it came to mean modern liberals, as opposed to the more conservative Tories. In the twentieth century *Whig* became a term of opprobrium for what the historian of science Herbert Butterfield called "the tendency in many historians to write on the side of Protestants and Whigs, to praise revolutions provided they have been successful, to emphasize certain principles of progress in the past and to produce a story which is the ratification if not the glorification of the present." Herbert Butterfield, *The Whig Interpretation of History* (New York: Scribner's, 1951), preface. This Butterfield deemed to be a bad thing, "the very sum and definition of all errors of historical inference. The study of the past with one eye, so to speak, upon the present is the source of all sins and sophistries in history." Ibid., 3–4. Although Butterfield's essay occasioned little discussion prior to his death in 1979, his banner was later taken up by postmodernist academics alarmed "whenever history is written either by, or on behalf of, a

triumphant elite." Adrian Wilson and T. G. Ashplant, "Whig History and Present-Centered History," *The Historical Journal*, 1988, vol. 31, no. 1, 3. In this rather circular way of thinking, to be "triumphant" means to argue that circumstances got materially better, and the "elite" is whoever made them better.

61 Newton met John Locke: There is some evidence that the two may have met previously, but in any event this was their first meeting subsequent to the publication of Newton's *Principia*, which Locke in exile had greeted with great enthusiasm.

61 "be courageous therefore and steady to the laws": James Gleick, *Isaac Newton* (New York: Pantheon, 2003), 143; Newton to John Covel, February 21, 1689.

62 "the only philosophy then known at Oxford": Bourne, *Life of John Locke* (New York: Harper, 1876), 48.

62 "the faithful product of observation": Maurice Cranston, *John Locke: A Biography* (New York: MacMillan, 1957), 92.

63 "Drawn to the dregs of a Democracy": Dryden, "Absalom and Achitophel."

63 "I appeal to the consciences of those that persecute, torment, destroy, and kill": Locke, *Letter Concerning Toleration,* text prepared by Garry Wiersema. http://odur.let.rug.nl/~usa/D/1651-1700/locke/ECT/toleraxx.htm (accessed November 16, 2004).

63 "no person whatsoever shall disturb, molest, or persecute": Bourne, *Life of John Locke,* 242.

63 a seditious manuscript: Wrote Locke, in just one of the many paragraphs that would, if discovered by the king's agents, have sufficed to cost him his life, "Despotical power is an absolute, arbitrary power one man has over another, to take away his life, whenever he pleases. This is a power, which neither nature gives, for it has made no such distinction between one man and another; nor compact can convey: for man not having such an arbitrary power over his own life, cannot give another man such a power over it; but it is the effect only of forfeiture, which the aggressor makes of his own life, when he puts himself into the state of war with another: for having quitted reason, which God hath given to be the rule betwixt man and man, and the common bond whereby human kind is united into one fellowship and society; and having renounced the way of peace which that teaches, and made use of the force of war, to compass his unjust ends upon another, where he has no right; and so revolting from his own kind to that of beasts, by making force, which is their's, to be his rule of right, he renders himself liable to be destroyed by the injured person, and the rest of mankind, that will join with him in the execution of justice, as any other wild beast, or noxious brute, with whom mankind can have neither society nor security." Locke, *Second Treatise of Government,* Sec. 172.

64 two retrograde regents: Both were enemies of religious tolerance and upholders of regal pomp. Philip, who instigated an Inquisition to suppress Calvinism in Holland, once said he would rather lose control of the Low Countries than see them lapse from Catholicism—which is just what happened. Louis, whose motto was "*L'état c'est moi*" (I am the state), was known to supplicants as the "Sun King" and to his detractors as "the new would-be Caesar." He spent millions building the Versailles Palace, an elegant machine for beguiling and preoccupying noblemen who might otherwise have worked to abridge his peerless power.

64 "not in the hope of some better fortune": Biographical note to Spinoza's *Ethics,* W. H. White, trans. (Chicago: University of Chicago Press, 1952), 352.

64 "the only certain and reliable criterion of truth": Spinoza, *Tractatus Theologico-Politicus*, S. Shirley, trans. (Leiden: Brill, 1989), 126.

64 "The purpose of the state is freedom": Matthew Stewart, *The Courtier and the Heretic* (New York: Norton, 2006), 102.

64 "Democracy is of all forms of government the most natural": Jonathan I. Israel, *The Dutch Republic: Its Rise, Greatness, and Fall 1477–1806* (Oxford: Clarendon, 1995), 787.

64 "The supreme mystery of despotism": Spinoza, *Tractatus Theologico-Politicus*; in Stewart, *Courtier and Heretic*, 100.

65 The Dutch innovated oil painting: The invention of oil painting as we think of it today is generally credited to the Flemish painter Jan van Eyck, who around 1410 developed a stable, oil-based varnish, although oil paints themselves came into use a bit earlier. As Prof. John H. Lienhard of the University of Houston puts it, "Van Eyck was to oil painting what Watt was to the steam engine." Lienhard, "Engines of Our Ingenuity," 809, http://www.uh.edu/engines/epi809.htm (accessed March 22, 2009).

66 "We murder to dissect": Wordsworth, "The Tables Turned" (1798); Blake, "The New Jerusalem," from his *Milton* (1804–1808). D. H. Lawrence, a reliable amplifier of romantic opinions, wrote in the twentieth century, "The dark, satanic mills of Blake / how much more darker and more satanic they are now!"

66 "Here were products of a major European civilization": Gary Schwartz, "Art in History," in David Freedberg and Jan de Vries, eds., *Art in History / History in Art* (Santa Monica, CA: Getty Center, 1991), 8.

67 "Visitors continually marveled": Israel, *Dutch Republic*, 1.

68 "I found myself in a storm": Cranston, *Locke*, 3.

69 "In an age that produces such masters": John Locke, *An Essay Concerning Human Understanding*, Peter H. Nidditch, ed. (Oxford: Clarendon Press, 1979), 10. Still smarting at his woeful attempts to absorb Aristotelian philosophy at Oxford decades earlier, Locke goes on to assert that science would already have advanced much further "if the endeavors of ingenious and industrious men had not been much cumbered with the learned but frivolous use of uncouth, affected, or unintelligible terms, introduced into the sciences, and there made an art of, to that degree that Philosophy, which is nothing but the true knowledge of things, was thought unfit or incapable to be brought into well-bred company and polite conversation. Vague and insignificant forms of speech, and abuse of language, have so long passed for mysteries of science; and hard and misapplied words, with little or no meaning, have, by prescription, such a right to be mistaken for deep learning and height of speculation, that it will not be easy to persuade either those who speak or those who hear them, that they are but the covers of ignorance, and hindrance of true knowledge. To break in upon the sanctuary of vanity and ignorance will be, I suppose, some service to human understanding."

69 "He everywhere takes the light of physics for his guide": Voltaire, François-Marie Arouet, *Letters on England*, Letter XIII, "On Mr. Locke" (London: Cassell & Company, 1894); etext prepared by David Price, http://isis.library.adelaide.edu.au/cgi-bin/pg-html/pg/etext00/lteng10.txt (accessed May 25, 2005).

69 "not able to endure much": *Oxford English Dictionary*, 2002.

69 "you endeavored to embroil me with women": Newton to Locke, September 16, 1693.

69 "I have been . . . so entirely and sincerely your friend": Locke to Newton, October 5, 1693.

70 "the greatest man in the world": John Dunn, J. O. Urmston, and A. J. Ayer, *The British Empiricists* (New York: Oxford University Press, 1992), 11.

70 small, informal gatherings to discuss science: The names of only two members of this little circle have come down to us. They are Dr. David Thomas, who had brought about Locke's introduction to Lord Ashley; and James Tyrrell, a college friend of Locke's and later an influential Whig thinker and writer.

70 "puzzled . . . perplexed": Locke, *Essay Concerning Human Understanding*, Epistle to the Reader, 7.

70 "To explicate by subtle hints": Regarding this poem A. E. Shipley observes, "Butler did not appreciate what even in these days is not always appreciated, that the minute investigation of subjects and objects which to the ordinary man seem trivial and vain often lead to discoveries of the profoundest import to mankind." A. E. Shipley, "The Progress of Science," in A. W. Ward and A. R. Waller, eds., *The Cambridge History of English and American Literature*, vol. 8 (New York: Putnam, 1907–21), online edition copyright 2000 Bartleby.com, Inc., http://www.bartleby.com/cambridge (accessed December 28, 2004).

71 "The stage is too big for the drama": Richard P. Feynman interviewed by Bill Stout, 1959, in Richard Feynman, *Perfectly Reasonable Deviations from the Beaten Track* (New York: Basic Books, 2006), appendix 1.

71 "philosophy is the misuse of a terminology": Eugene P. Wigner, "The Unreasonable Effectiveness of Mathematics in the Natural Sciences," in *Communications in Pure and Applied Mathematics*, vol. 13, no. I (February 1960). Wigner attributes it to W. Dubislav, *Die Philosophie der Mathematik in der Gegenwart* (Berlin: Junker und Dunnhaupt Verlag, 1932), 1.

71 "We should not then perhaps be so forward": Locke, *Essay*, IV, 22.

72 "All true science begins with empiricism": *Nature*, no. 615 (1881), 343.

72 "undervalue the advantages of our knowledge": Ibid., 46.

73 "What is not surrounded by uncertainty cannot be the truth": Richard P. Feynman to the editor of the *California Tech*, February 27, 1976.

73 "The extent of knowledge . . . is so vast": Locke, journal entry made in France in 1677; in Bourne, *Life of Locke*, vol. 1, 360. An entry in Locke's commonplace book dated 1671 gives an impression of the slipperiness of the term "ideas": "I imagine that all knowledge is founded on, and ultimately derives itself from, sense or something analogous to it, and may be called sensation, which is done by our senses conversant about particular objects, which gives us the simple ideas or images of things, and thus we come to have ideas of heat and light, hard and soft, which are nothing but the reviving again in our minds these imaginations which those objects, when they affected our senses, caused in us, whether by motion or otherwise it matters not here to consider; and thus we do when we conceive heat or light, yellow or blue, sweet or bitter. And therefore I think that those things which we call sensible qualities are the simplest ideas we have, and the first objects of our understanding." Ibid., vol. II, 89

73 rights such as liberty: Locke writes variously of "life, liberty, or estate," and "life, liberty, health, and indolency of body; and the possession of outward things, such as money, lands, houses, furniture, and the like." Jefferson's phrase "life, liberty, and the pursuit of happiness" also echoes Locke, who referred to the "pursuit of our happiness."

73 "the power a man has to do or forbear doing": Locke, *Essay*, XXI, 15.

73 "Our understanding and reason were given us": Locke, *Essay*, XXI, 24.

73 "As far as this power reaches": Locke, *Essay*, XXI, 8.

73 "It is lawful for the people . . . to *resist*": Locke, *Two Treatises of Government*, II.xix.232: 437, Peter Laslett, ed. (New York: Cambridge University Press, 1967).

74 "I have seen parents so heap rules on their children": Locke, *Some Thoughts Concerning Education*, 1692, sec. 65.

74 "blank slate": If, for instance, we envision the brain as akin to a binary computer, then our neuroanatomy might be compared to the central processing unit (CPU) and the firmware in a personal computer when we first take it out of the box: The computer's subsequent "experiences" soon make it unique, but its CPU and

its firmware determine the fundamentals of what it can do. For a discussion see, e.g., Steven Pinker, *The Blank Slate: The Modern Denial of Human Nature* (New York: Penguin, 2002).

74 as natural as fish find water: As John Stuart Mill observed, "The book which has changed the face of a science, even when not superseded in its doctrines, is seldom suitable for didactic purposes. It is adapted to the state of mind, not of those who are ignorant of every doctrine, but of those who are instructed in an erroneous doctrine. So far as it is taken up with directly combating the errors which prevailed before it was written, the more completely it has done its work, the more certain it is of becoming superfluous. . . ." Mill, "Professor Sedgwick's Discourse on the Studies of the University of Cambridge," *London Review*, April 1835.

74 "the Age of Paine": John Adams to Benjamin Waterhouse, October 29, 1805.

74 "the friend of human rights": James Monroe to Thomas Paine, September 18, 1794.

74 "free America without her Thomas Paine": *Thomas Paine Readers Newsletter*, vol. 1, no. 1, Fall 1997, 5.

74 "Is it a greater miracle": William Blake, *Annotations to "An Apology for the Bible in a Series of Letters Addressed to Thomas Paine by R. Watson,"* in Geoffrey Keynes, ed., *Blake: Complete Writings* (London: Oxford University Press, 1972), 391.

75 "It was read by public men": John Keane, *Tom Paine: A Political Life* (Boston: Little, Brown, 1995), 112, who references George W. Corner, *The Autobiography of Benjamin Rush* (Princeton: Princeton University Press, 1949),114–15; and see Rush to James Cheetham, Philadelphia, July 17, 1809.

75 "Every living man in America . . . who could read": Elbert Hubbard, "A Little Journey to the Home of Thomas Paine," in Daniel Edwin Wheeler, ed., *Life and Writings of Thomas Paine* (New York: Vincent Parke, 1908), vol. 1, 312.

75 "in a rage when our affairs were at their lowest ebb": Keane, *Paine*, 143, referencing the *Pennsylvania Journal;* and the *Weekly Advertiser*, December 19, 1776.

75 "I think the game is pretty near up": Kenneth Nebenzahl, ed., and Don Higginbotham, *Atlas of the American Revolution* (Chicago: Rand McNally, 1975), 99.

75 "Victory or Death": Willard Sterne Randall, *George Washington: A Life* (New York: Henry Holt, 1997), 322.

76 "These are the times that try men's souls": Paine, *The American Crisis*, Number 1, December 19, 1776, 1.

76 "as much effect as any man living": Jefferson to Paine, March 18, 1801.

76 "never been grasped in their full complexity": Eric Foner, *Tom Paine and Revolutionary America* (New York: Oxford University Press, 1976), xii.

77 "Love abhors clamor and soon flies away": Paine, "Reflections on Unhappy Marriages," *Pennsylvania Magazine*, June 1775.

77 "very well recommended to me as an ingenious worthy young man": Benjamin Franklin to his son-in-law Richard Bache, September 30, 1774, in Keane, *Paine*, 84.

77 "I am led to this reflection by the present domestic state of America": Paine (writing as "Atlanticus"), "Useful and Entertaining Hints," *Pennsylvania Magazine*, February 1775; italics in original.

78 "all destroyed—at least all which the General had": Moncure Daniel Conway, *The Life of Thomas Paine, with a History of His Literary, Political and Religious Career in America, France, and England* (New York: Putnam, 1893), vol. 1, xix–xx, http://www.thomaspaine.org/bio/ConwayLife.html.

79 "the natural bent of my mind was to science": Paine, *Age of Reason*, Part 1, Section 11.

79 "by unwearied application (without a master)": E. Henderson, *Life of James Ferguson, F. R. S.* (London: A. Fullerton, 1867), 450.

79 "grounded on demonstration": Samuel Johnson, *A Dictionary of the English Language* [London, 1755] (Delray Beach, FL: Levinger, 2002), 455.
79 "I purchased a pair of globes": W. E. Woodward, *Tom Paine: America's Godfather 1737–1809* (New York: Dutton, 1945), 30.
79 "the use of the Globes . . . is the first consideration": Benjamin Martin, "The Description and Use of Both the Globes, the Armillary Sphere, and Orrery" (London: Self-published, 1758).
80 "like a bubble or a balloon in the air": Paine, *The Age of Reason*, Part 1, Eric Foner, ed., in *Paine: Collected Writings* (New York: Library of America, 1995), 705.
80 "as our system of worlds does round our central Sun": Ibid., 708.
80 "every tree, every plant, every leaf": Ibid., 704.
80 "There is room for millions of worlds": Ibid., 705.
80 "The same universal school of science presents itself to all": Ibid., 709; see also Marjorie Nicolson, "Thomas Paine, Edward Nares, and Mrs. Piozzi's Marginalia," Huntington Library Bulletin, no. 10 (1936), 113.
80 "no disposition for what is called politics": Ibid., 701.
80 "I took the idea of constructing it from a spider's web": Paine to Sir George Staunton, in Wheeler, *Life and Writings*, vol. 10, 231.
81 "Each of us had a roll of cartridge paper": Paine, "The Cause of the Yellow Fever," in Wheeler, *Life and Writings*, vol. 1, 310–11.
81 "breeding grounds for a new radical politics": Keane, *Paine*, 45.
82 "a silly, contemptible thing": Paine, *Rights of Man*, in Wheeler, *Life and Writings*, vol. 1, 260.
82 "The state of a king shuts him from the world": Paine, *Common Sense,* in Wheeler, *Life and Writings*, vol. 1, 8.
82 "a banditti of ruffians": Paine, *Rights of Man*, 2:2, in Foner, *Paine,* 556.
82 "never traces government to its source": Paine, *Rights of Man*, in Wheeler, *Life and Writings*, vol. 4, 147.
82 "to trump up some superstitious tale": Paine, *Common Sense*, in Wheeler, *Life and Writings*, vol. 2, 21.
82 "All hereditary government is in its nature tyranny": Paine, *Rights of Man*, *Part 2*, in Foner, *Paine*, 559.
82 "As the republic of letters brings forward the best": Paine, *Rights of Man*, *Part 2*, in Foner, *Paine*, 563.
83 "Such governments consider man merely as an animal": Paine, *Agrarian Justice*, in Foner, *Paine*, 410. The phrase italicized by Paine is a quotation from a speech to Parliament by Bishop Samuel Horsley, a Tory.
83 "No one man is capable . . . of supplying his own wants": Paine, *Rights of Man*, Part 2, 1, in Foner, *Paine*, 551.
83 "All the great laws of society are laws of nature": Paine, *Rights of Man*, in Foner, *Paine*, 553.
83 "liberally opened a temple where all may meet": Paine, "Letter to the Abbe Raynal," 1782, http://ideas.repec.org/h/hay/hetcha/paine1908-99.html.
84 "make the best of life": Ibid.
84 "the quiet field of science": Paine to George Clymer, December 13, 1786.
84 "an ingenious, honest man": Keane, *Paine*, 271.
84 "every thing in it that constitutes the idea of a miracle": Paine, *Age of Reason, Part I*, in Foner, *Paine,* 714.
84 "Every thing is a miracle": Paine, *Age of Reason, Part I*, in Foner, *Paine,* 713.
84 "You see the absurdity of monarchical governments": Keane, *Paine*, 313.
85 "mistakes are likely enough to be committed": Keane, Ibid., 315.
85 "I saw my life in continual danger": Paine to Samuel Adams, January 1, 1803; in Wheeler, *Life and Writings*, vol. 6, 301.

85 "running headlong into atheism": Ibid.
85 "postpone it to the latter part of my life": Keane, *Paine*, 390.
85 "To believe that God created a plurality of worlds": Paine, *Age of Reason*, *Part 1*, in Foner, *Paine*, 704.
85 "in an endless succession of death": Paine, *Age of Reason*, *Part 1*, in Foner, *Paine*, 710.
85 "how happened it that he did not discover America?": Paine, *Age of Reason*, *Part 1*, in Foner, *Paine*, 716.
86 "revolutions cannot be made with rosewater": Keane, *Paine*, 381.
86 "I have been less fortunate, but not less innocent": Ibid., 408.
86 "what rendered the scene more horrible": Ibid., 412.
86 "a very funny, witty, old man": Ibid., 449.
87 "I have lived an honest and useful life": Ibid., 533.
87 "The principles of science lead to this knowledge": Paine, *Age of Reason*, in Wheeler, *Life and Writings*, vol. 6, 277.
87 "The Bible represents God to be a changeable, passionate, vindictive Being": Paine, "Letters Concerning the 'Age of Reason': An Answer to a Friend," Paris, May 12, 1797, in Wheeler, *Life and Writings*, vol. 6, 290.
87 "That which is now called natural philosophy": *Age of Reason: Part I*, in Foner, *Paine*, 691.
88 "The study of theology . . . is the study of nothing": *Age of Reason*, in Wheeler, *Life and Writings*, vol. 6, 276–77.
88 a "better Christian": William Blake, *Milton* E140; and in Geoffrey Keynes, ed., *The Complete Writings of William Blake* (London: Nonesuch, 1957), 396.
88 "filthy little atheist": Theodore Roosevelt, *Gouverneur Morris* (Boston: Houghton Mifflin, 1888), 289; in Keane, *Paine*, 323.

CHAPTER FIVE: AMERICAN INDEPENDENCE

90 "a science of the very first order": Thomas Jefferson to David Williams, November 14, 1803.
90 American Indian vocabularies: Jefferson viewed the American Indians as living in a state of nature and having "no law but that of Nature"; Jefferson to John Manners, June 12, 1817. He wondered whether studying Indian societies might adduce evidence supporting the liberal assertion that human equality is based in nature's laws. Jefferson divided societies into three kinds—those governed by force; by the people, as in England and the United States; and those "without government as among our Indians"; Jefferson to James Madison, January 30, 1787. He frequently expressed sympathy for the Indians ("those much-injured people") and ranked the oratory of their greatest speakers with that "of Demosthenes and Cicero"; *Notes on the State of Virginia*, in Merrill D. Peterson, ed., *Jefferson: Writings* (New York: Library of America, 1984), 188. The bulk of Jefferson's papers on American Indian vocabulary were, however, destroyed by thieves who ransacked his baggage.
90 "I would wish to form them into a knot on the same canvas": Jefferson to Richard Pric, January 8, 1789.
90 "sure knowledge": Jefferson, *Notes on the State of Virginia*, Query VI.
90 the solar eclipse of September 17, 1811: Jefferson to Paine Todd, October 10, 1811. Computer analysis by T.F., September 16, 2004.
91 "Science is my passion, politics my duty": Silvio A. Bedini, *Thomas Jefferson: Statesman of Science* (New York: Macmillan, 1990), 1. By the word "science," Jefferson and his contemporaries meant something both broader and rather different from what we mean by the word today, though the differences are not so

daunting as a few scholars have claimed. It is true that Jefferson sometimes employed the term "science" to mean any sort of well-founded knowledge, as when he wrote, "All the branches, then, of useful science, ought to be taught in the general schools, to a competent degree, in the first instance. These sciences may be arranged into three departments, not rigorously scientific, indeed, but sufficiently so for our purposes. These are, 1. language; 2, mathematics; 3, philosophy." But more often he meant what he called "real science," such as the study of "arithmetic, geometry, astronomy, and natural history." Saul K. Padover, *A Jefferson Profile as Revealed in His Letters* (New York: John Day, 1956), 239, 304.

91 "the most flattering incident of my life": Jefferson to the secretary of the APS, January 28, 1797. The French naturalist Georges Cuvier later credited Jefferson with discovering this ice-age ground sloth, known today as *Megalonyx jeffersonii.*

91 "abandoning the rich and declining their dinners": Jefferson to Martha Jefferson Randolph, 1800; in I. Bernard Cohen, *Science and the Founding Fathers* (New York: Norton, 1995), 63–4. Jefferson kept to this resolve so assiduously that twelve years later, at an inaugural ball honoring the induction of James Madison as his successor to the presidency, Jefferson felt obliged to inquire whether he'd arrived too early and to otherwise ask advice on how properly to behave.

91 "attacked . . . by your excellency's ram": William Keough to Thomas Jefferson, February 15, 1809.

91 "Nature intended me for the tranquil pursuits of science": Thomas Jefferson to Pierre Samuel Du Pont de Nemours, March 2, 1809.

92 "laws of nature": The term "law" became a scientific metaphor (planets "obey" Newtonian "laws" of motion) and then came full circle when Enlightenment thinkers argued that scientifically identified natural laws revealed a universal order to which human society could conform.

92 "sacred and undeniable": John Adams also reviewed Jefferson's draft, but the handwriting of this editorial alteration appears to be Franklin's.

92 science . . . founded on "self-evident principles": Joseph Priestley's *A Course of Lectures on Oratory and Criticism* was not published until 1777, but he had been delivering the lectures since 1762, and his friend Jefferson would have been familiar with his thinking in this regard. John Harris's *Lexicon technicum, or an universal English Dictionary of Arts and Sciences*, the first alphabetical encyclopedia in English, was originally published in London in 1704; this excerpt appears in Cohen, *Science and the Founding Fathers*, 123.

92 "every single Argument should be managed as a Mathematical Demonstration": For Locke on the self-evident, see his *An Essay Concerning Human Understanding*, Chapter VII.

92 "turned to neither book nor pamphlet": For Locke see William Samuel Howell, "The Declaration of Independence and Eighteenth-Century Logic," *The William and Mary Quarterly*, 3rd Ser., vol. 18, no. 4 (October 1961), 483. Jefferson at age 77 recalled that "Richard H. Lee charged it as copied from Locke's treatise on Government. . . . I know only that I turned to neither book nor pamphlet while writing it." Carl Becker, *The Declaration of Independence: A Study in the History of Political Ideas* (New York: Harcourt, Brace, 1922), 25. Regardless of whether this recollection, decades after the fact, is entirely accurate, Jefferson did draw—whether from the page or from his prodigious memory—on a number of sources, ranging from the ancient Greeks to Thomas Paine. Clear influences include Jefferson's own draft for the Virginia constitution, which he had just written and from which he cribbed many of the grievances against King George, and the Virginia Declaration of Rights, adopted by the Virginia Constitutional Convention on June 12, 1776. Written by the Virginia planter George Mason and destined to form the basis of the Bill of Rights, the Virginia Declaration expresses similar

Lockean sentiments, in something like the same language that Jefferson employed in Philadelphia. It begins by asserting "that all men are by nature equally free and independent and have certain inherent rights . . . namely, the enjoyment of life and liberty, with the means of acquiring and possessing property, and pursuing and obtaining happiness and safety." Jefferson esteemed Mason as "one of our really great men, and of the first order of greatness" (Jefferson to Augustus B. Woodward, April 3, 1825) and never claimed that anything in the Declaration of Independence was original: "I did not consider it as any part of my charge to invent new ideas altogether and to offer no sentiment which had ever been expressed before." Becker, ibid.

93 "a man profound in most of the useful branches of science": Jefferson, *Autobiography*, third paragraph. Hired to teach natural philosophy, which he defined as being made up of "Physicks, Metaphysicks, and Mathematicks," Small was also enlisted to teach moral philosophy, which the college defined as comprised of "Rhetorick, Logick, and Ethicks." He applied modern scientific and mathematical precepts to both, taking to heart the William and Mary charter, which advised that the faculty need not follow "Aristotle's *Logick and Physicks*, which reigned so long alone in the Schools" but were free to "teach what Systems . . . they think fit." William Samuel Howell, "The Declaration of Independence and Eighteenth-Century Logic," *The William and Mary Quarterly*, 3rd Ser., vol. 18, no. 4 (October 1961), 472.

93 "Every creature possessed of reason and liberty is accountable for his actions": William Duncan, *The Elements of Logic* (New York: Nichols, 1802), 126, 131.

93 "given added persuasive power": William Samuel Howell, "The Declaration of Independence and Eighteenth-Century Logic," *The William and Mary Quarterly*, 3rd Ser., vol. 18, no. 4 (October 1961), 465.

94 "applicable to all men and all times": Lincoln, address in Independence Hall, Philadelphia, February 22, 1861.

94 the Library of Congress: Jefferson was paid only a fraction of the books' worth by Congress. His comments on the "sublime luxury" of reading, e.g., "Homer in his own language" are from a letter to Joseph Priestley, January 27, 1800.

94 submarines armed with torpedoes: The world's first military submarine, the *Turtle*, was constructed as a school science project at Yale by the inventor David Bushnell and deployed against the British flagship *Eagle* during its blockade of New York harbor in August 1776. The one-man craft successfully navigated to a point beneath the *Eagle* but was unable to affix a bomb to the ship's copper-shielded hull.

94 "degrading the Almighty": Paine, *The Age of Reason*, Part I; in Eric Foner, ed., *Paine: Collected Writings* (New York: Library of America, 1995), 715.

94 "a delicious luxury indeed": Jefferson to Thomas Lomax, March 12, 1799.

94 "the most extraordinary collection of talent": *Washington Post*, April 30, 1962, B5; *Public Papers of the Presidents of the United States: John F. Kennedy* (New York: Harper and Row, 1965), 384. The dinner was held on April 29, 1962. As a few academics have fallen victim to a habit of disparaging Kennedy's intellect generally, and of assuming that this toast in particular is to be credited to his speechwriters, it should be noted that the original, preserved at the National Portrait Gallery in Washington, D.C., shows Kennedy's own hand crossing out the toast written for him and writing two discarded alternates—presumably on his knee while at the banquet—before inscribing the one he delivered.

95 "Is it best to have the sliders . . . a little circular?": Garry Wills, *Cincinnatus: George Washington and the Enlightenment* (Garden City, NY: Doubleday, 1984), 196.

95 "I knew there was as much love of science . . .": Cohen, *Science and the Founding Fathers*, 196.

95 "Oh that I had devoted to Newton": John Adams to Thomas Jefferson, February 3, 1812.

95 "a man who never said a foolish thing in his life": http://www.rogersherman.net (accessed August 31, 2004).

95 Franklin . . . soothed Jefferson's temper: As Jefferson recalled it, "I was sitting by Dr. Franklin, who perceived that I was not insensible to these mutilations. 'I have made it a rule,' said he, 'whenever in my power, to avoid becoming the draughtsman of papers to be reviewed by a public body. I took my lesson from an incident which I will relate to you. When I was a journeyman printer, one of my companions, an apprentice hatter, having served out his time, was about to open shop for himself. His first concern was to have a handsome signboard, with a proper inscription. He composed it in these words, "John Thompson, Hatter, makes and sells hats for ready money," with a figure of a hat subjoined; but he thought he would submit it to his friends for their amendments. The first he showed it to thought the word "Hatter" tautologous, because followed by the words "makes hats," which show he was a hatter. It was struck out. The next observed that the word "makes" might as well be omitted, because his customers would not care who made the hats. If good and to their mind, they would buy, by whomsoever made. He struck it out. A third said he thought the words "for ready money" were useless, as it was not the custom of the place to sell on credit. Every one who purchased expected to pay. They were parted with, and the inscription now stood, "John Thompson sells hats." "*Sells hats!*" says his next friend. "Why nobody will expect you to give them away, what then is the use of that word?" It was stricken out, and "hats" followed it, the rather as there was one painted on the board. So the inscription was reduced ultimately to "John Thompson" with the figure of a hat subjoined.' " Jefferson to Robert Walsh, December 4, 1818.

96 "Liberty is to faction what air is to fire": Madison, *Federalist, No. 10*, 55.

96 "The Parliament is the heart": James Harrington, *Oceana*, http://www.constitution.org/jh/oceana.htm (accessed September 6, 2004).

96 "North America is become a new primary planet": Thomas Pownall to Adam Smith, September 25, 1776; see Pownall, *The Administration of the Colonies* (London: Wilkie, 1764), 32–33.

96 "The progress of science": Jefferson to John Adams, June 11, 1812.

97 social mobility: Inasmuch as some revisionist historians having sought to portray the American founders as English-style aristocrats who were hypocritical in proclaiming a love of learning and liberty, it may be worth pointing out that, as the historian Gordon S. Wood notes, "By English standards American aristocrats like Washington and Jefferson, even with hundreds of slaves, remained minor gentry at best [while] lawyers like Adams and Hamilton were even less distinguished." From an English standpoint America resembled Scotland, in that both states "tended to be dominated by minor gentry—professional men and relatively small landowners—who were anxious to have their status determined less by their ancestry or the size of their estates and more by their character or their learning." As Wood notes, "Almost all the revolutionary leaders—even including the second and third ranks of leadership—were first-generation gentlemen. That is to say, almost all were the first in their families to attend college, to acquire a liberal arts education, and to display the new eighteenth-century marks of an enlightened gentleman. Of the ninety-nine men who signed the Declaration of Independence or the Constitution only eight are known to have had fathers who attended college." Gordon S. Wood, "The Greatest Generation," *New York Review of Books*, March 29, 2001, 17–22. See also "England's Cultural Provinces: Scotland and America," *William and Mary Quarterly*, 3rd Ser., vol. 11, April 1954, 200–213.

97 "Once works of the intelligence became sources of power": Alexis de Tocqueville,

Democracy in America, introduction, Arthur Goldhammer, trans. (New York: Library of America, 2004), 5.

97 "Although democracy does not encourage men to cultivate science for its own sake": Ibid., 527.

98 slavery . . . was explicitly protected by the Constitution: The Enumeration Clause of the Constitution apportioned congressional representatives under a formula, negotiated between northern and southern states, that counted each slave as three-fifths of a person. Article 1, Section 9 barred Congress from prohibiting the importation of slaves until the year 1808. The Fugitive Slave Clause (IV, 2) required that persons "held to service in one state, under the laws thereof . . . shall be delivered up on the claim of the party to whom such service or labor may be due."

98 "How is it . . . that the loudest YELPS for liberty come from the drivers of Negroes?": James Boswell, *The Life of Samuel Johnson*, Chapter XLII, 1777–1778. Many Americans expressed similar sentiments at the time. John Jay, a *Federalist* author, noted in 1786 that "to contend for our own liberty, and to deny that blessing to others, involves an inconsistency not to be excused." Patrick Henry in 1773 expressed his hope that "an opportunity will be offered to abolish this lamentable evil." Shortly after signing the Constitution, Oliver Ellsworth wrote that "all good men wish the entire abolition of slavery."

98 "revengeful and cruel": Franklin, "A Conversation on Slavery," *Public Advertiser*, January 30, 1770.

98 "Why increase the sons of Africa by planting them in America?": Franklin, "Observations on the Increase of Mankind," 1751, in Walter Isaacson, *Benjamin Franklin: An American Life* (New York: Simon & Schuster, 2003), 152.

98 "a higher opinion of the natural capacities of the black race": Franklin to John Waring, December 17, 1763.

99 those "who alone in this land of freedom are degraded": Pennsylvania Society for the Abolition of Slavery, Petition to Congress, February 12, 1790. The elder Franklin wrote biting attacks on slavery; see, e.g., his "An Address to the Public from the Pennsylvania Society for Promoting the Abolition of Slavery, and the Relief of Free Negroes Unlawfully Held in Bondage" (1789); "Plan for Improving the Condition of the Free Blacks" (~1789); and "Sidi Mehemet Ibrahim on the Slave Trade" (1790). The last of these, a satire published as an anonymous letter to the editor of the *Federal Gazette*, invents a Muslim slaveowner of old who protests that if the Christian slaves of Algiers were freed, the economy would suffer and the slaves themselves would be deprived of proper religious instruction.

99 "vehement philippic against Negro slavery": Carl Becker, *The Declaration of Independence: A Study in the History of Political Ideas* (New York: Harcourt, Brace, 1922), 212.

99 "I tremble for my country when I reflect that God is just": Jefferson, *Notes on the State of Virginia,* Query 18. Jefferson adds, a few lines later, "I think a change already perceptible, since the origin of the present [i.e., American] revolution. The spirit of the master is abating, that of the slave rising from the dust, his condition mollifying, the way I hope preparing, under the auspices of heaven, for a total emancipation, and that this is disposed, in the order of events, to be with the consent of the masters, rather than by their extirpation."

99 "I am happy to be able to inform you": Jefferson to Condorcet, August 30, 1791. Jefferson's political opponents assailed both his "thus fraternizing with Negroes," as one put it, and his expressions of sympathy for the Indians: "When we see him employed . . . to establish that the body of the American savage is not inferior in form or in vigor to the body of an European . . . how much more solicitous may we suppose him to have been to prove that the mind of this savage was also [well formed]?" sputtered Luther Martin, the attorney general of Maryland. Then as

now, those determined to fault Jefferson can always find reasons to call him a racist when he displayed anything less than a modern liberal sensibility, while branding him a hypocrite when he was being fair and open-minded. See Bedini, *Jefferson*, 224, 278–79.

99 "We have the wolf by the ear": Jefferson to John Holmes, April 22, 1820.

100 "another Peloponnesian war": Jefferson to John Adams, January 22, 1821.

100 "the best man living": *Pennsylvania Journal,* February 19, 1777, in David Hackett Fischer, *Washington's Crossing* (New York: Oxford University Press, 2004), 428. Jefferson said of Washington, with whom he had quarreled bitterly, "His person, you know, was fine."

100 "I wish from my soul": Washington to Lawrence Lewis, his future son-in-law, August 1797.

100 "to see some plan adopted": John Rhodehamel, *The Great Experiment: George Washington and the American Republic* (New Haven: Yale University Press, 1999), 97.

100 "This may seem a contradiction": John Bernard, *Retrospections of America, 1797–1811* (New York: Harper, 1887), in Richard Norton Smith, *Patriarch: George Washington and the New American Nation* (Boston: Houghton Mifflin, 1993), 305–6.

101 "and I do moreover most pointedly": Washington, Last Will and Testament, July 9, 1799.

102 "No experiment can be more interesting": Jefferson to John Tyler, 1804.

102 "arises as well from the object itself": James Madison, *Federalist No. 37, No. 38*, January 12, 1788, in Jacob E. Cooke, ed., *The Federalist* (Middletown, CT: Wesleyan University Press, 1961), 235, 241.

102 "planets . . . fly from their orbits": I. B. Cohen notes that, while the Constitution can be regarded as a Newtonian document, the founders and some of the latter-day academics who have pursued this line of inquiry have, through an inadequate command of Newton's work, got it a bit muddled. For a discussion see Supplement 2 to Cohen's *Science and the Founding Fathers,* 283ff.

102 a botanical analogy: The duality between those who view the Constitution as comparable to mechanistic physics and those who prefer to think of it in terms of the life sciences has proved enduring. James Russell Lowell in 1888 described the founders as having "invented a machine that would go of itself" and Woodrow Wilson in 1912 compared the Constitution to a mechanical model of the solar system. On the other side the journalist Amos K. Fiske contributed a long essay to the *New York Times* in 1900 titled, "The Constitution: An Organism Not a Mechanism," Justice Oliver Wendell Holmes in 1914 declared that "the provisions of the Constitution . . . are organic living institutions," and Felix Frankfurter in 1915 asserted that "the Constitution is an organism." Inevitably, some writers mixed their metaphors. In the 1880s the prominent attorney George Ticknor Curtis referred to the "necessity of organic laws to supply the machinery of the new government," and James Fenimore Cooper, Mark Twain's favorite example of a writer whose tangled style confuses even himself, managed in a single sentence to envision the "fruits" politicians could produce were the nation not in a "harness" that had something to do—Cooper becomes a bit obscure here—with advancing "the great national car." For a discussion, see Michael Kammen, *A Machine That Would Go of Itself: The Constitution in American Culture* (New York: Knopf, 1986).

103 "Without the Constitution and the Union": Lincoln, unpublished fragment, in Kammen, *A Machine That Would Go of Itself*, 16.

103 "I confess that there are several parts of this Constitution which I do not at present approve": Franklin to the Constitutional Convention, September 17, 1787; text from James Madison's notes. Italics added. Franklin, over eighty years old and

only two years from the end of his life, was too feeble to speak himself, so his remarks were read on his behalf by his fellow Pennsylvania delegate James Wilson.

103 "Science is far from a perfect instrument of knowledge": Carl Sagan, *The Demon-Haunted World: Science as a Candle in the Dark* (New York: Ballantine, 1996), 27.

104 "Science is a kind of open laboratory for a democracy": Lee Smolin, "Loop Quantum Gravity," *Edge*, 112, http://www.edge.org/3rd_culture/smolin03/smolin03_index.html, posted February 24, 2003 (accessed September 20, 2004).

104 "I tried to explain that . . . there are supposed to be votes": Richard P. Feynman, lecture at the University of Washington, 1963, published as *The Meaning of It All: Thoughts of a Citizen Scientist* (Reading, MA, Addison-Wesley, 1998), 101. As the text is based on the transcription of an extemporaneous lecture, its punctuation has been modified here for clarity's sake.

104 "The government of the United States was developed": Ibid., 49.

105 "If he does that, he will be the greatest man in the world": This story comes from the American artist Benjamin West, who counted among his students Washington's best-known portraitist, Gilbert Stuart. West claimed that King George, believing that West understood Washington's character, asked him what Washington would do if the Americans won the war, to which West responded by predicting that Washington would relinquish all his political power. This story has never been verified and in some respects sounds too good to be true, but the often ridiculed George III was capable of penetrating insights. In the audience he gave on June 2, 1785, to John Adams, formerly condemned as a treasonous traitor but now welcomed as the former colonies' first ambassador to England, George said, according to Adams: "I wish you Sir, to believe, and that it may be understood in America, that I have done nothing in the late Contest, but what I thought myself indispensably bound to do, by the Duty which I owed to my People. I will be very frank with you. I was the last to consent to the Separation, but the Separation having been made and having become inevitable, I have always said, as I say now, that I would be the first to meet the Friendship of the United States as an independent Power." See David McCullough, *John Adams* (New York: Simon & Schuster, 2001), 335–337.

105 "the success of the experiment": Washington, State of the Union address, December 7, 1796.

105 "Is there a doubt? . . . Let experience solve it": Washington's Farewell Address was published in *Claypoole's American Daily Advertiser* and the *Gazette of the United States* on September 19, 1796. Based on a draft written by Alexander Hamilton four years earlier, it reads today as perhaps a bit prolix, weighing in at over six thousand words and ignoring its own advice, in the sixth paragraph, that, "Here, perhaps, I ought to stop." Nevertheless it became an article of democratic scripture, read aloud annually in the House of Representatives for two centuries thereafter. As was usually the case with Washington, though, what mattered were less his words than his deeds.

105 "newspapers without a government": Jefferson to Edward Carrington, January 16, 1787.

106 the emergence of political parties: The two principal parties of the day were the Federalists, centered on Alexander Hamilton and favoring a strong national government, and the Anti-Federalists (sometimes called Republicans), who rallied around Jefferson's vision of an agrarian union where the states held most of the power. Weakened by Hamilton's death in a duel with Aaron Burr in 1804, the Federalist party had petered out by about 1825, while the Republicans, confusingly enough, evolved into today's Democratic Party. The modern Republican Party originated in 1854, in part as an alliance of abolitionists and suffragettes, and won the White House with Abraham Lincoln's election in 1860.

106 "fathered children by one of his slaves": One of Jefferson's libelers—Harry Croswell, of Hudson, New York, whose *Wasp* specialized in lampooning the powerful—was put on trial for printing that Jefferson had paid another journalist to accuse George Washington of treason, thievery, and perjury. Alexander Hamilton defended Croswell by pleading that evidence bearing on the truth or falsehood of the libel should be introduced at the trial. It was not, but Hamilton's plea was so effective that the New York state legislature promptly enacted a statute to that effect, establishing the "Hamilton doctrine"—that truth is an absolute defense against libel actions. While in Albany for the trial, Hamilton in the course of a casual conversation made derogatory remarks about Aaron Burr; this led to the duel in which Hamilton was killed. Theirs was one of several "newspaper duels," in that each combatant was championed by his own newspaper—Burr by Washington Irving's New York *Morning Chronicle*, Hamilton by the *Evening Post*—a circumstance that enhanced their notoriety while turning them into what amounted to action figures in a lurid melodrama played out for circulation dollars. The charge that Jefferson fathered children by Sally Hemings continues to resurface, bolstered by a 1998 DNA test indicating a link between the Jefferson and Hemings family lines. For a recent study of this fascinating question, see Annette Gordon-Reed, *Thomas Jefferson and Sally Hemings: An American Controversy* (Charlottesville: University Press of Virginia), 1998.

106 "I have lent myself willingly as the subject of a great experiment": Jefferson to Thomas Seymour, February 11, 1807.

107 "The experiment has been tried": Jefferson, Second Inaugural Address, March 4, 1805.

107 "The press, confined to truth": Ibid.

107 "I offer to our country sincere congratulations": Ibid. Similarly, Abraham Lincoln based the Gettysburg Address on the concept that liberal democracy was an experimental system now undergoing a particularly agonizing empirical test: "Now we are engaged in a great civil war, testing whether that nation or any nation so conceived and so dedicated can long endure."

107 "every man of science feels a strong and disinterested desire of promoting it": Jefferson to Marc Auguste Pictet, February 5, 1803. As Alexis de Tocqueville observed, "The Americans did not rely on the nature of the country to counter the dangers arising from their constitution and political laws. To ills that they share with all democratic peoples they applied remedies that had previously occurred to no one else; and, though they were the first to try those remedies, they succeeded." Tocqueville, *Democracy in America*, vol. 1, part 2, chapter 9, Goldhammer trans.

107 "An insurrection has consequently begun": Jefferson to John Adams, October 28, 1813.

107 "shrink from the advance of truth and science": Jefferson to John Manners, 1814.

107 "There is not a truth existing which I fear": Jefferson to Henry Lee, 1826.

108 "All eyes are opened": Jefferson to Roger C. Weightman, June 24, 1826.

108 "I love you with all my heart": Jefferson to Adams, August 15, 1820.

CHAPTER SIX: THE TERROR

110 "Madame Deficit": The government's financial distress was due not to the profligacy of its queen but to its having financed the war entirely through debt. The one thing that students still learn about Marie Antoinette—that when informed that the peasants were starving because they could no longer afford to buy bread, she flippantly replied, "Let them eat cake"—is false. The tale appears in Jean-Jacques

Rousseau's *Confessions*, where it is attributed to "a great princess," but Rousseau was writing in 1766, when Marie was ten years old and living in Austria. A teenager when she married and came to France, Marie wrote to her mother, Empress Maria Theresa of Austria, that on her entrance into Paris on June 14, 1773, "What touched me most . . . was the tenderness and earnestness of the poor people, who, in spite of the taxes with which they are overwhelmed, were transported with joy at seeing us. When we went to walk in the Tuileries, there was so vast a crowd that we were three-quarters of an hour without being able to move either forward or backward. The dauphin and I gave repeated orders to the Guards not to beat any one, which had a very good effect. . . . What a happy thing it is for persons in our rank to gain the love of a whole nation so cheaply. Yet there is nothing so precious; I felt it thoroughly, and shall never forget it."

111 "the most unexampled success . . . this great crisis being now over": Jefferson to Madison, May 11, 1789; Jefferson to Jay, June 29, 1789.

111 "all danger of civil commotion here is at an end": Jefferson to John Trumbull, June 29, 1789.

111 "Tranquility is well established in Paris": Jefferson to Paine, September 13, 1789. One wonders what so clouded Jefferson's crystal ball. It may be that he was isolated from contact with more than a small circle of French friends. Although he read and wrote French, Jefferson did not speak it with the eloquence to which he was accustomed, and he seems to have shrunk from the raised eyebrows that his unpolished spoken French might have aroused; nor was he much given to society gatherings in any language. Absent such conversations he evidently relied excessively on the advice of his friend General Lafayette, who shared in the prevailing American optimism about the French revolutionary cause.

112 "to change the general way of thinking": Arthur M. Wilson, "*Encyclopédie,*" in Paul Edwards, ed. *The Encyclopedia of Philosophy* (New York: Macmillan, 1967), vol. 2, 506.

112 "All things must be examined": Diderot, encyclopedia entry for the word *Encyclopédie,* in Isaac Kramnick, ed., *The Portable Enlightenment Reader* (New York: Penguin, 1995), 18.

113 "Philosophers were engaged in a winning battle": William Doyle, *The Oxford History of the French Revolution* (Oxford: Clarendon Press, 1989), 52.

113 "I have not the least doubt": Paine to Washington, May 1, 1790.

113 "the most famous and the most respected of antiquity": Keane, *Tom Paine*, 250.

114 quadrupled the annual rate of science book publication: The modern French historian Daniel Roche calls eighteenth-century popularization of science "the most important intellectual phenomenon of the time." Roche, *France in the Enlightenment*, Arthur Goldhammer, trans. (Cambridge, MA: Harvard University Press, 1998), 508.

114 The marvelous mathematical clockwork: Véron de Forbonnais, for one, attempted in the *Encyclopédie* to draw God, nature, and human sociability together into one pleasing web: "Infinite Providence . . . sought, through the variety it established in nature, to make men dependent on one another. The Supreme Being bound nation to nation in such a way as to preserve the peace and promote love for one another, and also, by filling the universe with marvels, to garner their praise by demonstrating his greatness and his love for all." Roche, *France in the Enlightenment*, 42.

115 "invented nothing . . . set everything on fire": Mme de Staël, in Will and Ariel Durant, *Rousseau and Revolution* (New York: Simon & Schuster, 1967), 887. And she admired Rousseau!

115 "Uneducated, he wrote the most influential book": John Herman Randall Jr., *The Career of Philosophy*, vol. 1 (New York: Columbia University Press, 1962), 964.

116 "the most sociable and loving of men": Rousseau, *Reveries of the Solitary Walker*, Peter France, trans. (New York: Penguin, 1979), 27.

116 "she was present in my mind": Bertrand Russell, *A History of Western Philosophy and Its Connection with Political and Social Circumstances from the Earliest Times to the Present Day* (New York: Simon & Schuster, 1954), 685. Russell observes that Rousseau "considered that he always had a warm heart, which, however, never hindered him from base actions towards his best friends."

116 "The physical wants which were satisfied": Rousseau, *Confessions*, Book IX. From 1755 Rousseau was unable to continue having sexual relations, owing to a disorder of the urinary tract. He saw the transformation as yet another tribute to his personal merit: "From that moment I became virtuous."

116 "deceitful, vain as Satan": Paul Johnson, *Intellectuals* (New York: Harper & Row, 1988), 27.

116 "an inconsequential poor pygmy of a man": Durant, *Rousseau and Revolution*, 165.

116 "Rousseau is . . . a Rascal": Adam Smith to David Hume, July 6, 1766.

117 "entertained no principle": Russell, *A History of Western Philosophy*, 690.

117 "my odd, romantic notions": Biographical note to Rousseau's *A Discourse on the Origin of Inequality* (Chicago: University of Chicago Press, 1952), 319.

117 "I see him satisfying his hunger at the first oak": Rousseau, *Origin of Inequality*, 336, 334. In fairness to Rousseau it should be noted than many other writers have fallen victim to bucolic romances—among them Virgil, who in his *Georgics* (2.458) describes "how lucky farmers are" since the earth provides them with "an easy living from the soil." From what is known of Virgil's life it appears unlikely that he ever worked on a farm for so much as a day in his life.

117 "Let us begin then by laying facts aside": Randall, *Career of Philosophy*, vol. 1, 968.

117 "I hate books": Jean-Jacques Rousseau, *Emile or On Education*, Alan Bloom, trans. (New York: Basic Books, 1979), 184.

117 "have true ideas": Rousseau, "The Social Contract and Discourses," in Randall, *Career of Philosophy*, vol. 1, 967.

117 "Man is naturally good": Rousseau, "Discourse on Inequality," 1754, in Russell, *A History of Western Philosophy*, 687.

118 "The first man": Rousseau, *Origin of Inequality*, 348.

118 "The worst mistake in the history of the human race": Jared Diamond, in "Noble or Savage?" *The Economist*, December 22, 2007, 129.

118 the results do not look like Eden: For critiques of the notion that early man was peaceable, see Steven A. LeBlanc with Katherine E. Register, *Constant Battles: The Myth of the Noble Savage and a Peaceful Past* (New York: St. Martin's, 2003) and Lawrence Keeley, *War Before Civilization* (New York: Oxford University Press, 1996). An ecological explanation of the high rate of violent mortality among youthful males may be that these societies lacked sufficient nutrition to otherwise support their expanding numbers. The Tasaday hoax was exposed in large part by scientists—among them an archeologist who saw from photos that the "stone age" people's stone axes were obvious fakes, and an anthropologist who noted that their alleged cave dwellings lacked "middens," or garbage heaps normally found in such situations if people have actually been living there.

119 indifferent to their environment: Cave painters in the south of France 32,000 years ago focused on rhinoceroses; 17,000 years later their subjects were bison, bulls, and horses—the rhinos presumably having all been eaten. This testimony would appear to be firsthand, if, as seems likely, the painters were mostly youthful males; see R. Dale Guthrie, *The Nature of Paleolithic Art* (Chicago: University of Chicago Press, 2005). Although climate change or some other phenomenon rather than hunting may have been responsible for these wildlife depredations, it remains the case that little evidence supports the romantic assumption that prehistoric

humans practiced much environmental stewardship. This is not to say that they did not love and revere the world or find it beautiful, but stewardship requires more than a motive. If, for instance, you wish to limit the hunting of the wildebeest, it is not sufficient that your one tribe does so: Neighboring tribes will simply kill the wildebeest you spared. So your campaign to save the wildebeest requires alliances with, or dominion over, your neighbors—whereupon you are on the road toward civilization.

119 "my face [is] almost as well known as that of the moon": Gordon S. Wood, "Founders & Keepers," *The New York Review of Books*, July 14, 2005, 35.

119 "a living advertisement for the virtues of Rousseauistic simplicity": Doyle, *Oxford History of the French Revolution*, 63.

119 "a scene of continual dissipation": Isaacson, *Benjamin Franklin*, 352.

120 "Throwing aside . . . all those scientific books,": Rousseau, *Origin of Inequality*, Preface; Randall, *Career of Philosophy*, vol. 1, 968.

120 "men like me . . . can no longer subsist": Randall, *Career of Philosophy*, vol. 1, 969.

120 "*forced to be free*": Rousseau, *Social Contract*, bk. 1, Chapter 7, "The Sovereign." Susan Dunn, *Sister Revolutions: French Lightning, American Light* (New York: Faber, 1999), 62. Italics added.

120 "Divine man!": Durant, *Rousseau and Revolution*, 890.

120 "the fundamental principle": Robespierre, "On the Moral and Political Principles of Domestic Policy," speech of February 5, 1794.

120 "What is the end of our revolution?": Robespierre, "Report upon the Principles of Political Morality," Paul Halsall, trans. Internet Modern History Sourcebook, http://www.fordham.edu/halsall/mod/1794robespierre.html (accessed February 17, 2005).

121 "*Truth* and *reason* alone": Robespierre, speech of May 16, 1791, in Dunn, *Sister Revolutions*, 117.

121 "I abhor any kind of government that includes factious men": Robespierre, July 14, 1791, in Dunn, *Sister Revolutions*, 88.

121 "The duty of journalists": Jean-Paul Bertaud, "An Open File: The Press Under the Terror," in Keith Michael Baker, ed., *The French Revolution and the Creation of Modern Political Culture* (New York: Elsevier, 1994), 301.

121 "to lead the people by reason": Robespierre, "On the Moral and Political Principles of Domestic Policy," Paul Halsall, trans. Modern History Sourcebook, from http://www.rjgeib.com/thoughts/french/french.html (accessed January 19, 2005).

122 "conspirators . . . assassins": Robespierre, "Report upon the Principles of Political Morality Which Are to Form the Basis of the Administration of the Interior Concerns of the Republic," Paul Halsall, trans. Modern History Sourcebook, from http://www.rjgeib.com/thoughts/french/french.html (accessed January 19, 2005).

122 "already been judged": Doyle, *Oxford History of the French Revolution*, 194.

122 "I now relinquish that hope": Paine to Jefferson, April 20, 1793.

122 "painted the charms of virtue in strokes of fire": Jean Starobinski, "Rousseau and Revolution," *New York Review of Books*, April 25, 2002, 59.

123 "to suspend a penalty over each action": Keith Michael Baker, introduction to Keith Michael Baker, ed., *The French Revolution and the Creation of Modern Political Culture*, vol. 4 (New York: Elsevier, 1994), xiiv.

123 "I met but outrages": Marat to Roume de Saint-Laurent, November 20, 1783.

123 "I am maybe the only author beyond suspicion": Marat, "Denunciation Against Mr. Necker, First Minister of Finances, made by Mr. Marat, the Friend of the People, in presence of the Public's Court," http://seni.club.fr/life_of_marat.html (accessed January 20, 2005).

124 "cowardly murderers": Ibid.

124 Law of Suspects: Doyle, *Oxford History of the French Revolution*, 251.

124 "the blood of those who had been executed": Ibid., 254.

124 "It only took them an instant to cut off that head": Ken Alder, *The Measure of All Things: The Seven-Year Odyssey and Hidden Error That Transformed the World* (New York: Free Press, 2002), 145.

125 "Let anyone be a savant": Charles Coulston Gillispie, "Science in the French Revolution." *Proceedings of the National Academy of Sciences (USA)*, vol. 45, issue 5 (May 15, 1959), 681.

125 "theory of madness": Dunn, *Sister Revolutions*, 35.

125 "reformers . . . divided into two groups": Randall, *Career of Philosophy*, vol. 1, 684–85.

126 "Time will tell": Starobinski, "Rousseau and Revolution," p 55.

126 "put his country's freedom on a sure basis": Vincent Cronin, *Napoleon* (New York: HarperCollins, 1994), 177.

126 "I haven't been able to understand": Ghita Ionescu, *Opposition: Past and Present of a Political Institution* (London: Watts, 1968), 59.

126 "If Washington had been a Frenchman": Napoleon to Benjamin Constant, shortly before his defeat at the battle of Waterloo, in J. Christopher Herold, ed. and trans., *The Mind of Napoleon: A Selection from His Written and Spoken Words* (New York: Columbia University Press, 1955), 276; and J. Christopher Herold, *The Age of Napoleon* (Boston: Houghton Mifflin, 2002), 123.

127 "Search the unknown forests": James Thomas Flexner, *George Washington: Anguish and Farewell (1793–1799)* (Boston: Little, Brown, 1972), 501. Chateaubriand was inclined to romanticize Washington a bit. The historian J. Christopher Herold notes that "when Chateaubriand saw Washington in Philadelphia in 1791, he was disappointed to see him riding in a coach; he had expected to find him behind a plow." Herold, *Age of Napoleon*, 11.

127 "Infallibility not being the attribute of Man": Washington, draft of his Farewell Address, in Dunn, *Sister Revolutions*, 127.

CHAPTER SEVEN: POWER

129 "Until they were satisfied that knowledge was money": Henry Adams, *History of the United States of America During the Administrations of Thomas Jefferson*, Earl N. Harbert, ed. (New York: Library of America, 1986), 53.

129 "Advances will be most frequent": H. B. Phillips, "On the Nature of Progress," *American Scientist*, October 1945, 253–59.

129 "be laughed at": Strutt to the novelist Maria Edgworth, in Margaret C. Jacob, *Scientific Culture and the Making of the Industrial West* (New York: Oxford University Press, 1997), 95.

130 "years after we had taken chemistry to guide us": William J. Bernstein, *The Birth of Plenty: How the Prosperity of the Modern World Was Created* (New York: McGraw Hill, 2004), 123. Carnegie's chemistry research enabled him to produce better steel. His principal management innovation was to keep his mills running around the clock even if to do so meant sometimes selling at a loss. "As manufacturing is carried on today, in enormous establishments with five or ten millions of dollars of capital invested and with thousands of workers, it costs the manufacturer much less to run at a loss per ton or per year than to check his production," he wrote. "Stoppage would be serious indeed. The condition of cheap manufacture is running full." Stuart Weems Bruchey, *The Wealth of the Nation: An Economic History of the United States* (New York: Harper, 1988), 123. Though reviled as a "robber baron," Carnegie was a lifelong liberal and science devotee who gave away vast sums, favored estate taxes as "of all forms of taxation . . . the wisest," and maintained that "the man who dies . . . rich, dies disgraced.'" Carnegie, *The*

Gospel of Wealth, 1889, http://www.fordham.edu/halsall/mod/1889carnegie.html. His critics perpetuated the myth that the public libraries Carnegie financed were required to hang his portrait on the wall, as one would expect of a royal patron; actually, his only such stipulation was that the libraries bear an image of the sun along with the maxim, "Let There Be Light."

131 "clamber over or plough through": David E. Nye, *America as Second Creation: Technology and Narratives of New Beginnings* (Cambridge, MA: MIT Press, 2003), 153.

131 "quite delightful and strange beyond description": Paul Johnson, *The Birth of the Modern: World Society 1815–1830* (New York: HarperCollins, 1991), 191.

131 "uneducated persons . . . It is not against railways": Wordsworth, letter to the editor of the *Morning Post*, 1844, reprinted as a pamphlet the following year; in W. J. B. Owen and Jane Worthington Smyser, eds., *The Prose Works of William Wordsworth* (Oxford: Clarendon, 1974), vol. 3; also see Ernest de Selincourt, ed., *Wordsworth's Guide to the Lakes* (Oxford University Press, 1970), 156. Wordsworth later protested that he had no desire "to interfere with the innocent enjoyments of the poor, by preventing this district becoming accessible to them by a railway." James Mulvihill, "Consuming Nature: Wordsworth and the Kendal and Windermere Railway Controversy." *Modern Language Quarterly* 56(3):305–26 (September 1995). But he felt that appreciation of nature was about as rare as appreciation of poetry, about which he advised his patron Lady Beaumont, "It is an awful truth that there neither is, nor can be, any genuine enjoyment of poetry among nineteen out of twenty of those persons who live, or wish to live, in the broad light of the world." Adam Kirsch, "Strange Fits of Passion," *New Yorker*, December 5, 2005, 92. Like Samuel Taylor Coleridge, who in his *Definitions of Poetry* of 1836 declared that "poetry is opposed to science," Wordsworth was not at his best when addressing technology. The politically reactionary William Butler Yeats would continue the tradition, dismissing scientific knowledge in unintentionally funny lines like these from his poem "The Song of the Happy Shepherd":

> Seek, then,
> No learning from the starry men,
> Who follow with the optic glass
> The whirling ways of stars that pass

131 "conquer space": John C. Calhoun, 1817; *Annals of Congress,* 14th Congress, 2nd Session, 851–960. Calhoun, a congressman and senator from South Carolina, served as vice president under John Quincy Adams and Andrew Jackson.

131 "Railroad iron is a magician's rod": Nye, *America as Second Creation*, 157. Railroad technology became a profitable export for Americans, who dispatched 162 locomotives and 2,700 freight and passenger cars to Russia alone. While the artist James McNeill Whistler was painting portraits of his mother, his father—an engineer—was working on the Russian railroads.

132 "I long to hear from you": Samuel Morse to Lucretia Morse, February 10, 1825, in Tom Standage, *The Victorian Internet: The Remarkable Story of the Telegraph and the Nineteenth Century's On-Line Pioneers* (New York: Walker, 1998), 25.

132 "severed the preexisting bond": Robert Lucky, "The Quickening of Science Communication." *Science*, vol. 289, issue 5477 (July 14, 2000): 259–64.

132 "Is it not a feat sublime?": Neil Baldwin, *Edison: Inventing the Century* (New York: Hyperion, 1995), 39.

132 "removing causes of misunderstanding": Standage, *Victorian Internet*, 90–91.

133 "know one another better": Ibid., 104.

133 little more than a few clapboard buildings and unpaved streets: One such town was Norman, Oklahoma, where my maternal grandfather, Nelson MacCandless Crowe (1853–1920), arrived around 1890 to assume a teaching position at the newly founded University of Oklahoma. Norman was a creation of the Atchison, Topeka & Santa Fe Railway, which in 1886–1887 built the town as an Indian Territory station site. Few lived there before the Great Land Run of 1889, when white settlers rushed in to claim "unassigned lands" in what had been Indian Territory. Crowe's hopeful letters to his bride, who remained back east with their young daughter (my mother), were accompanied by photos of Norman's few buildings, between which could be seen miles of starkly empty prairie receding to the horizon.

133 "in the West the railroad itself builds cities": Horace Greeley et al., *The Great Industries of the United States,* 1872, in Nye, *America as Second Creation*, 157. Greeley, an abolitionist journalist, is best remembered for the injunction, "Go west, young man," which he often repeated but did not invent.

133 each community regulated its own clocks: Solar observers being in short supply, methods for determining the local time were often rather informal, as expressed in the old joke about a small-town telephone operator who gets a call every weekday morning for years from a man asking for the correct time. Eventually, out of curiosity, she inquires why he keeps calling. "I'm the superintendent at the mill," he informs her, "and I have to keep our clock accurate in order to blow the factory whistle at noon." The operator laughs. "That's funny," she says. "For years, I've been setting my clock by your whistle."

134 "There isn't time": *High Noon*, written by John W. Cunningham and Carl Foreman and directed by Fred Zinnemann, 1952.

134 "chagrined a little": Franklin to Peter Collinson, April 29, 1749.

135 "my desire to escape from trade": Michael Faraday, *Life and Letters* 1:54; from http://www.woodrow.org/teachers/chemistry/institutes/1992/Faraday.html (accessed January 30, 2005).

135 "ALL THIS IS A DREAM": James Hamilton, *A Life of Discovery: Michael Faraday, Giant of the Scientific Revolution* (New York: Random House, 2004), 334.

136 "No experiments are useless": Neil Baldwin, *Edison: Inventing the Century* (New York: Hyperion, 1995), 51.

136 "You have the thing. Keep at it!": Steven Watts, *The People's Tycoon: Henry Ford and the American Century*. New York: Knopf, 2005, 42.

136 "You can't have that much and drive a Ford": Ibid., 259.

136 "Edison was enraptured": Harold Evans, *They Made America* (New York: Little, Brown, 2004), 159.

136 "suddenly came to me": Ibid., 159–60.

137 "I speak without exaggeration": Ronald W. Clark, *Edison: The Man Who Made the Future* (New York: Putnam's, 1977), 90.

137 "In this business": Joseph Priestley, *Experiments and Observations on Different Kinds of Air* (London: J. Johnson, 1776) http://web.lemoyne.edu/~giunta/priestley.html (accessed December 22, 2005).

137 "to make the dynamos, the lamps, the conductors": Mitchell Wilson, *American Science and Invention* (New York: Simon & Schuster, 1954), 291.

137 "had to walk the streets": Matthew Josephson, *Edison: A Biography* (New York: McGraw-Hill, 1959), 252.

137 "Sundown no longer emptied the promenade": Robert Louis Stevenson, "A Plea for Gas Lamps," in his *Virginibus Puerisque*, http://robert-louis-stevenson.classic-literature.co.uk/virginibus-puerisque/ebook-page-65.asp (accessed October 12, 2005).

137 "a symbol of infinity": Henry Adams, "The Dynamo and the Virgin," in his *The*

Education of Henry Adams (New York: Library of America: 1983), 1067. Adams adds that without the help of a technologically adept friend, he "might as well have stood outside in the night, staring at the Milky Way"—a touching reminder that in 1900 one could still look up from a city street in Paris and *see* the Milky Way. Electric lighting would soon put an end to that, at least until the day when campaigns against "light pollution" succeed in returning to cities the splendors of the night sky.

138 "Mr. Tesla may accomplish great things": Nikola Tesla, *My Inventions* (Zagreb, Yugoslavia: Skolska Knjiga, 1977), ch. 3; http://modena.intergate.ca/personal/marcop/tesla/tesla.htm (accessed October 13, 2005). This odd little book was originally published in *Electrical Experimenter*, May, June, July, October, 1919.

138 "a fly alighting on a table": Ibid.

138 "I must return to work": John J. O'Neill, *Prodigal Genius: The Life of Nikola Tesla* (New York: McKay, 1944), 49. For more on Tesla see Jill Jonnes, *Empires of Light: Edison, Tesla, Westinghouse, and the Race to Electrify the World* (New York: Random House, 2003), ch. 4.

139 Einstein regarded authority: This is not to say that Einstein romanticized revolutionaries. Widely regarded as a "Red" in Germany following World War I, Einstein in November 1918 was invited to consult with leftist students who had formed a "soviet" and imprisoned the rector and deans at the University of Berlin. He declined, saying, "I have always thought that the German universities' most valuable institution is academic freedom, whereby the lecturers are in no way told what to teach, and the students are able to choose which lectures to attend, without much supervision and control. Your new statutes seem to abolish all this and to replace it by precise regulations, I would be very sorry if the old freedom were to come to an end." Max Born, *The Born-Einstein Letters*, Irene Born, trans. (New York: Walker, 1971), 150. For context see Thomas Levenson, *Einstein in Berlin* (New York: Bantam, 2003), ch. 12.

139 "Something deeply hidden": Paul Arthur Schilpp, *Albert Einstein: Philosopher-Scientist* (La Salle, IL: Open Court, 1969), vol. 1, 9.

140 "twins paradox": For further discussion see "The Twin Paradox" Web site of the physicist John Baez, from whom I have borrowed the names of Terrence for the earthbound twin and Stella for the astronaut: http://math.ucr.edu/home/baez/physics/Relativity/SR/TwinParadox/twin_paradox.html.

141 "clarify . . . 'time' ": Einstein, "On the Thermodynamics of Moving Bodies." *Annalen der Physik* 17 (1905) 891–921, Anna Beck, trans., in John Stachel, ed., *The Collected Papers of Albert Einstein* (Princeton, NJ: Princeton University Press, 1989), English translation supplement., vol. 2, 140.

141 "Of the gods we believe": Robert B. Strassler, ed., *The Landmark Thucydides: A Comprehensive Guide to the Peloponnesian War* (New York: Free Press, 1996), 5.105.2.

142 "The persistent effort of Europeans": David B. Abernethy, *The Dynamics of Global Dominance* (New Haven, CT: Yale University Press, 2000), 10. Abernethy adds that "to the extent that non-European societies did not possess or value the explore-control-utilize syndrome, they were at a power disadvantage when encountering people who did. Where institutions and norms did not support scientific investigation and technological development, it was practically impossible to resist invaders who could call upon the latest round of advances generated by their home societies." Ibid., 39.

142 "*a right to interfere*": Daniel R. Headrick, *The Tools of Empire: Technology and European Imperialism in the Nineteenth Century* (New York: Oxford University Press, 1981), 28. Emphasis added.

142 "accompanied by misgivings": Evelyn Baring, the first Earl of Cromer, *Ancient*

and Modern Imperialism, reprint of 1910 edition (Honolulu: University Press of the Pacific, 2001), 19–20.

143 the British Empire alone dominated 444 million people: Great Britain in 1913 controlled nearly 32 million square kilometers of colonial lands, more than the next nine largest colonial powers combined and ten times that of the United States. As the political scientist David Abernethy notes, "Two-thirds of the United Nations' member states as of January 2000—125 of 188—consisted of territories outside of Europe which at one time were governed by Europeans. Three-fifths of the world's population live in countries whose entire territory has at one time been claimed by a European states. If one includes states portions of whose current territory were under the legal jurisdiction of Europeans—notably China, with its treaty ports—then in excess of 80 percent of human beings now living inhabit states that experienced some version of formal European rule. That rule lasted for more than 250 years in 37 U.N. member states and for more than a century in 60." Abernethy, *Dynamics of Global Dominance*, 12–13. At the peak of the colonial era there were only 59 independent nations in the world; today, colonialism having waned, their number is approaching 200. Some 88 nations today, with a combined population of 2.3 billion, "list a west European tongue as an official language." Ibid., 15. This is not to say that substantial numbers of the subjugated populations typically spoke the languages of the imperialist powers: prior to 1914, to take just one historical snapshot, it is estimated that only half of 1 percent of the world's 300 million colonials were literate in a European language. Eric Hobsbawm, *The Age of Empire, 1875–1914* (New York: Random House, 1989), 202.

143 "Raw power . . . made a virtue in itself": Norman Davies, *Europe: A History* (New York: Oxford University Press, 1996), 759.

144 an enterprise that had begun in freedom devolved into freedom's opposite: Americans, having begun as colonials themselves, were skeptical about colonialism. John Quincy Adams recalled expressing his exasperation at British colonial ambitions to Stratford Canning (1786–1880), the British minister to Washington, in 1821. "You claim India," Adams fumed. "You claim Africa. You claim—" to which Canning urbanely interjected, "Perhaps a piece of the Moon?" "No, I have not heard that you claim exclusively any part of the Moon," Adams replied. "But there is not a spot on this habitable globe that I could affirm you do not claim." *The Diary of John Quincy Adams 1794–1845*, Allan Nevins, ed. (New York: Scribner, 1951), in Johnson, *Birth of the Modern*, 47.

144 turning trade into tyranny: Examples of old-style invasions of inlands as well as coastal cities included Cortés's conquest of the Aztecs in 1519 and Pizarro's of the Incas in 1531, Spanish colonization of Luzon Island in the Philippines, the Dutch of Batavia in Java, Dutch and French Huguenots migrating into the interior of South Africa, and tax collection in Bengal by agents of the English East India Company. The more common pattern of controlling only coastal enclaves was found, among other places, in the trading ports of Luanda, Mombassa, and Malindi in Africa, and of Hormuz, Goa, Pondicherry, Madras, Calcutta, and Macao in Asia.

144 "How did a people who thought themselves free": Simon Schama, "A History of Britain," BBC television series, http://www.bbc.co.uk/history/programmes/hob/index.shtml (accessed November 21, 2005).

145 "The Battas are not a bad people": Johnson, *Birth of the Modern*, 345, 348–9. Raffles later expanded on his respect for the Battas, observing in his memoirs that they "have many virtues. I prize them highly. However horrible eating a man may sound in European ears, I question whether the party suffers so much, or the punishment itself is worse than the European tortures of two

centuries ago. I have always doubted the policy, and even the right, of capital punishment among civilized nations; but this once admitted, and torture allowed, I see nothing more cruel in eating a man alive than in torturing him for days with mangled limbs and the like. Here they certainly eat him up at once, and the party seldom suffers more than a few minutes. It is probable that he suffers more pain from the loss of his ear than from what follows: indeed he is said to give one shriek when that is taken off, and then to continue silent till death." C. E. Wurtzburg, *Raffles of the Eastern Isles* (New York: Oxford University Press, 1984).

145 "Racism reared its head": Although many European attitudes toward "exotic" peoples from distant lands were mind-bogglingly dehumanizing—as when, in 1904, several Samoan women were exhibited in the Hamburg zoo—others ranged from bewilderment to naïveté. Black Elk of the Oglala Sioux, who performed for Queen Victoria as part of Buffalo Bill's Wild West Show in 1887, reported that she requested an audience with the Sioux following the performance, and told them, "Today I have seen the best-looking people I know. If you belonged to me, I would not let them take you around in a show like this." John G. Neihardt, *Black Elk Speaks* (New York: Washington Square Press, 1959), 188.

145 "Westward the course of empires": Berkeley, "Verses on the Prospect of Planting Arts and Learning in America," in A. Fraser, ed., *The Works of George Berkeley, D.D.* (New York: Oxford University Press, 1901), vol. 4, 364. The technological historian David Nye notes that Berkeley's dictum had a visual counterpart in an enormously popular Currier and Ives lithograph, "Across the Continent: Westward the Course of Empire Takes Its Way," published in 1868, when the first transcontinental railroad was nearing completion. "The relationship between the railroad and the landscape is unambiguous," Nye writes. "The railroad is rapidly developing the American West. The track ahead is still being laid, but the area already served by the railroad contains a prosperous new community and a prominent public school. The power to move across the land is translated into expansion and settlement. Human creations transform the landscape, which in the distance remains vague and unformed, whereas in the foreground a new town has been impelled into life by the railroad . . . . The Currier and Ives image was both a vision of the recent past and a prediction of the future. It used space to represent time: the new community in the foreground was the present; the empty land ahead of the train was the future, which extended to the vanishing point of perspective. The land ahead was presented as empty space awaiting the coming of white civilization. A few Native Americans on horseback were literally on the margins, watching, with the smoke from the train blowing over them and obscuring their view of the land, and thus of the future." Nye, *America as Second Creation*, 158–59.

145 "The continent lay before them": Henry Adams, *History of the United States*, 1300.

146 "By his invention every river is laid open to us": Daniel R. Headrick, *Tools of Empire*, 17.

146 "gunboat diplomacy": The Chinese, forced to trade Indian opium for tea, were further humbled when Western gunboats in 1842 defeated a fleet of paddle-wheel warships based on those with which the Sung dynasty had prevailed *six hundred years earlier*—against pirates in 1132 and Digunai forces in 1161. The fact that Western paddle-wheel steamboats were based on the old Chinese design "was to haunt China in later centuries when her innovative spirit had flagged and her technology was surpassed by that of the Western barbarians," writes Daniel Headrick, who adds that "in the case of gunboats, we cannot claim that technological innovation caused imperialism, nor that imperialist

motives led to technological innovation. Rather, the means and the motives stimulated one another in a relationship of positive mutual feedback." Headrick, *Tools of Empire*, 53–4.

146 "the most signal triumph": Twenty British and twenty Egyptian troops died on one side, eleven thousand Dervishes on the other. As Churchill added, in what became a colonialist trope, "The strongest and best-armed savage army yet arrayed against a modern European Power had been destroyed and dispersed, with hardly any difficulty, comparatively small risk, and insignificant loss to the victors." Churchill, "The River War: An Account of the Reconquest of the Sudan," in Headrick, *Tools of Empire*, 118–9.

147 exporting the very gifts . . . that their conduct defied: The absurdity of the colonialist position encouraged absurdities on opposing sides, as when Mohandas Gandhi asserted, at the outset of World War II, that Great Britain and the United States "lack the moral basis for engaging in this war unless they put their own houses in order . . . They have no right to talk about protecting democracies and protecting civilization and human freedom, until the canker of white superiority is destroyed." Abernethy, *Dynamics of Global Dominance*, 145. Gandhi makes a stirring point but it is hardly the case that nations have "no right to talk" about reform until they themselves have fully reformed. The English prohibition against slavery at home was a good start even though England still permitted slavery overseas, for a time, and it would hardly have been a good idea for the Allied powers to have spent years "putting their houses in order" before confronting the Nazis.

147 "the very idea of distant possessions will be even ridiculed": Priestley, *Letters to the Right Honorable Edmund Burke*, 1791, in Kramnick, *The Portable Enlightenment Reader*, 670. In 1768, Priestley put the case more broadly: "Knowledge will be subdivided and extended; and knowledge, as Lord Bacon observes, being power, the human powers will, in fact, be increased; nature, including both its materials, and its laws, will be more at our command; men will make their situation in this world abundantly more easy and comfortable; they will probably prolong their existence in it, and will grow daily more happy, each in himself, and more able (and, I believe, more disposed) to communicate happiness to others. Thus, whatever was the beginning of this world, the end will be glorious and paradisiacal, beyond what our imaginations can now conceive." Priestley's effusions about "happiness" increasing to a "paradisiacal" state may have been fulsome but his predictions otherwise proved to be admirably accurate.

147 "a history that placed its own progress at the heart of the story": John Darwin, *After Tamerlane: The Global History of Empire Since 1405* (New York: Bloomsbury, 2008), 11.

CHAPTER EIGHT: PROGRESS

149 "No one talks seriously about 'progress' ": Hayek, *Constitution of Liberty*, 39; Tom Bethell, *The Noblest Triumph: Property and Prosperity Through the Ages* (New York: St. Martin's, 1998), 18.

150 "the time when . . . he finds life a burden": Marie Jean Antoine Nicolas Caritat, marquis de Condorcet, "The Future Progress of the Human Mind," in Paul Halsall, ed. Internet Modern History Sourcebook, http://www.fordham.edu/halsall/mod/condorcet-progress.html (accessed June 13, 2005).

150 "the prejudices that have established inequality": Ibid.

150 "speed up the advances of . . . sciences": Condorcet, "Sketch," in Steven Kreis, *The History Guide: Lectures on Modern European Intellectual History*, http://www.historyguide.org/intellect/lecture10a.html (accessed June 13, 2005).

150 "the perfectibility of man": Ibid.

150 "capable of more than they have ever so far achieved": John Passmore, *The Perfectibility of Man* (Indianapolis, IN: Liberty Fund, 2000), 512.

151 "we the middling people": Franklin, "Plain Truth," November 17, 1747.

152 "We must, indeed, all hang together": Isaacson, *Benjamin Franklin*, 313.

152 "The rapid Progress *true* science now makes": Franklin to Joseph Priestley, February 8, 1780.

152 "Furnished as all Europe now is with academies of science": Franklin to Joseph Banks, July 27, 1783.

152 He literally bet on the future: Believing that one should not profit from holding office in a democratic state, Franklin bequeathed his salary as president of Pennsylvania, some £2,000, to charity. That was a lot of money at the time, but Franklin wanted to let it accrue interest until it could make a greater impact. He stipulated that it first be loaned out, at 5 percent interest, to "young married artificers" who sought to establish their own businesses, and that a century later Boston and Philadelphia each draw £100,000 for "public works, which may be judged of most general utility to . . . make living in the town more convenient to its people, and render it more agreeable to strangers resorting thither for health or a temporary residence." The remainder was to continue earning interest for *another* century, whereupon each city could employ the funds as it wished. Boston's share of this legacy funded the Benjamin Franklin Institute of Technology, while Philadelphia's established the Franklin Institute. In other provisions of the will Franklin left "my fine crab-tree walking stick, with a gold head curiously wrought in the form of the cap of liberty" to George Washington; £2,173 to his son-in-law, Richard Bache, on condition that he "immediately after my decease manumit and set free his Negro man Bob"; and his best telescope to the astronomer David Rittenhouse, "for the use of his observatory."

152 "the humanity of the United States can never reach the sublime": Keats, letter to his brother and sister-in-law, October 14–31, 1818.

152 "something of the pettiness and materiality of his first occupation": Leigh Hunt, *Autobiography* (New York: Harper, 1850), I, 130ff; in Lemay and Zall, *Benjamin Franklin's Autobiography*, 267.

152 "Our inventions are wont to be pretty toys": Thoreau, *Walden*, "Economy," Frederick P. Rose, ed. (New York: Library of America, 1985), 363–65. As dismissive of science and political liberty as he was of technology, Thoreau compared voting in democratic elections to playing board games, and argued that students would be better off learning how to manage their own financial accounts than studying the economics of Adam Smith and David Ricardo. Franklin's biographer Walter Isaacson notes that although Franklin's *Autobiography* was the only book Davy Crockett took with him to the Alamo, "a backwoodsman as refined as Thoreau had no place for it when heading off the Walden Pond." Isaacson, *Franklin*, 479.

153 "'My religion is to serve my fellow men'": Sinclair Lewis, *Babbitt*, ch. 16, section 3.

153 life expectancy had more than doubled: United Nations Population Division, 2004; World Federation of UN Associations Millennium Project *2007 State of the Future* report.

153 influenza epidemic . . . and the HIV/AIDS epidemic: James C. Riley, *Rising Life Expectancy: A Global History* (New York: Cambridge University Press, 2001), 1; United Nations Secretariat, Department of Economic and Social Affairs, Population Division, *World Population Prospects: The 1998 Revision* (New York: 1998).

154 personal energy levels increased by roughly 50 percent: Robert William Fogel, *The Escape from Hunger and Premature Death, 1700–2100: Europe, America, and the Third World* (New York: Cambridge University Press, 2004); and William H. McNeill, "Bigger and Better?" *New York Review of Books*, October 21, 2004, 61.

154 malnutrition in the developing world dropped: *The Economist*, August 4, 2001, 63–65.

154 "the hope that all people will enjoy a long and healthy life": Riley, *Rising Life Expectancy*, 221.

154 Regarding wealth . . . extreme poverty has dropped: *World Bank Atlas*, 2004; World Bank *World Development Indicators*, 2005, http://www.worldbank.org/data. These figures are expressed in terms of purchasing power parity (PPP), defined as the number of units of a nation's currency required to buy the same amount of goods and services domestically as a U.S. dollar would buy in the United States. PPP is regarded by economists as a somewhat more accurate way than GDP to compare the wealth and poverty of nations.

154 "We are all illiterate here": Patrick E. Tyler, "In China's Outlands, Poorest Grow Poorer," *The New York Times*, October 26, 1996, A1.

155 In 1970, 37 percent of all people over age fifteen were illiterate: Amartya Sen, "To Build a Country, Build a Schoolhouse," *The New York Times*, May 27, 2002, A13; UNESCO, *2007 State of the Future* report.

155 "What will save me?": Celia W. Dugger, "In Africa, Free Schools Feed a Different Hunger," *The New York Times*, October 24, 2004, section 1, column 1.

155 Wider-reaching indicators of happiness . . . Happier than their grandparents: *The Economist Pocket World in Figures* (London: Profile, 2003); Ronald Inglehart and Hans-Dieter Klingemann, "Genes, Culture, Democracy, and Happiness," in Ed Diener and Eunkook M Suh, eds., *Culture and Subject Well-Being* (Cambridge, MA: MIT Press, 2000); Angus Maddison, *The World Economy: A Millennial Perspective* (Paris: OECD, 2001); Hadley Cantril, *The Pattern of Human Concerns*, (Piscataway, NJ: Rutgers University Press, 1966); William J. Bernstein, *The Birth of Plenty: How the Prosperity of the Modern World Was Created* (New York: McGraw Hill, 2004), 326–27.

156 the number of liberal-democratic nations: Freedom House, "Freedom in the World 2008," http://www.freedomhouse.org (accessed July 21, 2008).

156 the least-free fifth . . . had a per-capita GDP of only $3,305: *Economic Freedom of the World 2007 Annual Report*, http://www.freetheworld.com/release.html (accessed July 21, 2008).

157 per capita GDP increased nearly tenfold: Angus Maddison, *Monitoring the World Economy, 1820–1992* (Paris: OECD, 1995), 347, 349; the GDP figures are in 1990 international dollars.

157 "Globalization has dramatically increased": Jay Mazur, "Labor's New Internationalism," *Foreign Affairs*, January/February 2000; in Michael Kremer and Eric Maskin, "Globalization and Inequality," Brookings Institution Globalization and Inequality Group meeting, June 17, 2002, http://www.brookings.edu/gs/research/projects/glig/glig_hp.htm (accessed June 20, 2005). "Already obscene": Kevin Watkins, "Making Globalization Work for the Poor," March 2002. "Inequality is soaring": Noam Chomsky, "September 11th and its Aftermath: Where Is the World Heading?" Chennai (Madras) India, November 10, 2001. "The attacks on the World Trade Center": *International Herald Tribune*, editorial, October 3, 2001. From Gary Burtless, "Is the Global Gap Between Rich and Poor Getting Wider?": The Brookings Institution, June 17, 2002.

157 global inequality overall ceased growing from 1980 to 2000 and in many respects began to shrink: A. Boltho and G. Toniolo, "The Assessment: The Twentieth

Century: Achievements, Failures, Lessons," *Oxford Review of Economic Policy,* vol. 15, no. 4. They write: "Worldwide divergence in per capita GDP increased steadily from the beginning of the century to the early 1980s. A turning point occurs, however, around 1980. The more rapid growth rates of India and, especially, China in more recent years have led to some modest convergence." "Nine measures of global income inequality . . . deliver the same picture: inequality declined substantially during the last two decades," notes Xavier Sala-i-Martin, "The World Distribution of Income," *NBER Working Paper* 8933. "When international inequality is appropriately measured on the basis of [PPP] . . . and countries are weighted according to the size of their populations, plausible measures of international inequality indicate that income convergence has taken place since the late 1960s," writes Arne Melchior: "Global Income Inequality: Beliefs, Facts and Unresolved Issues," *World Economics*, vol. 2, no. 3, July-September 2001.

157 income inequalities tend to be largest within the nations least touched by globalization: *Economic Freedom of the World 2007 Annual Report,* http://www.freetheworld.com/release.html (accessed July 21, 2008).

158 "notably in India and China": For an analysis see "More or Less Equal?" *The Economist*, March 13, 2004, 69.

158 "Nor was this just a brief spurt": IMF World Economic Outlook database, May 2001.

158 the Gini ratio: Bob Sutcliffe, "A More or Less Unequal World? World Income Distribution In The 20th Century," Political Economy Research Institute, Working Papers wp54. (Amherst: University of Massachusetts, 2003), http://www.umass.edu/peri/pdfs/WP54.pdf (accessed June 18, 2005).

158 "the most lucrative new products": Nathan Rosenberg and L. E. Birdzell Jr. *How the West Grew Rich: The Economic Transformation of the Industrial World.* New York: Basic Books, 1986, 26.

159 "All institutions of freedom are adaptations": Friedrich A. Hayek, *The Constitution of Liberty* (Chicago: University of Chicago Press, 1960), 30, 38.

159 "The Western scientific community": Rosenberg and Birdzell, *How the West Grew Rich*, 255.

160 "'Bureaucratic feudalism'": Needham, *Grand Titration*, 197.

160 "Results! Why, man, I have gotten a lot of results!": Frank Lewis Dyer, *Edison: His Life and Inventions* (New York: Harper, 1929), 615–616. The physicist Pierre Duhem relates a similar story, told him by an assistant to Pasteur: "Pasteur arrived at the laboratory having in his head what Claude Bernard called a preconceived idea, that is, a proposition which he wished to submit to experimental verification. His assistants, under his direction, prepared experiments which, according to that preconceived idea, ought to produce certain results. Most of the time the expected results were not those which the experiment in fact produced. The experiments were then repeated from the beginning, and with greater care. Another failure. A third attempt was then launched, only to result in a further failure. My friends the laboratory assistants were often astonished at the obstinacy of 'the boss,' intoxicated by erroneous preconception. Finally a day came when Pasteur announced an idea different from the one which experiment condemned. One then realized with admiration that none of the contradictions to which the latter had led had been in vain; for each of them had been taken into account in the formation of the new hypothesis. It, in its turn, had its corollaries put to the test of facts, and, quite often, the process resulted in new failures. However, in the process of destroying this hypothesis, these failures prepared the conception of a new idea. So, little by little through this sort of struggle between preconceived ideas that suggested experiments, and the

experiments that constrained the preconceived ideas to be transformed, a hypothesis was formed which conformed perfectly with the facts." Pierre Duhem, *German Science: Some Reflections on German Science and German Virtues*, John Lyon, trans. (La Salle IL: Open Court, 1991), 22.

161 People say that they hate their mobile phones: 2004 Lemelson-MIT Invention Index study. http://web.mit.edu/invent/n-pressreleases/n-press-04index.html (accessed August 18, 2009).

161 The number of urban dwellers quadrupled: Most of the recent urbanization has occurred in Africa (whose urban population increased ninefold), and in Asia, Latin America, and the Caribbean (up fivefold each). Meanwhile North American urbanization more than doubled, from 110 million to 243 million. See "The Prospects for World Urbanization and Rural Population Growth," UN Population Division, *World Urbanization Prospects: The 2001 Revision*. http://www.un.org/esa/population/publications/wup2001/WUP2001_CH2.pdf (accessed June 28, 2005).

162 put the brakes on population growth: Ibid.

162 "perhaps the most notable effect": Carroll Pursell, *The Machine in America: A Social History of Technology* (Baltimore, MD: Johns Hopkins University Press, 1995), 131–32. "If the American city by 1920 was much more than a machine for making money," Pursell observes, "it was also unthinkable without that machine."

162 urbanites rose from 10 percent of the population: Hobsbawm, *Age of Empire*, 343.

162 Life expectancy at birth: Tristram Hunt, *Building Jerusalem: The Rise and Fall of the Victorian City* (London: Weidenfeld & Nicholson, 2004), and reviews thereof: John Brewer, "City Lights," *New York Review of Books*, May 11, 2006, 18; and Kate Colquhoun, "Among the Satanic Mills," *The Telegraph*, July 6, 2004, at www.telegraph.co.uk.

163 "FAMINE, FILTH AND DISEASE": Ackroyd, *London: The Biography*, 573.

163 "Rich London is the creature of slum-London": William Morris, *Pall Mall Gazette*, September 4, 1888, 1–2, http://www.marxists.org/archive/morris/works/1888/ugly.htm (accessed June 28, 2005).

164 "rogues, vagabonds, and thieves": D. C. Coleman, *The Economy of England, 1450–1750* (New York: Oxford University Press, 1977), 18–19.

164 "they excel . . . in policy and industry": Hernando de Soto, *The Mystery of Capital: Why Capitalism Triumphs in the West and Fails Everywhere Else* (New York: Basic Books, 2000), 99.

165 two hundred bureaucratic steps: Hernando de Soto, interviewed for the PBS series *Commanding Heights*, http://www.pbs.org/wgbh/commandingheights/shared/minitextlo/int_hernandodesoto.html (accessed June 28, 2005). In Manila, de Soto found, the same process would take at least 24 years. In Egypt, getting permission "to build a house on a sand dune" would take "something like 17 years."

165 Peru "had become two nations": http://www.aworldconnected.org/article.php?id=309&print=1 (accessed June 28, 2005).

165 "paper wall": de Soto, *Commanding Heights* interview.

165 "The bell jar": de Soto, *Mystery of Capital*, 159, 67.

165 Peru . . . Russia . . . Latin America . . . Mexico: de Soto, interviewed by Dirk Verhofstadt, http://www.stephenpollard.net/002060.html (accessed June 29, 2005).

166 "In every country we have examined": de Soto, *Mystery of Capital*, 34.

166 "The life of the law has not been logic": Ibid., 153.

166 "the principles upon which the landed property of this country should . . . be established": Harold Perkin, *Origins of Modern English Society* (London: Routledge, 1999), 452.

167 "Capitalism is in trouble if it doesn't adapt": de Soto, *Commanding Heights* interview; quotation slightly edited for clarity.

167 "Einstein taught us": de Soto, *Mystery of Capital*, 40.

CHAPTER NINE: THE SCIENCE OF WEALTH

168 "no real English gentleman": *The First Edinburgh Reviewer*, 1855, 1965:324. Bagehot was an influential early editor of the *Economist* who felt that economists were unpopular mainly because, as Francis Bacon said of unsentimental political analysts like Machiavelli, they "write what men do, and not what they ought to do." (Bacon, *Advancement of Learning*, 1605, 1963:201).

168 Freud . . . discovered nothing and cured nobody: Freud's hyperbolic career is wryly critiqued by the Berkeley English professor Fred Crews in his *The Memory Wars* (New York: New York Review of Books, 1990).

169 "gigantic inevitable famine": Thomas Malthus, *An Essay on the Principle of Population*, 1798, preface.

169 the world population climbed: Colin McEverdy and Richard Jones, *Atlas of World Population History* (London: Penguin and Allen Lane, 1978).

170 The world's three hundred richest individuals have more money than the entire bottom half: *The Economist Pocket World in Figures* (London: Profile, 2003).

170 "the number of people in the developing world living on less than a dollar a day has been cut almost in half": *The Economist*, June 28, 2003, 80.

171 "anarchic, formless, and appallingly unjust": William Manchester, *A World Lit Only by Fire: The Medieval Mind and the Renaissance* (Boston: Little, Brown, 1993), 1, 15.

171 Grimm's Fairy Tales: " 'Do give me one night's lodging, and a little to eat and drink,' said he to her, 'or I shall starve' ": ("The Blue Light"). " 'We must get supplies for the winter,' said the cat, 'or else we'll starve' ": ("The Companionship of the Cat and the Mouse"). " 'I have no money,' thought he, 'I have learnt no trade but that of fighting, and now that they have made peace they don't want me any longer; so I see beforehand that I shall have to starve' ": ("Bearskin"). "Next to a great forest there lived a poor woodcutter who had come upon such hard times that he could scarcely provide daily bread for his wife and his two children. Finally he could no longer even manage this, and he did not know where to turn for help": ("Hansel and Gretel"). "Hunger tormented her . . . She thought, 'Ah, if I were but inside, that I might eat of the fruit, else must I die of hunger!' Then she knelt down, called on God the Lord, and prayed. And suddenly an angel came towards her, who made a dam in the water, so that the moat became dry and she could walk through it. And now she went into the garden and the angel went with her. She saw a tree covered with beautiful pears, but they were all counted. Then she went to them, and to still her hunger, ate one with her mouth from the tree": ("The Girl Without Hands").

171 "*nothing would ever change*": Manchester, *A World Lit Only by Fire*, 27. Italics in original.

172 "The increase of any estate": Jan de Vries, *Economy of Europe in an Age of Crisis, 1600–1750* (New York: Cambridge University Press, 1976), 177.

172 to lend money at interest was a sin: The prohibition against usury—often conflated with any lending that involved interest—originated with Exodus 22:25: "If thou lend money to any of my people that is poor by thee, thou shalt not be to him as a usurer." The ban was enthusiastically upheld by Saint Augustine ("Business is itself an evil") and Saint Jerome ("A man who is a merchant can seldom if ever please God"). The Catholic Church forbade clerics from loaning money in 325 and extended the prohibition to the laity in 850, provisions not lifted until

the Fifth Lateran Council, in 1571, by which time the Vatican had become one of Europe's major debtor states. Jews earned their reputation as moneylenders during this period: it helped that they could not be excommunicated, since they were not members of the Church in the first place.

172 "masters and possessors of nature": Descartes, *Discourse on Method*, F. E. Sutcliffe, trans. (Harmondsworth: Penguin, 1968), 78.

172 The English "gentleman": Macaulay describes the old-style lord as typically a parochial boor, who had little education or polish, whose "oaths, coarse jests, and scurrilous terms of abuse were uttered with the broadest accent of his province" and whose "chief pleasures were commonly derived from field sports and from an unrefined sensuality." A seventeenth-century English gentleman, on the other hand, "generally receives a liberal education . . . [and] has seen something of foreign countries. A considerable part of his life has generally been passed in the capital; and the refinements of the capital follow him into the country." Lord Macaulay, *The History of England* (London: Penguin, 1986), 56–57. In her *Scientific Culture and the Making of the Industrial West* (New York: Oxford University Press, 1997, 87), Margaret C. Jacob notes that with the rise of science in the eighteenth century "a new *persona* emerged, first in England and then in western Europe: the literate gentleman who read the periodical press, attended literary and philosophical lectures or clubs for the purpose of being cultured, and remained vaguely Christian, generally Protestant, but explained his beliefs in terms of the order and harmony of creation. He might be a merchant of the city or a landed gentleman of the country; he might even be a shopkeeper, a doctor, or a lawyer. He believed in educating his children; his wife, although generally more pious than he, was almost certainly literate and a reader of books, especially novels. By the 1720s, particularly in England, such a gentleman or merchant could have increasingly easy access to applied science as taught by the Newtonian lecturers." According to one study, the ranks of the gentry increased from a quarter of the British population in 1436 to half in 1790, during which time those of "church and crown" declined from 35 to 10 percent. George Mingay, *The Gentry* (London: Longman, 1976), 59, Table 3.1. Although the gentry had their share of complaints, and like other rising classes could simultaneously "grab and weep," their circumstances were widely admired. "Could humanity ever attain happiness," wrote Hume, referring to the half-century leading up to the English civil wars of 1642–1651, "the condition of the English gentry at this period might merit that appellation." R. H. Tawney, "The Rise of the Gentry," in E. M. Carus-Wilson, ed., *Essays in Economic History* (London: Edward Arnold, 1954), 175.

173 "This new unwanted society": Robert L. Heilbroner, *The Worldly Philosophers: The Lives, Times, and Ideas of the Great Economic Thinkers* (New York: Simon & Schuster, 1999), 33. Nostalgia for the old order survives even today. One hears outcries against wealth as being inherently evil—it must have been accumulated at the expense of the environment, if not of the poor—and can visit a Federal Reserve vault deep beneath Wall Street where splay-fingered laborers heft bars of gold from one cage to another to reflect computer-generated, satellite-relayed transfers among nations that no longer even bother to keep their gold within their borders. This vault holds a quarter of the world's known gold reserves.

173 "nothing which requires more to be illustrated": Boswell, *Life of Johnson* (New York: Random House, 1952), 581. Boswell was surprised that Johnson approved of Smith's book, since the sole meeting between the two great thinkers had been disastrous. According to Sir Walter Scott, Johnson, encountering Smith at a London club to which both belonged, attacked a statement Smith had made praising David Hume. Smith said that his remark was accurate.

"You lie!" exclaimed Johnson. Smith, wounded, replied, "You are a son of a bitch!" One wishes their dialogue had been loftier in tone, but it is doubtful that the revolutionary economist and the deeply conservative Johnson would have found much common ground for agreement in any event. (For a critique of this anecdote as possibly unfounded, see the editor's appendix to the 1889 Bell edition of Boswell's *Life of Johnson,* 371, available online at http://books.google.com/books?id=ayhG0LE6_l4C&dq=%22Sir+Walter+Scott%22+%22Samuel+Johnson%22+%22Adam+Smith%22+%22you+lie%22&source=gbs_summary_s&cad=0.

173 "as dull a dog as he had ever met": Johnson quoted by Boswell, in James Buchan, *The Authentic Adam Smith* (New York: Norton, 2006), 89.

173 "by far the most useful, and therefore as by far the happiest . . . period of my life": Adam Smith to Dr. Archibald Davidson, Edinburgh, November 16, 1787.

174 "seems very likely to become": Heilbroner, *Worldly Philosophers,* 50.

174 "his happy talent of throwing light": John J. Lalor, *Cyclopædia of Political Science, Political Economy, and the Political History of the United States by the Best American and European Writers* (New York: Maynard, Merrill, 1899), III.196.10.

174 "fix the attention of public opinion": Ibid., III.196.15.

174 "We are all your scholars": Heilbroner, *Worldly Philosophers,* 74. For a skeptical view of the reliability of this anecdote, see Buchan, *The Authentic Adam Smith,* 136–37.

175 "does not perhaps contain a word of truth": Adam Smith, *Lectures on Rhetoric and Belles Lettres* (London: Thomas Nelson, 1964), 139–40.

175 "supercilious and ignorant contempt": Adam Smith, *History of Astronomy,* Dugald Stewart, ed. (London: Cadell & Davies, 1811), vol. 5, V, 115–16.

175 "superlative words and intellectual arguments": William Petty, *The Economic Writings*, C. H. Hill, ed. (London: Cambridge University Press, 1899), vol. 1, 244. As the economic historians Harry Landreth and David C. Colander note in their *History of Economic Thought* (Boston: Houghton Mifflin, 2002), 51, "Petty's seminal insight that ideas should be expressed in terms of numbers, weight, and measure and that only arguments that have visible foundations in nature should be accepted is the cornerstone of modern thinking in economics." For a discussion of quantitative thinking in Smith's day, see, e.g., Tore Frängsmyr, John L. Heilbron, and Robin E. Rider, eds., *The Quantifying Spirit in the Eighteenth Century* (Berkeley: University of California Press, 1990).

176 "perhaps not one pin in a day": Adam Smith, *Wealth of Nations*, vol. 1, bk. 1, ch. 1. (Chicago: University of Chicago Press, 1975), 3. One way to appreciate how the division of labor improves productivity is to think of it in terms of broadening the job pool: If a plant manager needs to hire someone who can skillfully perform every aspect of making a pin (or an automobile engine), he or she is likely to find few if any workers up to the task, whereas many more are available who know or can be trained how to attach a pinhead or a set of piston rings. On the other hand, specialization can make work drearily repetitious. Smith realized this; his remedy was to promote universal public education, including what today is called job training, to keep workers stimulated and help them move on to other tasks.

176 "beautiful machines": Ibid., 4.

176 "It was probably a farmer": Smith, *Lectures on Justice, Police, Revenge and Arms* (Oxford: Clarendon Press, 1896), II, ii, 167–78.

177 "continually gravitating . . . towards the natural price": Roy Porter, *Enlightenment: Britain and the Creation of the Modern World* (London: Penguin, 2000), 387.

177 "It is not from the benevolence of the butcher": Smith, *Wealth,* bk. 1, ch 2, 7.
177 "invisible hand": Ibid., bk. 4, ch. 2, 194.
177 "If a foreign country can supply us with a commodity": Ibid., 194.
178 "Opulence and freedom": Porter, *Enlightenment*, 389.
178 "It is impossible indeed": Smith, *Wealth*, bk. I, ch. 10, 55.
178 "Though the interest of the laborer is strictly connected": Ibid., 110.
179 "Their superiority over the country gentleman": Ibid., 110.
179 "The inland corn dealer": Ibid., bk. IV, ch. 5, 229.
179 "nothing seems more likely to establish this equality": Ibid., bk. IV, ch. 7, 271–72. Emphasis added.
180 "If natural philosophy . . . shall at length be perfected": Newton, *Opticks*, III, 1, qu. 3.
180 "an immense chain of the most important and sublime truths": Smith, *History of Astronomy*, 189–90.
181 Charles Darwin's first musings: Darwin learned a lot from reading the economists, especially Adam Smith (see Silvan S. Schweber, "The Origin of the *Origin* Revisited." *Jour. History Biol.* 10:229–316). Stephen Jay Gould suggested that Smith's free-market economics, though too cruel for a society to endure if completely unfettered, is just how natural selection works: "The theory of natural selection lifts [Smith's] entire explanatory structure, *virgo intacta*, and then applies the same causal scheme to nature—a tough customer who can bear the hecatomb [sacrifice] of deaths required to produce the best polity as an epiphenomenon. Individual organisms engaged in the 'struggle for existence' act as the analog of firms in competition. Reproductive success becomes the analog of profit—for, even more than in human economies, you truly cannot take it with you in nature." Gould, *The Structure of Evolutionary Theory* (Harvard University Press, 2002), 123.
181 "I never loved anybody . . . so much": Heilbroner, *Worldly Philosophers*, 88.
181 "I was never a boy": W. L. Courtney, *Life of John Stuart Mill* (London: Walter Scott, 1889), 40.
181 "a certain evil": J. S. Mill, *Principles of Political Economy*, bk. V, ch. 11, 7 (Amherst, NY: Prometheus Books, 2004), 865.
181 "There are some things with which government ought not to meddle": Ibid., 834, 857.
181 ability to distinguish: Mill was so scrupulous that he even remained open-minded about communism, pending the outcome of experiment: "It is yet to be ascertained whether the Communistic scheme would be consistent with that multiform development of human nature, those manifold unlikenesses, that diversity of tastes and talents, and variety of intellectual points of view, which not only form a great part of the interest of human life, but by bringing intellects into a stimulating collision, and by presenting to each innumerable notions that he would not have conceived of himself, are the mainspring of mental and moral progression." (*Principles of Political Economy,* 219.) The experiment now having been made, with hideous consequences, Mill's doubts appear to have been well-founded.
182 "misleading and disastrous": John Maynard Keynes, *The General Theory of Employment, Interest, and Money* (Amherst, NY: Prometheus, 1997), 3.
182 "In the long run we are all dead": John Maynard Keynes, *A Tract on Monetary Reform* (London: Macmillan, 1923), 80.
182 "We are all Keynesians now": The journalist Howard K. Smith reported on ABC News in January 1971 that Nixon had made this remark to him privately some weeks earlier, in late 1970. The context is unclear but Nixon most likely was defending deficit spending, anathema to a classical conservative but business as usual for a Keynesian.

183 "I know of no example in time or place": Friedman, *Capitalism and Freedom* (Chicago: University of Chicago Press, 2002), 9, 15.

183 "government . . . threatens freedom": Friedman, *Capitalism and Freedom*, 5, 39. Friedman approvingly quotes his fellow economist Joseph Schumpeter's objection, in 1954, that "as a supreme, if unintended compliment, the enemies of the system of private enterprise have thought it wise to appropriate its label" by calling themselves liberals.

184 "the beliefs of the great majority": F. A. Hayek, *The Road to Serfdom* (Chicago: University of Chicago Press, 1994), 19.

185 "the American economy . . . has performed better under Democratic than Republican administrations": U.S. Census Bureau; Larry M. Bartels, *The New York Times*, 31 August 2008, BU 1,7; Bloomberg Financial Markets, in Tommy McCall, "Bulls, Bears, Donkeys and Elephants," *The New York Times*, October 14, 2008, A27. (The calculations of stock market investments, if normalized for the Great Depression by eliminating Herbert Hoover's presidency, change the Republican return to $51,211 rather than $11,733 as against a yield of $300,671 under the Democrats.) Like all time-sensitive data in this book, these figures are of course adjusted to correct for inflation.

186 "The constancy of the laws of nature": Malthus, *An Essay on the Principle of Population* (London: J. Johnson, 1798), Chapter XVIII, par. 10.

186 "looking at the faces of the poorest people": A. C. Pigou, *Memorials of Alfred Marshall* (London: Macmillan, 1925), 10.

187 "All you needed was $10,000 and a satellite phone": Paul Collier, *The Bottom Billion: Why the Poorest Countries Are Failing and What Can Be Done About It* (New York: Oxford University Press, 2007), 21.

187 "the quack remedy . . . of populism": Ibid., 85.

187 "We need stronger and fairer globalization": "Paul Collier: Reforming Agriculture," *Economist's View*, http://economistsview.typepad.com/economistsview/2008/05/paul-collier-re.html (accessed June 17, 2008).

187 "Market capitalism is the best economic system": Felix Rohatyn, "From New York to Baghdad," *New York Review of Books*, September 21, 2002, 4.

188 "Economic growth makes a society more open": Benjamin M. Friedman, *The Moral Consequences of Economic Growth* (New York: Knopf, 2005), 15.

188 "all human progress": Ibid., 19, 15.

188 "The hope that poverty and ignorance may gradually be extinguished": Ibid., 67.

189 Cab drivers: Colin Camerer, et al., "Labor Supply of New York City Cabdrivers: One Day at a Time," *The Quarterly Journal of Economics*, May 1997, 407–41.

CHAPTER TEN: TOTALITARIAN ANTISCIENCE

191 "killed a hundred million people": The toll includes approximately 62 million in the USSR, 35 million in communist China, and 21 million in Nazi Germany.

191 "The physicists have known sin": J. Robert Oppenheimer, MIT lecture of November 25, 1947, published as *Physics in the Contemporary World* (Portland, ME: Anthoensen, 1947), 11. The complete sentence reads, "In some sort of crude sense which no vulgarity, no humor, no over-statement can quite extinguish, the physicists have known sin; and this is a knowledge which they cannot lose." But at what point did the physicist come to "know sin"? Certainly not when first learning about nuclear reactions: no knowledge is sinful, and the pioneering physicists saw no evident application of their work to weapons. (Einstein, whose equation $e=mc^2$ stands at the headwaters of nuclear research, long assumed that nuclear weapons were impossible.) Nor is it evident that the Manhattan Project physicists became sinners by working on the bomb, considering the prospects

for a world in which the Nazis or communists had it and the free world did not. That leaves the decision to drop it, which was not made by scientists. One can feel strongly that the bomb ought not to have been dropped on Japanese cities—as does the present author—without agreeing that President Truman's action turned the atomic scientists into sinners.

191 "an evil thing": Oppenheimer, statement to a joint meeting of the American Philosophical Society and the National Academy of Sciences shortly after the atomic bombings in Japan, in Jennifer Caron, "Biology and 'The Bomb,'" *Engineering & Science* 2004, no. 2, 17.

191 "a latent but growing feeling that science is somehow turning evil": Philip Morrison, "The Atomic Bomb and the Teacher of Science," *Science Education*, February, 1946, 8. Note that Morrison, unlike Oppenheimer, couched his remarks in terms of popular apprehensions rather than his own views as a scientist and scholar.

192 "the destructive potentialities of the scientific outlook": Michael Polanyi, "Why Did We Destroy Europe?" in R. T. Allen, ed., *Society, Economics & Philosophy* (New Brunswick NJ: Transaction, 1997), 111.

192 scientists ought to be held accountable: Such proposals dated back to the thirties, when the British baron and banker Josiah Stamp proposed a moratorium on research and invention and twenty-two Cambridge University scientists urged their colleagues to refuse work on national defense. Similar campaigns have popped up ever since. Recently, both right-wing conservatives and left-wing progressives have sought to curb stem-cell research—the former on religious grounds, the latter for fear that biotechnology would be used to create "simps," a science-fiction genetic blend of humans and chimpanzees intended for blue-collar work in hazardous environments such as cleaning up nuclear weapons laboratories.

192 "It was not that Bismarck lied": William Shawcross, *Sideshow: Kissinger, Nixon and the Destruction of Cambodia* (New York: Simon and Schuster, 1979), 304.

192 "mediaeval altogether": Princess Victoria to her mother, 1875, in Erich Eyck, *Bismarck and the German Empire* (New York: Norton, 1964), 199.

192 "eats too much, drinks too much": Prince Alexander Gortchakoff (1798–1883), in Eyck, *Bismarck and the German Empire*, 223.

192 "I want to make music in my own way": Michael Stürmer, *The German Empire: A Short History* (New York: Modern Library, 2002), 17.

193 "In exchange for lavish trinkets": Jonah Goldberg, *Liberal Fascism: The Secret History of the American Left from Mussolini to the Politics of Meaning* (New York: Doubleday, 2007), 240.

194 "those who would give up essential liberty": This quotation is customarily attributed to Franklin, in *An Historical Review of the Constitution and Government of Pennsylvania*, 1759, although his authorship is disputed. Many variations exist, one of which is inscribed on the steps of the Statue of Liberty.

194 "Instruction in the sciences": Adolf Hitler, *Mein Kampf*, James Murphy, trans., chapters 2, 4, http://www.nalanda.nitc.ac.in/resources/english/etext-project/Biography/hitler/intro.html (accessed January 2006).

194 "glacial cosmogony": The book *Glacial Cosmogony*, by Philipp Fauth and Hanns Hörbiger, was published in 1913. For a discussion see Robert Bowen, *Universal Ice: Science and Ideology in the Nazi State* (London: Bellhaven, 1993); Nicholas Goodrick-Clarke, *The Occult Roots of Nazism* (New York: Tauris Parke, 2004); and Robert Reeves, "Nazi Astronomy" (*Amateur Astronomy*, Fall 1995), 13. Heinrich Himmler, creator of the SS units which ran the death camps, was a vegetarian teetotaler and a devotee of homeopathy; he ordered that medicinal herbs be grown in the camps. Another crank theory popular with the Nazis—that

the earth was hollow and its (occupied) interior accessible through holes at the poles—is discussed in Walter Kafton-Minkel's *Subterranean Worlds* (Port Townsend, WA: Loompanics, 1989).

194 "The totalitarian belief that everything is possible": Hannah Arendt, *The Origins of Totalitarianism* (New York: Schocken, 2004), 591.

195 "Jewish physics": Albert Speer, *Inside the Third Reich*, Richard and Clara Winston, trans. (New York: Macmillan, 1970), 228.

195 "trying to buy the corpse of science": Bronowski, *Science and Human Values*, 70.

195 "flying in the face of established scientific fact": Joseph Needham, *The Nazi Attack on International Science* (London: Watts, 1941).

195 "the German forces were greatly superior": Alan J. Levine, "Was World War II a Near-run Thing?" in Loyd E. Lee, ed., *World War II: Crucible of the Contemporary World* (Armonk, NY: Sharpe, 1991), 365.

196 "the pesticide Zyklon B": Zyklon B was developed in part by Fritz Haber, a chemist whose career illustrates the pitfalls involved in passing moral judgments on scientists based on the consequences of their research. Haber worked on poison gas during World War I and directed the first release of chlorine gas against enemy troops—at Ypres, France, on April 22, 1915—for which he was declared a war criminal by the Allies. (His wife, the chemist Clara Immerwahr, shot herself in the garden behind their home—dying in the arms of her son, Hermann, who would himself commit suicide thirty years later.) Haber's synthesis of ammonia from nitrogen and hydrogen in the air, although discovered with little thought of its practical applications, provided a significant source of high explosives, the war minister telling Haber that but for his contribution Germany would have been defeated within a matter of months. So evidently we have in Haber a certified war criminal who fed the German munitions industry and personally oversaw a poisonous gas attack. On the other hand, the Haber process produced artificial fertilizers that saved millions from starvation worldwide, and when Haber was ordered to purge Jewish scientists from the Kaiser Wilhelm Institute for Physical Chemistry in Berlin, he refused and instead resigned—on April 30, 1933, objecting, "My tradition requires that I select the staff for a scientific post based only on their qualifications and character without asking about race." Haber to the Minister for Science, Art, and Folk Culture, 30 April 1933, in Kristie Macrakis, *Surviving the Swastika: Scientific Research in Nazi Germany* (New York: Oxford University Press, 1993), 54. Exiled, Haber died in Switzerland the following year. He was, by the way, a Jew who had converted to Christianity, and a lifelong personal friend of Einstein's. A number of his relatives perished in Nazi concentration camps, some perhaps murdered by Zyklon B. Nitrogen fertilizer continues to feed hundreds of millions of people, but is also a major source of global water pollution. Imagine trying to predict all this in, say, 1910, and you have an idea of the difficulties involved in attempting to regulate scientific research for social betterment.

197 "contrary positions and voices were simply eliminated": "Three Q's," *Science*, February 15, 2008, 883.

197 "it appears almost certain": Spencer R. Weart and Gertrud Weiss Szilard, eds. *Leo Szilard: His Version of the Facts* (Cambridge, MA: MIT Press, 1978), 95. Szilard escaped from Germany just one day before the authorities began interrogating every passenger on the same train. "This just goes to show that if you want to succeed in this world you don't have to be cleverer than other people," he reflected. "You just have to be one day earlier than most people" (Ibid., 14).

197 "we shall do without science for a few years!": Edward Y. Hartshorne, *The German Universities and National Socialism* (Cambridge, MA: Harvard University Press, 1937), 112. Although this statement is often attributed to a meeting between Hitler and Max Planck, it does not appear in Planck's short notes on their conversation. "It could be that Hitler made this or a similar statement to Carl Bosch, who also lobbied Hitler on behalf of Jewish scientists," writes Kristie Macrakis, in her *Surviving The Swastika*, 223.

197 "Bernhard Rust, a former provincial schoolmaster": Macrakis, *Surviving the Swastika*, 78.

197 "Suffered? . . . It doesn't exist anymore!": Cornwell, *Hitler's Scientists*, 198. Rust, who had been an inmate in an insane asylum before Hitler put him in charge of German science, became a lightning rod of criticism from scientists (Max Planck: "It is the fault of Rust") and eventually from prominent Nazis like Goebbels, who said, "You just can't let an absolute nitwit head German science for years and not expect to be punished for such folly." *Science*: *New Series*, February 4, 1949, 122.

198 "A few first-rate scientists did remain": Major figures in addition to those mentioned in this paragraph include the physicists Max von Laue, who sought to avoid having to make the heil-Hitler salute by always carrying a parcel in both hands when out in public; Otto Hahn, who was unable to accept the 1944 Nobel Prize awarded him because he was still being detained by the British authorities who debriefed him about nuclear fission; his fellow detainees Karl Wirtz and Carl-Friedrich von Weizsäcker; the biochemist Adolf Butenandt, the chemist Richard Kihn, and the biologist Fritz von Wettstein.

198 "If today thirty professors get up and protest": Cornwell, *Hitler's Scientists*, 136.

198 "The outside world is something independent": Max Planck, *Scientific Autobiography, and Other Papers* (London: Williams & Norgate, 1949), 13.

198 "Why is the one reflected in the many": Werner Heisenberg, *Across the Frontiers*, Peter Heath, trans. (New York: Harper & Row, 1974), 34.

198 "The world out there is really ugly, but the work is beautiful": David C. Cassidy, *Uncertainty: The Life and Science of Werner Heisenberg* (New York: Freeman, 1992), 330.

198 "Oh, what idiots we have been!": Ruth Moore, *Niels Bohr* (New York: Knopf, 1966), 226. This conversation took place on January 3, 1939. Bohr remained in Copenhagen until 1943, when he fled by boat to Sweden after being warned that the Nazis intended to arrest him. Churchill, alerted that Nazi agents in Sweden were hot on Bohr's trail, had him flown to Scotland in the bomb compartment of a Mosquito aircraft. Bohr met with Churchill and then went to America, where he became involved in the Manhattan Project—although the project leader, General Leslie Groves, told Meitner at a cocktail party in the States a couple of years later that Bohr had been no help at all with the bomb. Meitner's role in the discovery of nuclear fission has long been underestimated. Einstein, when called "the father of the bomb," replied that the bomb had no father but did have a mother, Lise Meitner.

199 "You all worked for Nazi Germany": Meitner to Hahn, June 27, 1945. Meitner continues, "Perhaps you remember when I was still in Germany (and I know today that it was not only stupid, but a great injustice that I didn't leave immediately), I often said to you, 'As long as just we and not you have sleepless nights, it won't get any better in Germany.' But you never had any sleepless nights: you didn't want to see—it was too disturbing." At the end of the war Hahn was astounded to learn that he had been awarded the Nobel Prize in physics, and that the Allies had produced nuclear weapons. At that point Hahn might have ameliorated the Nobel committee's oversight by giving Meitner credit for her part in

discovering nuclear fission, but he did nothing of the sort. For a discussion see Ruth Lewin Sime, "The Politics of Memory: Otto Hahn and the Third Reich," *Physics in Perspective*, March 2006, 3; Sime, *Lise Meitner: A Life in Physics* (Berkeley: University of California Press, 1996), 305.

200 "Flying won't be fun anymore": Cornwell, *Hitler's Scientists*, 267. Blamed for losing the Battle of Britain, Udet shot himself on November 17, 1941. The Nazis issued a statement claiming that he had died while testing an aircraft.

201 "Hey, don't let the cops get a hold of that": Jennet Conant, *Tuxedo Park: A Wall Street Tycoon and the Secret Palace of Science That Changed the Course of World War II* (New York: Simon & Schuster, 2002), 177.

201 "no time to proceed by ordinary channels": Winston S. Churchill, *The Second World War* (Boston: Houghton-Mifflin, 1949–54), vol. II [*Their Finest Hour*], 16.

201 "a war of masses of men": Cornwell, *Hitler's Scientists*, 273–74.

201 "Unless British science had proved superior": Churchill, *The Second World War*, 381–2.

201 "Would it help, sir": Cornwell, *Hitler's Scientists*, 273–74.

202 "he valued science and technology": R. V. Jones, *Most Secret War* (London: Hamish Hamilton, 1978), 106.

202 Historians differ sharply: See Thomas Powers, *Heisenberg's War* (New York: Knopf, 1993); Jeremy Bernstein, *Hitler's Uranium Club* (New York: Copernicus, 2001); and Michael Frayn's play *Copenhagen* (London: Methuen, 2003).

202 "Had I never lived": Owen Gingerich, ed., *The Nature of Scientific Discovery: Symposium Commemorating the 500th Anniversary of the Birth of Nicolaus Copernicus.* (New York: Braziller, 1975), 496. Heisenberg knew Beethoven's Opus 111 well enough to play it in full from memory. His motives aside, he may simply have been the wrong man for the job, in that his creative brilliance was not matched by administrative talent; as an American Nobel laureate once put it, Heisenberg "was a *great* physicist, but not a *good* one."

203 "In the early days of the war the world was amazed": Joseph Needham, "On Science and Social Change," written in 1944, first published in *Science and Society,* 1946, 10:225, and reprinted in his *The Grand Titration*, 146–47. Needham's staunch antifascism prompted him to imagine that communism could be combined with democracy to create a more cooperative society suitable to big science, where "an ever higher degree of collaboration and cooperation has become necessary." (Ibid., 134) "Though capitalism and democracy grew up together," he wrote, "we approach the time when either one or the other of them must go. And since science is indispensable to all future human civilization it is capitalism that is doomed, and not democracy." (Ibid., 146) Sometimes even Homer nods.

204 Some feared that the poor might vote to abrogate property rights: It remains a matter of debate why American and European majorities have so often restrained themselves from voting against property rights. One theory is that, in the United States at least, widespread ownership of stocks came to the rescue by giving Americans a sense of themselves as propertied. Another possibility is that majorities have more good sense than they're given credit for. Pollsters find that in national elections, Americans more often vote for what they see as the general good than out of narrow self-interest.

204 "A specter is haunting Europe": Karl Marx and Frederick Engels, *Manifesto of the Communist Party*, Samuel Moore and Frederick Engels, trans., 1847, 1888 (Moscow: Progress Publishers, 1969). From http://www.marxists.org/archive/marx/works/download/manifest.doc (accessed March 7, 2006).

205 "dragged from their squalid beds": Karl Marx, *Capital*, Samuel Moor and Edward Aveling, eds. (Chicago: University of Chicago Press, 1952), 117.

205 "the frightful condition of the working people's quarters": Frederick Engels, *The Condition of the Working Class in England* (Moscow: Progress Publishers, 1973), 313.

205 "Just as Darwin discovered": Engels, speech at the graveside of Karl Marx, 1883, in Eugene Kamenka, ed., *The Portable Karl Marx* (New York: Penguin, 1983), 69.

205 "with iron necessity": Marx, *Capital*, v. 1, Preface to the First German Edition, 1867.

206 no inevitable . . . direction: Even today, many imagine that evolution must have some sort of progressive vector that produces ever-better species—as when engineering improvements to an automobile or refrigerator are described in advertisements as "evolutionary." But while evolution can and does improve the survival prospects of particular species in a given environment, it is blind to the future and cannot intentionally produce a species better fit for some future environmental change.

206 "predicted that historic necessity would destroy": Michael Polanyi, *The Logic of Liberty* (Chicago: University of Chicago Press, 1980), 105.

206 "the clumsy and stupid Hegel": Arthur Schopenhauer, *The World as Will and Idea,* R. B. Haldane and J. Kemp, trans. (London: Kegan Paul, 1896), vols. 2, 8, 31, 22.

207 "Nature is the proof of the . . . dialectic": Frederick Engels, *Landmarks of Scientific Socialism* (Chicago: Charles Kerr, 1907), 44. Italics added. Communist presuppositions of this sort extended to virtually all fields. Maurice Merleau-Ponty expressed the position of many twentieth-century intellectuals when he declared that "Marxism is not a philosophy of history, it is *the* philosophy of history and to renounce it is to dig the grave of Reason in history. After that there can be no more dreams or adventures." The Polish philosopher Leszek Kolakowski characterized Marxism as "an instrument that made it possible to master all of history and economics without actually having to study either." Tony Judt, "Goodbye to All That?" *New York Review of Books*, September 21, 2006, 88.

207 "This book does not try to add": Karl Popper, *The Open Society and Its Enemies* (Princeton, NJ: Princeton University Press, 1950), 5.

207 change course when the experiments failed: The literature of dialectical materialism is all but devoid of provisions for testing its own validity. As the philosopher Sidney Hook noted in 1953, "If nature is a test of dialectics, then the principles or laws of dialectics must be something more than general summaries of results already won in the past. They must assert something testable about the future. They must function as hypotheses, no matter how broadly conceived. We are therefore justified in asking: What would the behavior of things have to be, what would we have to discover, what would we in principle have to observe, in order to reach the conclusion that dialectics *fails* to pass the test of nature? Dialectical materialists have never been able to tell us, because no matter what is observed in nature, no matter what happens, they can square it as easily with the so-called laws of dialectic as a pious believer can square any event that occurs with his belief that everything happens by the will of God or that it fulfills some Providential cosmic plan. Since no specific conditions are indicated under which the laws of genuine dialectic would be refuted or abandoned there is no test—illustrations of the laws are confused with proofs so that actually they do not function as hypotheses at all." Michael Polanyi, et al., *Science and Freedom. The Proceedings of a Conference by the Congress of Cultural Freedom and Held in Hamburg on July 23rd–26th, 1953* (Boston: Beacon, 1955), 185.

208 "I believed because I wanted to believe": Lev Kopolev, *The Education of a True Believer* (New York: Harper, 1980), 42.

209 "that science whose people . . . have the boldness": Joseph Stalin, toast given May 17, 1938, in David Joravsky, *The Lysenko Affair* (Cambridge, MA: Harvard University Press, 1970), 105.

209 "the rupture between intellectual and physical labor": Nils Roll-Hansen, "The Lysenko Effect: Undermining the Autonomy of Science." *Endeavour*, December 29, 2005, 143–47; http://www.ncbi.nlm.nih.gov/entrez/query.fcgi?cmd=Retrieve&db=PubMed&list_uids=16271764&dopt=Abstract (accessed April 16, 2006).

210 "When we put forward a measure": Lysenko, September 16, 1935, in Joravsky, *The Lysenko Affair*, 92.

210 "beacon-lights": In 1935, the Commissariat of Agriculture ordered that more than five hundred southern farms begin summer planting in accordance with Lysenko's instructions. When reports came back showing that most had failed, Lysenko made public only the attainments of the fifty "best" farms. This is a classic example of the pseudoscientific practice of restricting one's attention to "false positives." If, for instance, one tests for evidence of extrasensory perception among, say, a thousand individuals, a few will show strongly positive results, since such spikes show up in any random distribution. If one then looks for additional positive results in a second series of tests, a few of the top performers may again do well—just as a few unfortunate golfers will manage to get struck by lightning twice. One can then report that these "beacon" performers have demonstrated remarkable ESP ability, guessing hundreds of times better than chance would predict. Yet the results for the original thousand subjects, taken as a whole, are indeed random.

210 "It would be hard to imagine a more effective way to make genuine science useless": Joravsky, *The Lysenko Affair*, 90.

210 "raise more decisively the question": Ibid., 68.

211 "bases its selection work wholly on Darwin's evolutionary teaching": Peter Pringle, *The Murder of Nikolai Vavilov: The Story of Stalin's Persecution of one of the Great Scientists of the Twentieth Century* (New York: Simon and Schuster, 2008), 241–42. This transcript may also be found in Philip Boobbyer, *The Stalin Era* (London: Routledge, 2000), 149. Emphasis added.

212 Six million people died: For a differing interpretation of Stalin's motives in facilitating famine among the Kazakhs and the Ukrainians, see Robert Service, *Stalin: A Biography* (Cambridge, MA: Harvard University Press, 2005), 326–27.

213 "the overall record of Soviet science": Loren Graham, *Science in Russia and the Soviet Union: A Short History* (New York: Cambridge University Press, 1993), 5.

214 "because of its importance for the development of the atomic bomb": Andrei Linde, "How Physics Fostered Freedom in the USSR," *Physics Today*, June 1992, 13.

214 "respect human rights": For a full text of the Helsinki Accords, see http://www.osce.org/documents/mcs/1975/08/4044_en.pdf and http://www.hri.org/docs/Helsinki75.html#Introduction.

214 "scientific research institute . . . and a large labor camp": Andrei Sakharov, *Memoirs* (New York: Knopf, 1990), 113.

214 Sakharov called for an end to nuclear testing: Sakharov, "Reflections on Progress, Peaceful Coexistence, and Intellectual Freedom," *The New York Times*, July 22, 1968.

216 "The Earth is the cradle of humanity": Konstantin Tsiolkovsky, 1899, in William K. Hartmann, *Astronomy: The Cosmic Journey* (Belmont, CA: Wadsworth, 1978), 49. Similar research was conducted independently by the engineer Yuri Kondratyuk, who came up with the idea of landing men on the moon by dispatching a landing craft from a lunar orbiter—the technique ultimately employed in the American Apollo missions. A deserter from both the Red and

White Russian armies circa 1917, Kondratyuk survived by carrying the papers of a dead man with a similar name, Yuriy Vasilyevich Kondrayuk. Fear that his secret would be found out kept him from joining the thriving Russian rocketry clubs, one of which in 1933 managed to launch the first Russian liquid-fueled rocket to an altitude of four hundred meters. Kondratyuk died defending Moscow against the Nazis in 1942.

216 "flying saucers are real": *The American Weekly*, October 24, 1954.

216 "Opening it . . . I was aghast": Michael J. Neufeld, *Von Braun: Dreamer of Space, Engineer of War* (New York: Knopf, 2007), 24. Neufeld notes that this anecdote was ghost-written by one Curtis Mitchell, which may account for its rather contrived style, but corroborating sources indicate that it is substantially authentic.

217 "We needed money": Neufeld, *Von Braun*, 4.

217 "the bull's eye is the safest spot on the map": Ibid., 181.

217 "We despise the French": Tim Radford, "Down to Earth with a Bump," *The Guardian* (London), April 16, 1993, Features section, 2. The quotation evidently originates in an interview conducted in 1945 by Richard Porter, an American engineer dispatched with the army into Germany for the purpose of locating Germans working on rockets.

217 "Sometimes I Hit London": Sahl's was the best among hundreds of jests portraying von Braun as an exemplification of Nazi moral opacity. However, the physicist Freeman Dyson writes that he admires von Braun for "using his God-given talents to achieve his visions, even when this required him to make a pact with the devil." Differentiating himself from those who would condemn von Braun for collaborating with the Nazi regime, Dyson writes that "war is an inherently immoral activity. Even the best of wars involves crimes and atrocities, and every citizen who takes part in war is to some extent collaborating with criminals. . . . In my work for the RAF Bomber Command I was collaborating with people who planned the destruction of Dresden in February 1945, a notorious calamity in which many thousands of innocent civilians were burned to death. If we had lost the war, those responsible might have been condemned as war criminals, and I might have been found guilty of collaborating with them. . . . In order to make a lasting peace, we must learn to live with our enemies and forgive their crimes. Amnesty means that we are all equal before the law. Amnesty is not easy and not fair, but it is a moral necessity, because the alternative is an unending cycle of hatred and revenge" (*New York Review of Books*, January 17, 2008, 8).

217 "coddled to some extent": Von Braun, "Scientist and Engineer in a Totalitarian State," in Neufeld, *Von Braun*, 174.

218 "Soviet rockets must conquer space!": Asif A. Siddiqi, *Sputnik and the Soviet Space Challenge* (Gainesville: University Press of Florida, 2003), 6.

218 "Maybe you're making rockets": Ibid., 15.

218 "We will all vanish without a trace": William E. Burrows, *This New Ocean* (New York: Random House, 1998), 63. Korolyov's chief competitor, Valentin Glushko, informed on Korolyov but suffered a similar fate, serving time in the Butyrka prison (as did the authors Isaak Babel, Bruno Jasieński, and Alexander Solzhenitsyn). Once freed, he was continually harassed by the secret police for real or imagined crimes against the state.

219 "We gawked at what he showed us": James Oberg, *Red Star in Orbit* (New York: Random House, 1981), 25.

219 "The freed and conscientious labor": Burrows, *This New Ocean*, 183.

220 "Confidence in American science . . . evaporated": James Killian Jr., *Sputnik, Scientists and Eisenhower: A Memoir of the First Special Assistant to the President for Science and Technology* (Cambridge, MA: MIT Press, 1977), 7, in Philip

Taubman, *Secret Empire: Eisenhower, the CIA, and the Hidden Story of America's Space Espionage* (New York: Simon & Schuster, 2003), 212–13.

220 "The communists have established a foothold": Burrows, *This New Ocean*, 189.

220 "It is to be hoped that the explosion . . . shattered a myth": Ibid., 191.

221 "We have the hardware on the shelf": Neufeld, *Von Braun*, 312.

221 "We liked the difference": Mechanical engineer Carl Raggio, in John Noble Wilford, "Remembering When U.S. Finally (and Really) Joined the Space Race," *The New York Times*, January 29, 2008. The nose cone containing the satellite had been "bootlegged" by JPL staffers—they'd built it off the books, without government authorization, in case the Vanguard project faltered—and was kept hidden away in a closet at the lab.

222 "The units . . . consumed the cargo allotment": Richard Rhodes, *Dark Sun: The Making of the Hydrogen Bomb* (New York: Simon & Schuster, 2005), 97; see also George Racey Jordan, *From Major Jordan's Diaries* (New York: Harcourt, Brace, 1952).

223 "Uranium 92": Rhodes, *Dark Sun*, 98.

223 "The entry of Soviet personnel": Ibid., 99.

224 "Victory shall be ours": Ibid., 137, 55, 90–91.

224 "we must steal it!": Ibid., 71, 177.

225 "guided by scientific theories": "Bid to Build Democracy Comes to Fruition," *China Daily*, October 20, 2005, http://www.chinadaily.com.cn/english/doc/2005-10/20/content_486358.htm (accessed October 20, 2005).

225 "It is apparent that the communists have little appreciation": Alfred Zee Chang, "Scientists in Communist China," *Science*, New Series, June 4, 1954, 785.

226 "the eternal glory of Academician Lysenko": Jasper Becker, *Hungry Ghosts: Mao's Secret Famine* (New York: The Free Press, 1996), 65–66.

226 "Science is simply acting daringly": Ibid., 61–63.

226 "I have experienced famine": Ibid., 88.

227 "to be rejoiced over": Jung Chang and Jon Halliday, *Mao: The Unknown Story*. (New York: Knopf, 2005), 437–39. Fertilizer was a favorite metaphor for Mao, who told intimates that the real intent of his "Let a hundred flowers bloom" policy, which seemed to condone criticism of the party, was "to let the poisonous weeds grow first and then destroy them one by one. Let them become fertilizer." John Lewis Gaddis, *The Cold War: A New History* (New York: Penguin, 2005), 111.

227 "When you eat pork": Becker, *Hungry Ghosts*, 89.

227 "a new stage": Central Committee of the CCP, "Decision Concerning the Great Proletarian Cultural Revolution" (also known as the "Sixteen Points"), August 8, 1966.

227 roughly one million people died: Andres G. Walder and Yang Su, "The Cultural Revolution in the Countryside: Timing and Human Impact." *The China Quarterly*, March 2003, 74–99.

228 "must be absolutely protected": Chang and Halliday, *Mao: The Unknown Story*, 419.

228 "dialectical materialism insists": V. I. Lenin, *Materialism and Empirio-Criticism: Critical Comments on a Reactionary Philosophy*, chapter 5, "The Recent Revolution in Natural Science and Philosophical Idealism," and chapter 2. "Matter Has Disappeared," 1908, http://www.marxists.org/archive/lenin/works/1908/mec/five2.htm (accessed January 10, 2007).

228 "We must ferret out": Liu Bowen, "The Idealistic Doctrine of a 'Finite Universe' Must Be Criticized," *Acta Physics Sinica* no. 4, 1976; reprinted in Fang Lizhi, *Bringing Down the Great Wall: Writings on Science, Culture, and Democracy in China* (New York: Norton, 1990), 309.

228 "Many of us who have been to foreign countries": Orville Schell, "China's Andrei Sakharov: The Origins of a Dissident," *The Atlantic*, May 1988.

229 "The universe has no center": Edward Kolb, "On the Oneness of Nature," in Stephanie Pace Marshall, Judith A. Scheppler, and Michael J. Palmisano, eds, *Science Literacy for the Twenty-First Century* (Amherst, NY: Prometheus Books, 2003), 76.

230 "an atmosphere of freedom": Xu Liangying, "Essay on the Role of Science and Democracy in Society," *Journal of Dialectics of Nature*, 1981, 1(1):3–6; English translation in Fan Dainian and Robert S. Cohen, eds., *Chinese Studies in the History and Philosophy of Science and Technology* (Dordrecht: Kluwer, 1996), 6–7.

230 "There is no such thing as Chinese or Indian science": Fang Lizhi and Perry Link, "The Hope for China," a review of H. Lyman Miller's *Science and Dissent in Post-Mao China, New York Review of Books*, October 17, 1996, 17–18.

230 "In science, we approach a situation": Fang Lizhi, *Bringing Down the Great Wall*, 246.

230 "Authoritarianism needs authoritative statements": Fang Lizhi and Perry Link, "The Hope for China," 15.

231 "Were it not for Copernicus": Fang Lizhi, *Bringing Down the Great Wall*, 106.

231 "Marxism-Leninism was formulated . . . on the basis of nineteenth-century science": Miller, *Science and Dissent in Post-Mao China*, 21.

231 "inherently antiauthoritarian": Ibid., 19–21.

232 stress applied science over pure research: Although pure research tends to be the most rewarding form of long-term scientific investment, the communists believed that it was more "public spirited" to channel science toward immediate goals: "The ultimate aim of all science is the satisfaction of the needs of society," declared the *Soviet Encyclopedia*. Michael Polanyi, who toured the Soviet Union in 1935 and was not charmed by what he saw there, summarized this argument as consisting of four assertions: (1) "Each step in the progress of science occurs in response to practical needs"; (2) "Hence there is no essential distinction between science and technology"; (3) "All scientific research is [therefore] to be organized in the direct service of industrial or other practical aims"; (4) "All research . . . must be directed centrally as part of the process of economic planning." "I regard each of these statements as absurd," Polanyi adds. Polanyi, et al., *Science and Freedom*, 1955. See also Loren Graham, *Science in Russia and the Soviet Union*, 1993.

233 "Through painstaking exploration and hard struggle": "Bid to Build Democracy Comes to Fruition," *China Daily*, October 20, 2005, from http://www.chinadaily.com.cn/english/doc/2005-10/20/content_486358.htm (accessed October 20, 2005).

234 "Why do we always lag behind the USA?": Sakharov, *Memoirs*, 146. Sakharov adds, "Only then, I think, did I realize that I was face to face with a terrifying human being."

234 "It was decided to carry out this Marxist experiment on us": Michael Mandelbaum, *The Ideas That Conquered the World* (New York: Public Affairs, 2002), 239.

234 "From the bottom of my heart, we didn't want to do any bad things": Robert Stone, "The Unconscionable War," review of Tom Bissell's *The Father of All Things: A Marine, His Son, and the Legacy of Vietnam*, *New York Review of Books*, November 22, 2007.

235 "superman": Richard Pipes, *Communism: A History* (New York: Modern Library, 2001), 68–69. The slogan of Mao's Great Leap Forward was, "We shall teach the sun and moon to change places. We shall create a new heaven and earth."

CHAPTER ELEVEN: ACADEMIC ANTISCIENCE

236 "the spirit of democracy": David Starr Jordan, "Democracy and Science." *The American Museum Journal*, February 1918, 125.

236 "impossible to establish a simple causal relationship": Henry E. Sigerist, "Science and Democracy," *Science and Society* 2 [1938] 291. See also his "The History of Science and Its Place in Contemporary Civilization," address to the Department of Agriculture Graduate School, Washington, D.C., April 4, 1939.

236 "there is a distinct connection": Needham, "On Science and Social Change," in his *The Grand Titration*, 166, 145. Needham's concept of democracy was, however, sufficiently elastic that he could refer, in 1937, to "the great democracy of the Soviet Union"! Joseph Needham, *Moulds of Understanding: A Pattern of Natural Philosophy* (New York: St. Martin's, 1976), 164.

237 "Science . . . is afforded opportunity for development": Robert K. Merton, *On Social Structure and Science* (Chicago, University of Chicago Press, 1942, 1996). 268.

237 "only in certain types of society": Bernard Barber, *Science and the Social Order* (Westport, CT: Greenwood, 1952), 60–6.

237 "Examination will show that certain 'liberal' societies": Ibid., 61.

237 "one among many truth games": Patti Lather, *Getting Smart: Feminist Research and Pedagogy with/in the Postmodern* (New York: Routledge, 1991), 104.

237 there *was* no objective reality: Students of Plato will recognize this as the philosophy of the rhetorician Gorgias, a cynic whose position may be summarized as (1) nothing exists; (2) if anything *does* exist, we cannot know it; and (3) if we did know it, we could not express it. On one level, the academic movements described in this chapter can be summed up as an assault by cynics and sophists against science and reason.

237 "Western-style liberal democracy": Carola Sachse and Mark Walker, "A Comparative Perspective," *Osiris* 2005, 20:1–20. The insertion of the modifier "Western-style," customary in such attacks, makes a straw man of the argument: At issue is the relationship between science and liberal democracy, not that the democracy has to evince a certain "style." Sachse and Walker also argue that because Italian fascists, German Nazis, and Soviet communists "all declared that their particular political ideologies and systems were best suited to foster modern science," the same claim cannot be valid if made for liberal democracies such as the United States or the United Kingdom. This is like saying that the Queen of England cannot correctly assert that she *is* the Queen, since inmates of insane asylums make the same claim.

238 "logocentric": The term "logocentrism" was coined in the 1920s by the reactionary German psychologist and philosopher Ludwig Klages, regarded as one of the founders of existentialism.

239 "Only in socialist society": Boris Hessen, *The Social and Economic Roots of Newton's "Principia"* (New York: Howard Fertig, 1971); also available in *Science at the Cross Roads: Papers Presented to the International Congress of the History of Science and Technology Held in London from June 29th to July 3rd 1931 by the Delegates of the USSR* (London: Cass, 1971), 149–212.

239 "the complete coincidence of the physical thematic": Ibid.

239 "he was in deep political trouble": Loren R. Graham, "The Socio-Political Roots of Boris Hessen: Soviet Marxism and the History of Science," *Social Studies of Science*, November 1985, 715.

240 "Comrade Hessen is making some progress": Ibid.

240 "The overwhelming impression I gain from the London paper": Ibid., 716.

241 scientific findings are politically contaminated: Since the alleged "cultural

conditioning" of science was, for the postmodernists, the manner through which existence of an objective reality was questioned or dismissed, it should be noted that while science *is* culturally conditioned, in many ways—from the choices made in government and private funding of scientific research to the simple fact that scientists are human beings—what distinguishes science is precisely that its results transcend social conditioning. Just as no scientist would find it necessary to declare his or her race or nationality in defending a scientific thesis, neither would any scientist's claim to have found evidence against a thesis be taken seriously if it depended on a cultural provincialism. In other words, to restrict one's attention in studying science to that which is culturally conditioned is to insist on examining what in science is but trivially true.

242 "A critique of what I do is . . . impossible": *The Economist*, October 23, 2004, 89.

242 "In a profession full of fakeness, he was real": David Lehman, *Signs of the Times: Deconstruction and the Fall of Paul de Man* (New York: Poseidon Press, 1991), 143, 191, 82.

243 "The concept of making a charge": Ibid., 251.

243 "It is always possible to excuse any guilt": Paul de Man, *Allegories of Reading: Figurative Language in Rousseau, Nietzsche, Rilke and Proust* (New Haven, CT: Yale University Press, 1979), 293.

243 "The story of the emperor's new clothes": Alan Sokal and Jean Bricmont, *Fashionable Nonsense* (New York: Picador, 1998), 186.

243 "reason, glorified for centuries": Richard Wolin, *The Seduction of Unreason: The Intellectual Romance with Fascism from Nietzsche to Postmodernism* (Princeton, NJ: Princeton University Press, 2004), 159.

243 "inner truth and greatness of National Socialism": Thomas Sheehan, "Heidegger and the Nazis," *New York Review of Books*, June 16, 1988. For a discussion see Alex Steiner, "The Case of Martin Heidegger, Philosopher and Nazi," on the World Socialist Web Site, http://www.wsws.org.

243 He saw to it: Heidegger dedicated his best-known book, *Being and Time*, to Husserl, but deleted the dedication when the book was republished during the Nazi era. Hence generations of students, studying their dog-eared copies of *Being and Time* and trying to make sense of such statements as, "Scientific research is neither the sole nor the most immediate kind of being of this being that is possible," were reading a work that lacks a dedication because the professor to whom it was originally dedicated was a Jew. You might think this elision of interest to the deconstructionists, who filled many pages with discussions of the "silences" created by what is *not* said in a given text, but few seem to have paid much attention to it. Martin Heidegger, *Being and Time*, Joan Stambaugh trans. (New York: State University of New York Press, 1996), 10.

244 "demonstrated publicly my attitude toward the Party": Steiner, "The Case of Martin Heidegger" on the World Socialist Web Site, http://www.wsws.org.

244 "battle . . . fought out of the strengths of the new Reich": Martin Heidegger, "The University in the New Reich," in Richard Wolin, ed., *The Heidegger Controversy: A Critical Reader* (MIT Press, 1998), 44–45.

245 liberal educators: John L. Childs and George S. Counts, *America, Russia, and the Communist Party in the Postwar World* (New York: John Day, 1943); Evron Kirkpatrick and Herbert McClosky, *Minnesota Daily*, December 14, 1944; in John Earl Haynes and Harvey Klehr, *In Denial: Historians, Communism & Espionage* (San Francisco: Encounter, 2003), 30–31.

246 "A counter-revolutionary extreme right": Charles A. Micaud, *Communism and the French Left* (New York: Praeger, 1963), 7.

246 "a strong disloyal opposition": Ibid., ix.

246 "Left and right alike": Wolin, *Seduction of Unreason,* 158.

246 "This psychological interpretation": Micaud, *Communism and the French Left*, 73.

247 "so often betrayed by so many": Popper, *The Open Society and Its Enemies*, 3. As a teenager Popper supported "scientific socialism" but began to wonder "whether such a calculation could ever be supported by 'science.' . . . I still hoped for a better world, a less violent and more just world, but I questioned whether I really *knew*—whether what I had thought was knowledge was not perhaps mere pretense. . . . I was shocked to have to admit to myself that not only had I accepted a complex theory somewhat uncritically, but I had also actually noticed quite a bit that was wrong, in the theory as well as in the practice of communism, but had repressed this—partly out of loyalty to my friends, partly out of loyalty to 'the cause,' and partly because there is a mechanism of getting oneself more and more deeply involved: once one has sacrificed one's intellectual conscience over a minor point one does not wish to give up too easily; one wishes to justify the self-sacrifice by convincing oneself of the fundamental goodness of the cause, which is seen to outweigh any little moral or intellectual compromise that may be required. With every such moral or intellectual sacrifice one gets more deeply involved." "Karl Popper: Autobiography," in Paul Arthur Schilpp, ed., *The Philosophy of Karl Popper* (La Salle, IL: Open Court, 1974), vol. 1, 25.

249 "pseudoscientific theories like Marx's": Popper, "Autobiography," in Schilpp, *Philosophy of Karl Popper*, 31. Quoting Popper more fully: "Early in this period [the early 1920s] I developed further my ideas *about the demarcation between scientific theories* (like Einstein's) *and pseudoscientific theories* [like Marx's, Freud's, and Adler's]. It became clear to me that what made a theory, or a statement, scientific was its power to rule out, or exclude, the occurrence of some possible events—to prohibit, or forbid, the occurrence of these events: *the more a theory forbids, the more it tells us.*"

249 "paradigm shifts": Thomas S. Kuhn, *The Structure of Scientific Revolutions* [1962] (Chicago: University of Chicago Press, 1970), 10.

249 "A celestial body that had been observed off and on": Ibid., 115–16.

250 accurate only insofar as it was trivial: Kuhn himself seems to have glimpsed the void looming beneath his thesis, and occasionally in his later career protested against some of the "absurd" and "vaguely obscene" claims being made by deconstructionists in his name. Yet he was sufficiently influenced by them to declare, in his 1992 Rothschild Lecture at Harvard, "I am not suggesting, let me emphasize, that there is a reality which science fails to get at. My point is rather that no sense can be made of the notion of reality as it has ordinarily functioned in the philosophy of science." Steven Weinberg, "The Revolution That Didn't Happen," *New York Review of Books*, October 8, 1998, 48.

251 "pure cowardice": Murray Gell-Mann, *The Quark and the Jaguar* (New York: Freeman, 1994), 179.

251 "a historiographic revolution": Kuhn, *Structure of Scientific Revolutions,* 3.

251 "a landmark in intellectual history": Nicholas Wade, "Thomas S. Kuhn: Revolution Theorist of Science." *Science* (July 8, 1977): 143–45.

251 "science provides truths": Holton, "How to Think About the 'Anti-Science' Phenomenon," 103–28.

251 "particularist, self-aware": Lather, *Getting Smart*, 102. The quotation in full reads: "The relentless undermining of the Enlightenment code of values . . . increasingly appears the key Western intellectual project of the late twentieth century. Those choosing to encourage rather than resist this movement are using it to stretch the boundaries that currently define what we do in the name

of science. A behavioral science governed by adherence to methods and standards developed in the natural sciences is being displaced by what [author and "singer/songwriter" Richard] Harland calls 'a science for philosophers.' . . . A 'narrative, semiotic, particularist, self-aware' science is emerging. . . ."

252 "complex, and deploying them is often . . . complicated": Robert Conquest, *Reflections on a Ravaged Century* (New York: Norton, 2000), 224.

252 "There is no scientific method": Paul Feyerabend, *Killing Time: The Autobiography of Paul Feyerabend* (University of Chicago Press, 1995), 88.

252 "Successful research does not obey general standards": Paul Feyerabend, *Against Method* (London: Verso, 1993), 1.

252 "without mentioning a single fact": Feyerabend, *Killing Time*, 67.

252 "We can of course imagine a world": Feyerabend, *Against Method*, 18, 272.

252 "I recommend to put science in its place": Ibid., 162.

253 "There is not one common sense, there are many": Feyerabend, *Killing Time*, 143.

254 "the 'progress of knowledge and civilization'": Feyerabend, *Against Method*, 3.

254 "It often happens that parts of science become hardened": Ibid., 37.

254 "making a plurality . . . of views possible": Ibid.

254 "Scientists are not content with running their own playpens": Ibid., 163.

255 "'New' points of view": Feyerabend, *Realism, Rationalism & Scientific Method: Philosophical Papers*, vol. 1 (New York: Cambridge University Press, 1993), 103.

255 "Science is not a process of discovering": Carolyn Merchant, *Radical Ecology* (New York: Routledge, 1992), 236.

255 "neither logic nor mathematics": Stanley Aronowitz, *Science as Power* (Minneapolis: University of Minnesota Press), 1988, 346.

255 "since the settlement of a controversy is *the Cause*": Bruno Latour, *Science in Action* (Cambridge, MA: Harvard University Press, 1987), 99. For a discussion see Paul R. Gross & Norman Levitt, *Higher Superstition: The Academic Left and Its Quarrels with Science* (Baltimore: Johns Hopkins University Press, 1994), 57–58.

256 "There are many natural scientists": Alan Sokal, "Transgressing the Boundaries: Towards a Transformative Hermeneutics of Quantum Gravity," *Social Text*, Spring/Summer 1996, 217–52.

257 "The conclusion is inescapable": Paul A. Boghossian, "What the Sokal Hoax Ought to Teach Us," in Norette Koertge, ed., *A House Built on Sand* (New York: Oxford University Press, 1998), 25–26.

257 "*knowingness*": Norman Levitt, "More Higher Superstitions: Knowledge, Knowingness, and Reality," *Skeptic* vol. 4, no. 4, 1996, 78.

257 "What eventually 'crawled out of the laboratory'": Carl Feldbaum, "Some History Should Be Repeated," *Science*, vol. 295, February 8, 2002, 975.

257 which facts . . . you would prefer had never been learned: When people are asked what existing scientific knowledge they would prefer had been repressed, they often answer by referring to nuclear weapons. But does this mean that the Swiss government ought to have prevented Einstein from publishing his special relativity paper in 1905, or that the British Parliament ought to have padlocked Rutherford's laboratory before he could infer the structure of the atomic nucleus? And even if the leaders of all the Western nations had mustered such astonishing prescience—which, of course, they could not have done—what would the Cold War have been like had the Soviets failed to demonstrate similar restraint? The point is that government restrictions on scientific research assume a 20/20 foresight that neither the government nor anyone else possesses. (For a discussion see Timothy Ferris, "Keep Up the Search," *New Scientist*, October 22, 2005, 51.)

258 "Globalization has produced": Steven Best and Douglas Kellner, *The Postmodern Adventure* (New York: Guilford, 2001), 1.

259 "I am not a conservative.": "Allan Bloom, 20 Years Later," http://guardian.co.uk, updated September 17, 2007 (accessed September 6, 2008).

259 "to make themselves slaves willingly": Galileo, *Dialogues Concerning the Two Chief World Systems—Ptolemaic and Copernican*, Stillman Drake, trans. (Berkeley, CA: University of California Press, 1953), 112.

259 "Epochs which are regressive": Robert Hughes, *Culture of Complaint: The Fraying of America* (New York: Oxford University Press, 1993), 10.

260 "We're in the same boat, brother . . .": Lyrics by E. Y. Harburg.

CHAPTER TWELVE: ONE WORLD

262 "God is able . . . to vary the laws of nature": Newton, *Opticks*, Query 31.

262 "the principle of the uniformity of nature": Edwin Hubble, "NGC6822, A Remote Stellar System," *Astrophysical Journal*, 62, 432.

263 stasis is an illusion: Montaigne had it right when he supposed that stability is only relative: "All things," he wrote, "are in constant motion—the earth, the rocks of the Caucasus, the pyramids of Egypt—both with common motion and with their own. Stability itself is nothing but a more languid motion." For a discussion see George Hoffmann, "The Investigation of Nature," in Ullrich Langer, ed., *The Cambridge Companion to Montaigne* (New York: Cambridge University Press, 2005), 173.

263 "There is no reason at all why *he* won": Daniel C. Dennett, *Darwin's Dangerous Idea: Evolution and the Meanings of Life* (New York: Simon & Schuster, 1995), 214.

264 "Man is naturally good": Rousseau, "Discourse on Inequality," 1754, in Russell, *A History of Western Philosophy*, 687.

264 "Instincts do not create customs": Sociologist Ellsworth Faris, 1927, in C. N. Degler, *In Search of Human Nature: The Decline and Revival of Darwinism in American Social Thought* (New York: Oxford University Press, 1991), 84.

264 "Give me a dozen healthy infants": Jonathan Gottschall and David Sloan Wilson, eds., *The Literary Animal: Evolution and the Nature of Narrative* (Chicago: Northwestern University Press, 2005), 14–15.

264 "how similar their minds were to ours": Darwin, *The Descent of Man* (London: John Murray, 1871), vol. 1, 232.

264 "The sterilization of failures . . . and other defectives": Jonah Goldberg, *Liberal Fascism*, 249–55.

265 a third of American teachers . . . science museums shy away: National Science Teachers Association 2005 survey, http://www.nsta.org/publications/surveys/survey20050324.aspx (accessed March 26, 2009); T.F. interviews, Science+Society conference, Boston, 2007.

265 "systematic nonintrusive observations of . . . bonobo": Gottschall and Wilson, *Literary Animal*, 11.

266 "the key principle governing brain organization is a populational one": Gerald M. Edelman, *Neural Darwinism: The Theory of Neuronal Group Selection* (New York: Basic Books, 1987), 4–5, 176.

267 "clash of civilizations": Samuel P. Huntington, "The Clash of Civilizations?" *Foreign Affairs,* Summer 1993; 72, 3; 22. The phrase originated with Bernard Lewis.

268 "significant contributions to mathematics . . . and astronomy": Islamic contributions to mathematics and science, though real enough, are often inflated to create the misleading impression that the Islamic world had a scientific establishment—with laboratories, journals, and university departments of

science—comparable to those later found in the West. It did not. Such exaggerations seem to have been fielded in order to counter a suggestion—that Muslims are incapable of doing science—that no real scholar would take seriously anyway. As in China we find, however, that Muslim scientists winning Nobel Prizes have been obliged to do their research in exile. As of 2009, there had been two Muslim Nobel Laureates in science—Abdus Salam in physics and Ahmed Zewail in chemistry.

268 "They showed their friendship": Albert Hourani, *A History of the Arab Peoples* (New York: Time Warner, 1991), 266.

268 "passed virtually unnoticed in the lands of Islam": Bernard Lewis, *What Went Wrong? Western Impact and Middle Eastern Response* (New York: Oxford University Press, 2002), 7.

268 "the necessary rights of freedom": Ibid., 57.

269 "Abdul Wahhab": Wahhabist mullahs came to run many of Saudi Arabia's public schools, teaching their students little except to memorize the Quran and hate the West. Saudi Arabia has a per capita GDP comparable to that of Chile and Poland, yet 70 percent of its public schools don't even have a building, and instead must hold classes in rented rooms.

270 "We were racist": Paul Berman, *Terror and Liberalism* (New York: Norton, 2003), 55. The affinity between fascism and Islamism lives on the careers of people like Albert Friedrich Armand Huber, a Swiss journalist, neo-Nazi, and convert to Islam who after 9/11—which he celebrated with a night of drinking with like-minded friends in Bern bars—kept a picture of Osama bin Laden hanging next to one of Adolf Hitler on the wall of his study.

270 "live, screaming temptations": Sayyid Qutb, "The America I Have Seen," 1951, http://www.bandung2.co.uk/books/Files/Education/The%20America%20I%20Have%20Seen%20-%20Sayyid%20Qutb.pdf (accessed February 13, 2009).

270 "The resurgence of science has . . . come to an end": Sayyid Qutb, *Milestones* (New Delhi: Islamic Book Service, 2008), 103, 105, 114, 110, 8.

271 "a mockery": Bruce Lawrence, ed., *Messages to the World: The Statements of Osama bin Laden* (London: Verso, 2005), 112.

271 "society whose legislation does not rest on divine law": Marc Erikson, "Islamism, Fascism and Terrorism," *Asia Times,* November 5, 2002, http://www.atimes.com/atimes/Middle_East/DK05Ak01.html, (accessed February 12, 2009).

271 "The line from the guillotine and the Cheka": Ladan Boroumand and Roya Boroumand, "Terror, Islam, and Democracy." *Journal of Democracy* vol. 13, no. 2 (April 2002).

271 "certain uncomfortable resemblances": Bernard Lewis, "Communism and Islam," International Affairs (Royal Institute of International Affairs 1944–), January 1954, 9–10.

272 "suffering from affliction": Berman, *Terror and Liberalism*, 69.

272 "Just as Islam teaches us how to pray": Gulam Sarwar, *Islam: Beliefs and Teachings*, in Husain, *The Islamist*, 21.

272 "My son, the Prophet is not our leader": Husain, *The Islamist*, 52.

273 "Democracy is *haram*!": Ibid., 78.

273 "The state is the march of God": Ibid., 162.

273 "Combat is today the . . . duty of every Muslim": Abu-Hafs al-Masri Brigades, statement of July 1, 2004, in Hillel Fradkin, Husain Haqqani, and Eric Brown, eds., *Current Trends in Islamist Ideology* (Washington, D.C.: Hudson, 2005), 42.

273 "Glory does not build its lofty edifice except with skulls": Berman, *Terror and Liberalism*, 119.

273 "The Children are the Holy Martyrs of Tomorrow": Ibid., 112.

273 "Islam wants the whole earth": Marc Erikson, "Islamism, Fascism and Terrorism," *Asia Times,* November 5, 2002, http://www.atimes.com/atimes/Middle_East/DK05Ak01.html (accessed February 12, 2009).

274 "we are totally dependent on the West": Omar Nasiri, *Under the Jihad: My Life with Al Qaeda, A Spy's Story*, quoted in Ahmed Rashid, "Jihadi Suicide Bombers: The New Wave," *New York Review of Books,* June 12, 2008, 22.

274 "Take a vast area": David Frum and Richard Perle, *An End to Evil: How to Win the War on Terror* (New York: Random House, 2004), in Thomas Powers, "Tomorrow the World," *New York Review of Books*, March 11, 2004, 5.

275 "Western science is inherently destructive": Steve Paulson, "The Religious State of Islamic Science," *Salon*, August 13, 2007, http://www.salon.com/books/feature/2007/08/13/taner_edis/ (accessed February 16, 2009).

275 "In *jihad* the blood of the Muslims must flow": Lewis, *What Went Wrong?* 107.

275 "You are an Algerian": Larry Diamond, Marc F. Plattner, and Daniel Brumberg, eds., *Islam and Democracy in the Middle East* (Baltimore: Johns Hopkins University Press, 2003), 255.

275 "How would you assess the state of scientific knowledge": Steve Paulson, "The Religious State of Islamic Science," *Salon,* August 13, 2007, http://www.salon.com/books/feature/2007/08/13/taner_edis/ (accessed February 16, 2009).

276 "always triggers thunderous applause": Mohamed Talbi, "A Record of Failure," in Diamond et al., *Islam and Democracy*, 10.

276 "The Middle East has become dominated": Colin Rubenstein, "What Is Really Irking the Radical Islamists?" *The Age*, July 14, 2005, http://www.theage.com.au/news/opinion/what-is-irking-the-radical-islamists/2005/07/13/1120934299264.html (accessed February 13, 2009).

276 "a majority of Arab youths . . . want to emigrate": UN 2002 Arab Human Development Report, http://www.arab-hdr.org/publications/other/ahdr/ahdr2002e.pdf (accessed August 20, 2009).

276 "science operates in a matrix of freedom": Abdolkarim Soroush, *Reason, Freedom & Democracy in Islam*, Mahmoud Sadri and Ahmad Sadri, ed. and trans. (New York: Oxford University Press, 2000), 46.

277 "the only legitimacy": Rachid al-Ghannouchi in Robin Wright, "Islam and Liberal Democracy," *Journal of Democracy* 7.2 (1996): 64–75. http://muse.jhu.edu/journals/journal_of_democracy/v007/7.2wright01.html (accessed September 27, 2004.

277 "American freedom and democracy": Zogby International, National Society of Public Opinion Studies, Gallup, World Values Survey, NFO Middle East, 2002; *The Economist*, October 19, 2002.

277 Can such sentiments be translated into genuinely democratic governments?: Iraq is a democracy, but also a unique case in the Middle East since it was invaded by the United States and its democratic system thus installed by force of arms. Opponents of the war often said at the time that it is impossible to impose democracy, as the United States was trying to do. But whatever one thinks of America's military adventure in Iraq, that argument is not conclusive. Following World War II the United States imposed a democratic government on Japan by force; the resulting democracy not only survived but seems to have encouraged the rise of other democratic states elsewhere in Asia. In the words of Kenneth M. Pollack of the Brookings Institution, "For the first time East Asians could look at Japan and say, 'That's the kind of state that I could imagine living in.' Before Japan, East Asians thought about democracy the same way that Arabs do now. They thought of it as being an American or a European thing. Those were the only examples they had, and they knew they didn't want that. But then Japan came along and proved that you could build a

democracy . . . consistent with Japan's values, traditions and history." Kenneth M. Pollack, "Draining the Arab Swamp," Foreign Policy Research Institute, January 5, 2004.

277 Foreign nationals were kidnapped and tortured: Writes Mohamed Farag Ahmad Bashmilah, a citizen of Yemen: "From October 2003 until May 2005, I was illegally detained by the U.S. government and held in CIA-run 'black sites' with no contact with the outside world. On May 5, 2005, without explanation, my American captors removed me from my cell and cuffed, hooded, and bundled me onto a plane that delivered me to Sanaa, Yemen. I was transferred into the custody of my own government, which held me—apparently at the behest of the United States—until March 27, 2006, when I was finally released, never once having faced any terrorism-related charges. Since my release, the U.S. government has never explained why I was detained and has blocked all attempts to find out more about my detention. What I do know is that the Jordanian government—after torturing me for several days—handed me over to a U.S. 'rendition team' in Amman, which then abducted me, forced me onto a plane, and flew me to Afghanistan. During this, and several other transfers between CIA prisons, I was . . . stripped naked, dressed in a diaper, shackled, blindfolded and hooded, and then boarded onto a waiting plane. . . . During my detention, I agonized constantly about my family back in Yemen, knowing they had no idea where I was. They never once received information about who had taken me, why I was taken, or even whether I was alive. They were never contacted by the U.S. government or the International Committee of the Red Cross. My mother and wife were in such anguish that they had to be hospitalized for illness, stress, and anxiety. My father passed away while I was disappeared and I am still distraught thinking that he died without knowing whether I was dead or alive. I continue to suffer from bouts of illness [and] my physical symptoms are made worse by the anxiety caused by never knowing where I was held, and not having any form of acknowledgment that I was disappeared and tortured by the U.S. government." (Huffington Post, February 20, 2009, http://www.huffingtonpost.com/mohamed-farag-bashmilah/disappeared-in-the-name-o_b_168200.html (accessed February 20, 2009). Supporters of such measures—when they were not endlessly repeating that terrorism posed a threat to Americans, as if they alone appreciated that fact—liked to cite as precedents the Alien and Sedition Acts of 1789, Abraham Lincoln's suspension of habeas corpus during the Civil War, and the espionage acts of 1917–18, which made it a crime to obstruct the military draft or to advocate "disloyalty." But these acts encouraged abuses (the socialist leader Eugene V. Debs was sentenced to ten years in prison for a 1918 speech criticizing the Espionage Act itself!) and seldom proved lastingly popular. Thomas Jefferson denounced the acts as unconstitutional, Lincoln's highhandedness cost his party heavy losses in the 1862 congressional elections, and three of the four provisions of the acts, including one making it a crime for anyone to publish "false, scandalous, and malicious writing" against the government or its officials, were repealed or allowed to expire within four years of their enactment.

277 "A fearful superpower": G. John Ikenberry, paraphrasing the views of James Traub in Traub's *The Freedom Agenda: Why America Must Spread Democracy (Just Not the Way George Bush Did)* (New York: Straus & Giroux, 2008) in *Foreign Affairs*, January/February 2009, 183.

278 "threaten community harmony": Irving N. Jones, on behalf of the British Secretary of State, letter to Geert Wilders, http://www.jihadwatch.org/archives/024785.php (accessed February 20, 2009).

278 "Freedom of speech should be protected": *The Economist*, October 13, 2007, 67; Associated Press, "UK bars anti-Islamic Dutch lawmaker," February 12, 2009.

278 "fight the battle for the Enlightenment all over again": Lisa Applignanesi, ed., *Free Expression Is No Offence* (London: Penguin, 2005), 8.

278 "If liberty means anything": Ibid., 1.

279 While only 5 percent of Americans: American Religious Identification Survey, 2008; Pew Research Values Study, 2007.

279 "crime correlates *inversely* with levels of religious conviction": Pew Research Values Study, http://pewforum.org/publications/reports/poll2002.pdf; Harris poll, http://www.harrisinteractive.com/news/allnewsbydate.asp?NewsID=1131; CUNY Graduate Center study, 2001, http://www.gc.cuny.edu/press_information/current_releases/october_2001_aris.htm; Denise Golumbaski, Research Analyst, Federal Bureau of Prisons, compiled March 5, 1997; Max D. Schlapp and Edward H. Smith, *The New Criminology* (London: Routledge, 1973); *The Economist*, November 3, 2007. "A Special Report on Religion and Public Life," 9. The American incarceration rate—more than 2.25 million persons in prisons and jails circa 2006, according to the U.S. Department of Justice, or 751 per 100,000 residents—constitutes one of the most illiberal blemishes on American democracy. With under 5 percent of the world's population, the United States has almost a quarter of the world's prisoners. Its incarnation rate is five times that of the United Kingdom, seven times Canada's, and eight times France's, ranking even higher than that of repressive states like Libya, Iran, and China. Historically, it was surpassed only by Stalinist Russia and apartheid South Africa.

280 people should behave at least as ethically without religion as with it: Since religiosity fails to correlate with individual morality, the Christian right has resorted to claiming that atheistic *nations* behave immorally even though the same is not true of individuals. Nazi Germany and communist Russia are the examples most often cited. But fascism was not irreligious—at least not in the considered view of the Vatican, which supported Hitler, or of the Nazi party, which in its charter stated that it "stands for Positive Christianity," or of Hitler himself, who claimed in 1938 that "by warding off the Jews I am fighting for the Lord's work." Communism, as many have observed, was its own religion. "A totalitarian state is in effect a theocracy," wrote George Orwell, "and its ruling caste, in order to keep its position, has to be thought of as infallible." Louis L. Snyder, ed., *Hitler's Third Reich: A Documentary History* (Chicago: Nelson Hall, 1981), 167; Christopher Hitchens, *God Is Not Great*. (Hachette, 2007), 232.

280 "losing at least 5 percent of global GDP": Nicholas Stern, *The Economics of Climate Change* (New York: Cambridge University Press, 2007), xv.

284 "Warming seemed like a good thing": Spencer R. Weart, *The Discovery of Global Warming* (Cambridge, MA: Harvard University Press, 2003), 7.

285 "We owe much to Callendar's courage": Spencer R. Weart, "The Carbon Dioxide Greenhouse Effect," http://www.aip.org/history/climate/co2.htm (accessed February 28, 2009).

285 "We do not yet know": Weart, *Discovery of Global Warming*, 44.

287 "most of the global average warming": IPCC Climate Change 2007 Synthesis Report, http://www.ipcc.ch/pdf/assessment-report/ar4/syr/ar4_syr_spm.pdf (accessed March 21, 2009).

287 "having a probability of over 90 percent . . . slow the pace of progress": Ibid.

287 "no judgments will be given as absolutes": Lyman Bryson, *Science and Freedom* (New York: Columbia University Press, 1947), viii.

289 "Act only on those maxims": Immanuel Kant, *Fundamental Principles of the*

Metaphysic of Morals, sec. 2. When Jesus of Nazareth was asked which were the most important of the Ten Commandments, he mentioned the first one and then simply imported his own philosophy—"Love your neighbor as yourself," a version of the Golden Rule—into the mix. (Matthew 22:36–40).

290 "To tell the truth": Sydney Brenner, "Humanity as the Model System," *Science*, vol. 302, (October 24, 2003): 533.

INDEX